Advances in Immigrant Family Research

Series Editor
Susan S. Chuang
University of Guelph
Guelph, ON, Canada

More information about this series at http://www.springer.com/series/8807

Deborah J. Johnson • Susan S. Chuang
Jenny Glozman
Editors

Re/Formation and Identity

The Intersectionality of Development, Culture, and Immigration

Springer

Editors
Deborah J. Johnson
Michigan State University
East Lansing, MI, USA

Susan S. Chuang
University of Guelph
Guelph, ON, Canada

Jenny Glozman
The Calm Collective Psychotherapy Clinic
Toronto, ON, Canada

ISSN 2625-364X ISSN 2625-3666 (electronic)
Advances in Immigrant Family Research
ISBN 978-3-030-86425-5 ISBN 978-3-030-86426-2 (eBook)
https://doi.org/10.1007/978-3-030-86426-2

This Springer imprint is published by the registered company Springer Nature Switzerland AG
The registered company address is: Gewerbestrasse 11, 6330 Cham, Switzerland

To the countless brave souls, youth and their families, and elders who seek change, betterment, freedom, and resist or are open to new identities to make the impossible possible.

—Deborah J. Johnson, Susan S. Chuang, and Jenny Glozman

To Lois and Richard Johnson who handed me the universe and chose not to repress my restless, inquisitive spirit; and to my husband Jimmy, who champions and inspires me.

—Deborah J. Johnson

To my father, Ching Chu Chuang, who continues to support me, wherever my path may go.

—Susan S. Chuang

To my parents, Alex and Marina, for your support and resilience, and to my husband, Jon, for always being there to catch me.

—Jenny Glozman

Acknowledgments

We gratefully and collectively acknowledge the thousands of study participants that shared their immigrant and identity stories, completed numerous questionnaires, and helped researchers interpret their lives. We thank the many contributors of this volume for sharing their expertise to this critical topic as their research gives these pages meaning. Special acknowledgment to Dr. Junghee Yoon (who supported the book process throughout) and Drs. Barbara Thelamour, and Andrea G. Hunter for their timely efforts in shaping the final introduction.

Contents

About the Editors

Deborah J. Johnson, Ph.D. is a Professor in the Department of Human Development and Family Studies at Michigan State University. She serves as the current Director of the Diversity Research Network, within MSU's Office for Inclusion and Intercultural Initiatives. Her research explores racially and culturally related development, parental racial socialization and coping, and cultural adjustment from early childhood through emerging adulthood, in both domestic and international children and youth. She is the recipient of numerous fellowships and awards, including the SSHD Senior Mentor Award (2019). Recent books address the global rights of children and private/charter school experiences of African American children and immigrant identity explorations. She is the author of numerous articles, chapters, and edited volumes. Recent edited volumes include *Children and Prejudice* (2019) and *Vulnerable Children: Global Challenges in Education, Health, Well-Being, and Child Rights* (2013).

Susan S. Chuang, Ph.D. is a Professor at the University of Guelph, Canada. Her research focuses on fathering, parenting, parent-child relationships of young children among immigrant and ethnic families in various sociocultural contexts (Asian: Canada, China, Hong Kong, Taiwan; Latino/a: Paraguay, USA; Black: Jamaica). Her second line of research examines families and immigration, collaborating with community organizations in Canada. She conducts numerous volunteer workshops for the local communities that focus on various topics (e.g., parenting, education, child/youth development, employment). Her third line of research explores intimate partner violence against men, false allegations, child custody and access, high conflict relationships, parental alienation and their impact on individual and family well-being. She also is the Series Editor for Springer's *Advances in Immigrant Family Research*, and the Associate Editor for the *Journal of Family Psychology* and the *Journal of Adolescent Research*. She has organized six international conferences on immigrant families.

Jenny Glozman, Ph.D. completed her doctoral work in Family Relations and Human Development at the University of Guelph. She holds a B.A.H. in Psychology from Queen's University, Ontario, and her M.A. in Marriage and Family Therapy from Syracuse University, New York. She currently works as a Registered Psychotherapist and Marriage and Family Therapist and is the Co-Founder of the Calm Collective Psychotherapy Clinic in Toronto, Ontario. Her research focuses on topics of immigration, particularly among youth and families.

About the Authors

Mona M. Abo-Zena, Ph.D. is an Assistant Professor in the Department of Curriculum and Instruction within the College of Education and Human Development at the University of Massachusetts Boston. On individual, social, and structural levels, she studies the role of religion and other intersecting dimensions of diversity in human development, with a focus on minoritized groups and experiences. Her applied scholarship includes an equity focus on positive youth development that balances an asset-based perspective of children, families, and communities while acknowledging contextual challenges.

Lorena Aceves, Ph.D. is a first-generation Latina scholar who recently completed her Ph.D. in Human Development and Family Studies at the Pennsylvania State University. Her research throughout graduate school focused on understanding how cultural, familial, and individual-level factors contributed to Latinx adolescent's academic achievement and educational attainment. She is currently serving as an AAAS/SRCD Federal Executive Branch Fellow at the Office of Head Start within the Administration for Children and Families. She plans to leverage developmental science within federal policy to promote equitable opportunities and resources for students, children, and families from underrepresented backgrounds.

Pankhuri Aggarwal, M.A. is a fourth-year graduate student in the clinical psychology doctoral program at Miami University, Ohio. She completed her bachelor's in psychology from Lady Shri Ram College for Women in Delhi, India, and her master's in clinical psychology from Tata Institute of Social Sciences in Mumbai, India. Her research focuses on culturally informed conceptualization and interventions for depression and suicide, and education and training experiences of mental health professionals and trainees.

Mayra Y. Bámaca-Colbert, Ph.D. is an Associate Professor in Psychological Sciences at the University of California, Merced. She earned a B.A. in Psychology at California State University, Northridge, an M.S. in Human and Community Development at the University of Illinois, Urbana-Champaign, and a Ph.D. in

Family and Human Development at Arizona State University. Her research lies at the intersection of developmental (e.g., puberty), cultural (e.g., acculturation), and contextual (e.g., parenting, friendships, neighborhood) influences on the behavioral and health outcomes of ethnic-racial minoritized youth, with an emphasis on Latina/o/x populations.

Meeta Banerjee, Ph.D. is an Assistant Professor in the Department of Psychology at the University of South Carolina. She received her Ph.D. in Ecological-Community Psychology with a specialization in Applied Developmental Science from Michigan State University in 2012. As a Jacobs Foundation Pathways to Adulthood postdoctoral fellow, she worked directly with Dr. Jacque Eccles at the University of Michigan and was also an NIH minority postdoctoral scholar. Prior to her Ph.D., she received her M.S.W. and B.A. in Psychology from the University of Michigan in 2003 and 2001, respectively. Her research employs both integrative and ecological frameworks to understand the influence of contextual factors on the developmental trajectories in underrepresented minority youth and families.

Julia Z. Benjamin, Ph.D. is a Pediatric Psychologist at American Family Children's Hospital in the School of Medicine and Public Health, University of Wisconsin—Madison. She completed postdoctoral work at Mayo Clinic, Rochester, Minnesota, and her predoctoral internship at Boston Children's Hospital, Harvard University. She earned a Ph.D. in Counseling Psychology from the University of Wisconsin—Madison and M.A. in Developmental Psychology from Columbia University Teachers College. Her current clinical work and research is focused on treatment of chronic pain in children and adolescents, as well as supporting the well-being of youth with diverse gender identities.

Laura Bates, M.A. now retired, was a senior research consultant in University Outreach and Engagement, at Michigan State University. Her research focus included resilience among refugee children and youth and other children raised in challenging circumstances.

Marsha T. Carolan, Ph.D., LMFT is an Associate Professor Emeritus in the Department of Human Development and Family Studies at Michigan State University. She served as Program Director and Clinic Director for the Couple and Family Therapy program and introduced a feminist multicultural framework into the curriculum and training. Her teaching and research interests focus on diverse families and communities, health disparities and trauma, and social justice in clinical training.

Jasmin E. Castillo, B.S. is a third-year doctoral student at the Pennsylvania State University in Human Development and Family Studies. Her research interests lie at the intersection of psychology, education, and criminal justice and include students' perceptions of authority, teacher-student relationships, and the development of socioemotional skills in children. Her work examines the ways in which researchers can optimize interventions that leverage socioemotional skill development in order to divert students' who might otherwise enter the school-to-prison pipeline and foster school environments where equitable discipline is the standard.

Haeyoon Chung, M.A. is a doctoral student in Educational Psychology at the University of Illinois Chicago (UIC). She received her bachelor's and master's degrees from Ewha Woman's University in South Korea. Before pursuing her graduate studies, she worked as a licensed mental health counselor for young adults. Her research interest focuses on immigrant-origin young adults' identity development. Specifically, she is interested in how Asian American emerging adults navigate their lives regarding identity and career. Her current work focuses on young Asian American female's civic engagement and the model minority myth.

Annette V. Clarke, Ph.D. is an adjunct faculty member at Palo Alto University where she earned her Ph.D. and M.S. degrees in Clinical Psychology with an emphasis in Diversity and Community Mental Health. She holds an M.A. in Psychology with a concentration in Mental Health and Substance Use Counseling from The New School for Social Research and a B.A. in Psychology from New York University. Her interests include Black mental health, identity development, immigration, acculturation, body positivity, mindfulness and meditation, cultural competency, and reducing stigma and mental health care disparities. A licensed psychologist, she currently works across the developmental lifespan in public mental health and private practice.

Fernanda L. Cross, Ph.D. is an Assistant Professor at the University of Michigan School of Social Work. Her research program utilizes a strength-based approach to examining family and cultural factors that promote healthy development and mitigate the risk of poor psychological and/or educational outcomes for Latinx adolescents and families. She is particularly interested in examining the role undocumented status plays in family processes and adolescents' outcomes, as well as the protective effects of ethnic-racial socialization and identity development.

Sophie Fajardo, M.A. is a Ph.D. candidate in the Department of Sociology. She earned her M.A. in Sociology at the University of Chicago in 2014 and a B.A. in sociology from Grinnell College in 2012. Her research interests include international migration, inequality, knowledge production, and qualitative methodology. Her dissertation research focuses on the ways mass migration shapes content, production, and institutionalization of political knowledge.

Steven J. Gold, Ph.D. is Professor and graduate program director in the Department of Sociology at Michigan State University. His scholarly interests include international migration, ethnic economies, and qualitative field methods. He is author, co-author, editor, or co-editor of nine books, including *The Israeli Diaspora* which received the American Sociological Association's International Migration Section award for the best book on international migration in 2003. Gold has published over

100 papers and has served as the chair of 22 Ph.D. dissertations. He received the Distinguished Career Award from the American Sociological Association's International Migration Section in 2019.

Melinda Gonzales-Backen, Ph.D. is an Associate Professor in the Department of Family and Child Science at Florida State University. She earned a B.A. in Psychology at the University of Arizona and a M.S. and Ph.D. in Family and Human Development at Arizona State University. Her research focuses on how cultural stressors (e.g., discrimination, acculturative stress), cultural strengths (e.g., ethnic identity, familial ethnic socialization), adolescent development, and family processes intersect to predict adolescent adjustment among Latinx youth.

Christopher G. Graziul, Ph.D. is a data scientist for the Urban Resiliency Initiative in the Department of Comparative Human Development at the University of Chicago. His research integrates diverse theoretical approaches and methodological techniques to better understand the mechanisms that structure social systems, especially the co-constitution of physical context and social behavior.

Vrinda Kalia, Ph.D. is an Assistant Professor in the Department of Psychology at Miami University. She studies goal-directed behavior and the role of contextual forces (i.e., family, culture, language environment, stress) that shape the development of cognitive and emotional processes that enable goal achievement. Her work is interdisciplinary intersecting educational psychology, cultural psychology, and cognitive psychology, and her studies often combine behavioral data with physiological measurements (i.e., EEG and fNIRS).

Dalal Katsiaficas, Ph.D. is an Assistant Professor of Educational Psychology at the University of Illinois at Chicago. Her current research focuses on the positive social development of immigrant-origin youth as they emerge into adulthood. Specifically, her research examines the social responsibilities of immigrant-origin youth as well as the impact of undocumented immigration status on educational experiences. She utilizes participatory action research as a framework to conduct studies with immigrant-origin emerging adults. Her research agenda is guided by a commitment to socially meaningful issues that face immigrant-origin adolescents and emerging adults.

Su Yeong Kim, Ph.D. is a Professor of Human Development and Family Sciences at the University of Texas at Austin. She examines the role of cultural and family contexts in shaping the development of adolescents in immigrant and minority families in the USA. Her program of research has been funded by the National Science Foundation, National Institutes of Health, Russell Sage Foundation, and Spencer Foundation. She is currently investigating the role of bilingualism, executive function, and health in Latinx adolescents from immigrant families.

Peter S. Y. Kim, M.S. is a dual-Ph.D. candidate in Human Development and Family Studies and Demography at Pennsylvania State University. His main research interest is the role of culture in development, an interest piqued by his Korean American upbringing. The foci of his research include (1) the influence of place and context on cultural processes, (2) family functioning, parent-adolescent relationship, and socioemotional adjustment of adolescents, (3) bicultural competence and ethnic identity, and (4) maladjustment and emergence of risk-taking behavior. His interdisciplinary training allows him to bring fresh perspectives to studying minority culture and youth development.

Lisa López Levers, Ph.D., LPCC-S, LPC, CRC, NCC is a Professor of counselor education and human development at Duquesne University. She held the Rev. Francis Philben Endowed Chair in African Studies (2012–2017) and was a Fulbright Scholar at the University of Botswana, 2003–2004. Prior to entering the academy, she worked in community mental health for 15 years, focusing primarily on trauma counseling. She has published books, scholarly chapters in textbooks, and articles in peer-reviewed journals regarding issues of relevance to African culture, trauma and disaster, disability issues, and clinical counseling.

Bronwyn Nichols Lodato is a Ph.D. candidate in the Department of Comparative Human Development at the University of Chicago. She pursues a theoretical and applied research agenda that examines adolescent development and the transition to adulthood, assessing resources in varied contexts that support diverse youth. Her current research includes a mixed methods investigation of the impact of the Great Recession on identity development and postsecondary experiences for diverse youths.

Griselda Martinez, M.A. is a doctoral candidate in the Department of Human Development and Family Studies at the Pennsylvania State University. She earned a B.A. in Psychology and Chicana/o Studies and a M.A. in General Experimental Psychology at California State University, Northridge. Her research focuses on adolescent health and development with an emphasis on Latinx adolescents. She examines the role of individual, family, and cultural level processes in relation to mental health and substance use. Her research also seeks to identify sources of resilience that contribute to adolescent health and well-being.

Stephanie Miller-Tejada, M.S. is a doctoral candidate in the Personality and Social Contexts program at the University of Michigan. Her research examines how sociocultural factors (e.g., family, neighborhoods, and sociopolitical climates) inform the ethnic-racial identity development of youth of color. Further, she is interested in how ethnic-racial messages contribute to adolescents' understanding and beliefs about racial (in)justice and systemic inequality.

Alfredo Mirandé, J.D., Ph.D. is a native of Mexico City and the father of three children. He is Distinguished Professor of Sociology and Ethnic Studies at the University of California, Riverside. He also taught at the Texas Tech University

School of Law. He received a B.S. in Social Science from Illinois State University, M.A. and Ph.D. degrees in Sociology from the University of Nebraska, and a J.D. from Stanford University. He was a National Research Council Fellow in Ethnic Studies at the University of California, Berkeley, a Rockefeller Fellow in Sociology at Stanford University, and was recently inducted into the Illinois State University Hall of Fame. He is the author of ten books and many journal articles on race, gender, and law and a practicing attorney.

Jessica P. Montoro, M.S. is a doctoral candidate in the Developmental Psychology program at the University of Michigan. Her research examines how family, school contexts, and socioeconomic factors influence Latinx adolescents' ethnic-racial identity and socialization. She is currently investigating resilience factors among Latinx adolescents and how they relate to their psychological, social, and academic adjustment.

Elias Mpofu, Ph.D., D.Ed., CRC, MAPS, FASORC is a Professor of health services research at the University of North Texas, honorary professor of clinical and rehabilitation sciences at the University of Sydney, and distinguished visiting professor of educational psychology and inclusive education at the University of Johannesburg, South Africa. He is Editor-in-Chief of the *Journal of Psychology* in Africa and associate editor of the *Australian Psychologist*. Elias has published many articles, book chapters, and books on indigenous community health, including the seminal edited book entitled, *Counseling People of African Ancestry* (Cambridge University Press, 2010).

Hadidja Nyiransekuye, Ph.D. is an Assistant Professor of Social Work at the University of North Texas. She has a M.S.W. and Ph.D. from the University of Denver, Graduate School of Social Work, and LCSW from Massachusetts. She is a native of Rwanda and a member of the Refugee Congress. Her research interests are in the areas of forced migration, issues of genocide and genocide prevention, and African indigenous healing practices.

Sita G. Patel, Ph.D. is an Associate Professor of Clinical Psychology at Palo Alto University, and the director of the Culture, Community, and Global Mental Health Research Lab. She earned her doctorate in clinical and community psychology from the University of California, Berkeley, and her bachelor's at Vassar College. Her research focuses on immigrant and refugee mental health, including risk and resilience among newcomer immigrant adolescents, as well as global mental health and international trauma.

Scott W. Plunkett, Ph.D. is a professor of psychology at California State University Northridge. His research examines neighborhood, family, peer, and school qualities related to academics and mental health of adolescents and emerging adults in disadvantaged and/or underrepresented groups (e.g., immigrant families, ethnic minorities, sexual minorities). In addition, he examines the psychometric properties of mental health scales in various groups. He also evaluates campus programs (e.g., peer and faculty mentoring, peer learning facilitators) and community programs (e.g., parenting education, community nutrition, book flood, sexuality education).

Celeste H. Poe, Ph.D., LMFT is a clinical psychology postdoctoral fellow at the Yale Child Study Center. She obtained her Ph.D. in clinical psychology from Palo Alto University, her Master's degree in clinical psychology from Pepperdine University, and her Bachelor's degree in psychology from Xavier University of Louisiana. Her clinical and research interests include the treatment of acute, chronic, and intergenerational trauma in ethnically diverse children and families, with specific interests in early childhood, maternal mental health, and pediatric behavioral health.

Desiree B. Qin, Ed.D. is Associate Professor of Human Development and Family Studies at Michigan State University. Her research, funded by the William T Grant Foundation and the Spencer Foundation, focuses on understanding how immigration, culture, gender, and ecological contexts, especially family, impact adolescent and emerging adult development. Drawing on mostly qualitative, longitudinal data, her research has highlighted many nuanced, complex family processes that have been overlooked in quantitative work on Asian immigrant families, e.g., emotional alienation, parent-child conflicts, communication challenges, and parent-child separation. Her findings also point to negative effects of tiger parenting in child/adolescent development. Her most recent research interests include non-self, mindfulness meditation, culture, and well-being.

Stephen M. Quintana, Ph.D. is a Professor and Chair in the Department of Counseling Psychology and lead partner in the Wisconsin Partnership Program Grant with Centro Hispano. He received his M.A., Ph.D. from the University of Notre Dame, completed his predoctoral internship at Pennsylvania State University, and his B.A. from Carleton College, Northfield Minnesota. He received a Ford Foundation postdoctoral fellowship as well as Gimbel Child and Family Scholar Award. His research investigates development of children's and youth's understanding of social status, with more recent investigation into the enactment and performance of identity in social and public spaces.

Vaishali V. Raval, Ph.D. is Professor of psychology and affiliate of global and intercultural studies at Miami University. Her major program of research focuses on cultural and contextual foundations of parenting and child and adolescent mental health among Asians and Asian Americans. Emerging lines of work include culturally competent and evidence-based mental health care for diverse societies and factors that facilitate and hinder mental health of diverse racial and ethnic groups in the USA.

Deborah Rivas-Drake, Ph.D. is a Professor of Education and Psychology at the University of Michigan. The overarching goal of her work is to illuminate promising practices that disrupt racism and xenophobia and help set diverse young people on trajectories of positive contribution to their schools and communities. Together with the Contexts of Academic + Socioemotional Adjustment (CASA) Lab (casalab.org), she examines how school, peer, family, and community settings can support adolescents in navigating issues related to race and ethnicity, and how these experiences inform young people's academic, socioemotional, and civic development.

Constance Rose, M.A. is a fourth-year Ph.D. candidate attending Palo Alto University's Clinical Psychology Program. She obtained her Master of Science degree from Palo Alto University, her Master of Arts degree from Teachers College, Columbia University, and her Bachelor of Science degree from the Pennsylvania State University. Her research interests include domestic/intimate partner violence, intergenerational trauma, police brutality, immigration, LGBTQ, and Black mental health. Her clinical work involves aiding minoritized and marginalized populations dealing with a wide variety of comorbid medical and mental health diagnoses including serious mental illness. She aims to continue her research and clinical work both within California and within her hometown of New York City.

Yishan Shen, Ph.D. is an Assistant Professor of Human Development and Family Sciences at the School of Family and Consumer Sciences at Texas State University. She examines the roles of intersecting contexts (e.g., social class, culture, immigration) in adolescents' socioemotional, behavioral, and academic well-being. In collaboration with colleagues, she is currently conducting two research projects: one project investigates the roles of social positions and national contexts in the development of ethnic Korean youth across the United States, China, and South Korea, while the other project investigates the roles of everyday family and school experiences in predicting Hispanic/Latinx adolescents' academic outcomes.

Margaret B. Spencer, Ph.D. is the Charles L. Grey Distinguished Service Professor in Comparative Human Development and the College Marshall Field IV Professor of Urban Education and Life Course Development at the University of Chicago. Her Phenomenological Variant of Ecological Systems Theory (P-VEST) guides her scholarship and provides an identity-focused cultural ecological perspective, which frames her gender, race, culture, and context-acknowledging program of human development research. Recognizing the universality of human vulnerability, the theory addresses resiliency, identity, and competence formation processes for diverse humans situated both in the USA and abroad.

Sailaja Subramaniam, Ph.D. is an Assistant Professor in the Couple and Family Therapy program in the Department of Human Development and Family Studies at Michigan State University. She is also a licensed Marriage and Family Therapist and an Approved Supervisor with the American Association of Marriage and Family Therapy. Her teaching and research interests are primarily focused on the intersections of immigration, ethnicity, aging and health, trauma, and resilience in families.

Lester Sim, M.Sc. is a psychology doctoral student in the Personality and Social Contexts area at the University of Michigan and an Overseas Ph.D. Scholar with Singapore Management University. He is particularly interested in studying close relationships, examining cross-cultural differences, and understanding how one's identity may influence these processes. He examines romantic and family relationship processes through a variety of dyadic, biological, longitudinal, and multimethod approaches.

Barbara Thelamour, Ph.D. is an Assistant Professor of Psychology at Swarthmore College where she runs the Identity, Culture, and Immigration Lab. She received her Ph.D. from Michigan State University in Educational Psychology. Her mixed methods research focuses on the socialization process and identity development of Black immigrant youth, particularly in relation to Black American culture as the receiving culture. In her second line of research, the emphasis is on the educational experiences and outcomes of immigrants and other students of color. Across these research endeavors, she has highlighted how relationships with others, particularly parents, peers, and teachers, facilitate or hinder their identities, acculturation, and learning processes.

Numan Turan, Ph.D. is a Lecturer of Education Sciences department at Istanbul Medeniyet University in Turkey. He earned a Ph.D. in Counseling Psychology from the University of Wisconsin—Madison and a postgraduate degree from Middle East Technical University, Turkey. He conducts research in adult attachment development, psychological adjustment of displaced persons, self-determination, and attitudes toward seeking psychological services.

Melissa M. Vélez, B.S. is a doctoral candidate in Counseling Psychology at the University of Minnesota. She examines how familial and sociocultural influences, such as cultural socialization and ethnic identity development, impact the mental health of racial and ethnic minority adolescents and young adults. She is currently exploring other cultural values and assets present among individuals of Mexican origin and/or Mexican descent to be able to develop cultural adaptations. She is working on creating a bilingual measure of a cultural value of creativity among Mexican/Mexican Americans.

Dawn P. Witherspoon, Ph.D. is the McCourteny Family Early Career Professor of Psychology at the Pennsylvania State University and Director of Parents And Children Together (PACT), a university-community partnership to enhance the well-being of diverse families in urban central Pennsylvania. Her research focuses on contextual influences of families' and youth behavior, particularly focusing on how neighborhood, family, and race/ethnicity-related factors affect adolescents' well-being in multiple domains. Much of her research focuses on the neighborhood context and how it shapes cultural experiences and identifies positive characteristics in multiple contexts that are related to adolescent well-being.

Junghee Yoon, Ph.D. is a postdoctoral researcher in the Department of Human Development and Family Studies (HDFS) at Michigan State University (MSU). She completed her doctoral work in HDFS at MSU in 2018. With a special focus on racial/ethnic minority youth and young adults, her research explores developmental processes of identity formation and purpose development during adolescence through to emerging adulthood in multicultural setting in the USA. At MSU, she engages in several research projects investigating processes of resilience, acculturation, identity re/formation, and purpose development focusing on South Sudanese refugee young adults, African American and Latinx female college students, and Asian/Asian American students from immigrant families.

Minyu Zhang, Ph.D. is a postdoctoral research associate of the Population Research Center, University of Texas at Austin. She examines how general and sociocultural stressors, such as economic stress, discrimination, and language brokering stress, impact ethnic minority adolescents' sleep, health, and psychosocial well-being. Collaborating with colleagues, she is currently working on three projects: (1) testing the potential phonemic discrimination in monolingual versus bilingual infants, (2) examining the relations among cultural stress, hope, mental health, and substance use with a sample of recent Latinx immigrant adolescents, and (3) studying the semantic memory and lexical access among a group of bilingual Latinx college students.

Chapter 1
Identity Formation and Re-Formation: Resistance, Growth, and Emerging Pathways for Immigrant Youth and their Families

Deborah J. Johnson, Susan S. Chuang, and Jenny Glozman

Nascent in our conception of immigrant identity are the crucial concepts of resistance and novel pathways. These concepts are integral to a framework focused on the dynamic and nuanced contextual influences, processes, and unacknowledged evolutions of immigrant identity development; particularly among those who must negotiate race, ethnicity, ethno-religion, and color in new ways. Of particular interest are the immigrant youth and families, many who originated in nations with poorer economies or are considered "of color" or "not white" by others within Western cognitive dimensions, and all that this entails in their acculturative process.

We also deeply consider the appropriate applications of ethnic-racial identity theories (e.g., Cross Jr, 1991; Fine, 1994; Phinney, 1996; Suárez-Orozco, 2004; Tajfel & Turner, 1986) and thus recognize that the experience of immigrant youth and families is not just one of identity formation but one of reformation of previously developed ethnic-racial and cultural identities that make use of knowledge and socialization from both origin and new homelands. *Identity reformation* refers to the numerous and dynamic transactions taking place to inform and reshape identity and self-systems in relation to new information ecologically and culturally generated in relational and community contexts. Is it not reform as in passive "give one's self over," but in selective adjustment judged best for managing those contexts and self-systems. We insist that the immigrant or "new shore" experience (Chuang & Moreno, 2008, 2011) is one of resilience, not only adaptation or acculturation to,

D. J. Johnson (✉)
Michigan State University, East Lansing, MI, USA
e-mail: john1442@msu.edu

S. S. Chuang
University of Guelph, Guelph, ON, Canada
e-mail: schuang@uoguelph.ca

J. Glozman
The Calm Collective Psychotherapy Clinic, Toronto, ON, Canada
e-mail: jenny@thecalmcollective.ca

D. J. Johnson et al. (eds.), *Re/Formation and Identity*, Advances in Immigrant Family Research, https://doi.org/10.1007/978-3-030-86426-2_1

but the development and use of new skills and identity-centered strategies to make sense of novel environments. These processes and pathways allow the formation of meanings not previously existing at the individual or the family level or as a member of the origin society.

As co-editors, our own positionalities shape our interpretive lens, as well as how we have situated and organized the book. Two of us are insiders of the immigrant experience, one first-generation, another as second-generation born of immigrants, and the third an outsider with no recent immigrant experience. The first-generation, immigrated from two countries before settling in Canada. We are African American, Taiwanese Canadian, and European Russian Canadian. Each of us has worked extensively with refugee and immigrant populations in research and public service for the balance of our careers. All of us are developmental psychologists from departments and programs where child development and ecological perspectives are prominent. We clearly view systems as predominant influencers on identity development (Bronfenbrenner, 1979, 1995; Spencer et al., 1997).

It is important to note that in this volume, readers will encounter a range of terms used for origin, immigrant, and second-generation groups to depict national origin, ethnicity/race, transitional identities, and adopted identities (e.g., Latinx, Latino/a, Latine, Mexican American, Mexican, Puerto Rican, African American, Black American, BIPOC, Caribbean American, Nigerian, Asian, Asian American, Asian Indian, Chinese, Taiwanese). We chose not to standardize this language because it is dynamic in nature, and the terms reflect the complexities of cultural communities and the self-determination of those communities. Agreeing on specific terms can sometime reduce the richness of evolving accepted terms and does not approximate agreement. Racial-ethnic terms within the same cultural community may lack correspondence depending on origin, age, and geography (northern, southern, western, eastern, coastal, or mountain regions), even to the level of neighborhoods. Within each generation and between generations, meaning around race, ethnicity, and national origin can shift dramatically or subtly. The collection of terms and how they are employed across chapters captures a critical range of this dynamic process essential to immigrant adjustment and identity construction.

Social Historical Context

The journey of this volume began several years ago under the historical retrograde of immigrant policy of the administration of the 45th President of the United States of American (USA). Despite the numbers that exist, this period has been referred to as the harshest climate for migrants in a generation (Edyburn & Meek, 2021). Anti-immigrant sentiment has long plagued Western society and influenced both assimilation and immigrant identity patterns (Massey & Sanchez, 2010) with its most consequential impact on youth and their families and their respective future trajectories. Each chapter offered in this volume was written amidst the backdrop of xenophobic hegemony and threats, even terrorism. While this volume landed in a particular moment and historical period, it now emerges in a new era where many of these policies and enactments (travel bans against Muslim nations, threats to Deferred Actions for Children Arrivals (DACA) programs) were rescinded by the Biden administration that followed and within its first 100 days. Despite these advancements, the residual of acculturative processes begun under deep anti-immigrant attitudes linger, and the future immigration climate is uncertain. Public opinion largely supports immigration reform but does not address the specifics of that reform, contributing further to the uncertainty that exists.

While awaiting progress in immigration approaches and improvement in human rights conditions for newcomers, first- and second-generation citizens, this volume's emergence now sits at a powerful crossroads. The immigration experiences and all the identity and acculturative processes associated, shape families and who their youth become, how they adjust and adapt, or not, into the fabric of western receiving nations. These ecological issues represent a juncture in the thinking on identity formation, as well as social and historical contextual influences on immigrant children, youth, and families.

The work presented here underscores ecological, intersectional, developmental, and contextually based growth and transformation influences of immigrant and refugee families from a plethora of homelands. Counter to many assimilationist perspectives (Gordon, 1964; Portes & Rumbaut, 1996), immigrant youth and family processes are multidirectional in the formation and reformation of identity, propelled forward by much of their own agency. Latina/o communities are particularly vulnerable to anti-immigrant sentiment across multiple peaks in Western history. In addition, and more recently during the COVID pandemic, Asian immigrant groups have endured heightened negative reactive sentiment, and even violence. Reported incidences swelled from about 2600 in 2019–2020 to over 3795 incidences (March 2020–February 2021) in the USA. (Jeung et al., 2021).

Moreover, we particularly attended to African diasporic communities to aid in correcting the invisibility within the current literature and the understanding of their unique experiences. Ethno-religious immigrants (i.e., Muslim, Sikh) suffer many of the same vehemence directed at these other groups and are vulnerable like other immigrants of color for their visible markers of identity (clothing, hair coverings, etc.) (Rana et al., 2019). Black immigrants rarely enjoy full societal integration, and their identities are not only shaped by this exclusion, but by the "outsiderness" and rejection that often continues into the next generation (Massey & Sanchez, 2010).

Recent Demographic Patterns: Canada and the USA

Immigration patterns in the USA and Canada have long historical ebbs and flows of open and closed immigration policies and racialized patterns of both exclusion and opportunity. Current patterns underscore exceptionally diverse immigrant populations in these two nations. For example, immigration patterns globally are among the highest in the USA and Canada. In 2019, an historic high of 44.9 million immigrants resided in the USA (13.7% of the population). American immigrants are among the most diverse nationally and ethnically. The greater portion of these newcomers are Spanish speakers at nearly 61%, the next largest groups speak languages such as Mandarin, Tagalog, and Vietnamese and to the surprise of some, African diaspora group languages are well represented, Haitian French Creole, West African Twi/Igbo/Yoruba, and North Afro-Asian speakers of Arabic languages number above 3 million within the USA. Canada welcomes more than 300,000 new immigrants each year. About eight million immigrants with permanent residence were living in Canada in 2019—nearly 21.5% of the total Canadian population, second largest immigrant population in the world (Statista, 2019). Children represent a significant portion of migrants in Canada and the USA.

Refugees are a significant subgroup of immigrants to the USA. In 2019, forced displacement took place less than every two seconds, 79.5 million migrants experienced forced migration, up from 68 million in 2017, many fleeing violence and political assault in their home countries (Albu, UNHCR Global Trends Report, 2019). Children and youth represent 40% of that number, about 30–34 million. The USA admitted 29,916 refugees in 2019, a 34% increase from the 22,500 refugees admitted in 2018 (Baugh, Annual Flow Report: Refugees and Asylees, Sept, 2020), a minute fraction of the roaming migrants of the world.

Racial and Ethnic Identity

The critical components of our self-systems include identity. The complex transactions between multiple contexts and the individual contribute to their development and adjustment over time. Accordingly, Quintana and Scull (2009) argue that ethnic identity development is a component of the acculturative process among immigrant groups as well. Racial-ethnic identity becomes a prophylactic imperative in American culture against xenophobia and hierarchies of values associated with some identities over others. Not all identities or identity associations are available to all immigrant groups, and the constraints are learned quickly (Quintana & Scull, 2009; Thelamour & Johnson, 2017). The values placed on those identities and their inherent limiting properties become multicultural forces that influence the development and direction of evolving identities. Of importance to the diverse populations in this volume is the explicit focus on support for formation of both heritage and destination-country identities and the ongoing negotiation of these identities within family, national, and community contexts. (Schwartz et al., 2018).

Acculturation, Assimilation, and Identity

What can researchers learn about the two poles of the acculturation continuum as well as the middle? Berry (1997, 2007) developed one of the most influential and commonly used models, the Bidimensional Acculturation Framework. Berry's models (1997, 2007) put forth a bipolar continuum conceptualized with each end representing one of two extremes, termed marginality and assimilation. The first dimension pertains to the extent to which immigrants adopt the host culture and the second dimension refers to the extent to which they preserve their heritage culture. Marginality, referred to the experience of individuals who maintained strong ties with their heritage culture and rejected the receiving culture. In contrast, assimilation, referred to the experience of individuals who created strong ties with the receiving culture and foreclosed on their heritage culture. It was assumed that assimilation was the ultimate goal leading to optimum adjustment including success outcomes. Within these assimilationist frameworks, the bicultural center of this continuum was not seen as a point of its own, but merely as a transitional state. This focus on the benefits of assimilation was rooted in a strong belief in the superiority of the mainstream host culture over the heritage cultures of most immigrants (Buriel & De Ment, 1997; Buriel et al. 2006).

More recently, researchers have fervently challenged and revised the multidimensionality perspectives of acculturation to reflect the complexities inherent in individuals' change and transformation (Chuang, 2019). For example, Ferguson and Bornstein (2014) proposed the compelling argument that the dynamic, multidimensional, and multipurposed identity is not solely for acculturation (Ferguson & Bornstein, 2014; Ferguson, Bornstein, & Pottinger, 2012). Bolstering these explorations is critical for setting a future course for identity theory and research. As Nguyen and Ferguson (2019) reported, racial, ethnic, and national identities that fused land of origin and receiving cultures, creating a fourth beneficial "hiphop" identity for youth. In this hiphop identity, youth were able to express efficacy and confidence more consistent with global youth culture borrowing from the bravado of African American youth culture wherein hiphop originates. These developments supported self-esteem, neighborhood and school integration, and were noted as a form of resistance in the new setting. This was critical as other identities associated with national origin or cultural origin more often constrained such features of identity (Nguyen & Ferguson, 2019). This work is one example of future formulations and what it may mean to embrace the concept of immigrant identity reformation.

Following resettlement on new shores in Canada and in the USA, immigrant and refugee youth and families often face new challenges and stressors that significantly impact a range of health outcomes. Changes in these and other outcomes may also force reorganization and expansion of their identities with the added potential for new modes and contexts of thriving. While this volume provides research and theoretical perspectives consistently grounded in the literature on immigrant identity and acculturation, most chapters break the boundaries of traditional approaches rooted in segmented assimilation theory (Portes & Zhou, 1993; 2005) and the narrow confines of bifurcated theories of immigrant identity (e.g., Berry et al., 2006; Berry & Sabatier, 2011) and definitions of health adjustment. Rather, many of the chapters presented here consider the multiplicity and fluidity of identity alignment, as well as purpose of identity development and dynamic reformation. From the ecological perspective the notion of place or geographic location as a critical identity anchor provides another mechanism for resistance to reformation. Under some circumstances, researchers have found that family socialization practices are oriented toward retaining strong identity and ties to the place of origin while embracing some necessary adaptations in the new homeland and arresting full assimilation. Several chapters suggest that this approach supports success and positive development. Nevertheless, success may come at the price of increased isolation as integration increases. In one study, Chinese immigrant students were found to do well academically but suffer from mental health challenges derived from stress and isolation (Qin, 2008). Many families understand the unspoken demand of western societies that they assimilate and find ways to make accommodations that often lead to breakdowns in cohesion and intergenerational cultural learning. As parents, Sudanese immigrant refugees emerging adults promoted their children's Americanization, sometimes against values of collectivism, language, and cultural understanding (Johnson et al., 2014; Luster et al., 2009). Oscar Lewis (1959) more than 60 years ago reported similar dilemmas among Mexican American families making sacrifices of their family systems at the cost of hard-won assimilation of their children.

These formulations may be associated with poor acculturation. However, contributors in this volume applying new ideas and formulations argue that it is not poor acculturation. Rather, these are conscious decisions about aligning with a set of values judged to be to those available in the new U.S. home setting. Moreover, collectively, these chapters question and perhaps redefine optimal and healthy acculturation in socially dynamic societies.

Learning from the Identity Framework

This volume intentionally focuses from an inclusive significant place for non-European, non-Western experiences of immigration upon which immigrant identities theories were not conceptualized and developed. While having some validity, more conventional theories clearly are not sufficiently flexible to encompass the full range of experiences among immigrant youth and families to illuminate their identity, socialization, and acculturation processes.

To date, several critical issues have been understated or overlooked in the literature. What has not been captured previously are: (1) standing ground in identity as a strength, (2) iterative identity processes, remaking of individuals from the culture of origin, (3) reformation is adaptation and not only accommodation as previously implied in the literature, (4) exploring the factors that fit, are maintained, and developed into new identity streams for adjustment and resilience, (5) moving beyond unidirectional identity reformation process to transactional process and is a continuous adaptive process (identity formation and reformation), (6) understanding how misidentification fits into immigrant identity experience, and (7) the narrow and repressive orientation of assimilationist orientations that impact the theoretical considerations in the field. It is necessary to reclaim the concept of reformation as action prompted by personal or family agency. Many of the volume's chapters move in these directions, interweaving acts or orientations of resistance and resilience that are necessary to succeed in adjustment.

The Organization and Content of this Volume

In this volume, we explore identity as it is shaped and reshaped by "new shore" experiences. The chapters address the demands on youth and their families, ecological and inter/intra-identity and acculturative processes, as well the pathways to positive development. There are five segments in this volume that provide a deep exploration into identity reformation among immigrant groups. While we address the common experiences of identity development among immigrant groups as a whole, we also elevate the experiences that vary among these populations.

Each section shares multiple cultural viewpoints on both unique and similar experiences. We begin where the current theory exists, in Part A: *Acculturation, Multidimensional Identity, and Family Socialization Processes*, where the critical connections among these concepts are advanced. Part B: *Language, Language Role Identity, and Literacy* underscores that for most groups, language is a key identity element and one upon which acculturation rests. From an ecological perspective, place is an important context of identity formation and also can form the basis for resistance to traditional notions of the "model" or "good" immigrant. Thus, Part C: *Transnational Place and Identity: Resistant Re/Formation* addresses these ideas. The functional purpose that identity and adjustment serve for immigrant families and their children are discussed in Part D: *Identity as a Protective Factor: Coping with Depression, Distress, and Adjustment Stress*. These chapters explore whether identity is protective for these groups. The final section, Part E: *Positive Youth and Emerging Adult Development*, delves further into the complexities of harnessing identity for positive development across a variety of national settings. Though organized into clear segments, many of our chapters have content that speaks to other segments of the book. Thus, this book should be taken as an elegant dinner served in courses, each course can be savored on its own, while the full course is to be relished for all its notes and flavors.

Acculturation, Multidimensional Identity, and Family Socialization Processes

The interconnections among acculturative, socialization, and identity formation processes are underscored in the diverse immigrant experiences of youth, emerging adults and their families in this opening collection of the volume. Single pathways or even dual pathways are shown to be antiquated understandings of the experiences and identity transformations occurring within contemporary immigrant families. Here, we are able to address many of the ecological factors central to immigrant identity development.

While there can be stressful or even negative forces that produce such development, researchers have found that family closeness and love are pathways leading to positive strengths and outcomes among first-and second-generation offspring. Focusing on families, South Asian youth develop their identities within the context of strong family socialization experiences and practices. Subramaniam and Carolan (Chap. 3) demonstrate the complex interplay of the rapid acculturation of grandchildren and the steadfast cultural anchoring of their immigrant Indian grandparents. Though cultural imperatives and demands of grandparents appear unidirectional, there is evidence that reformation, identity shifts, and accommodations take place among all generations of the family system. Katsiaficus and Chung (Chap. 5) expound on the critical nature of families' socialization, obligation, and identity shaping function. Among the Latinx youth in this study, the mutual obligation and identity using the hyphenated framework are discussed. Mutually supportive obligations provided youth with new opportunities to contribute to family transactions that round out their cultural and personal identities.

Understanding and accepting obligations are a harmonious aspect of how families shape identity. Kalia, Aggarwal, and Raval (Chap. 6) describe a range of processes influencing identity development and experiences. Where South Asian emerging adults saw conflict between the two cultures through the lens of their families seems to challenge their academic and well-being outcomes. The ability to find bicultural harmony and the absence of perceptions of conflict allowed for the balance or melding and led to greater adjustment. All occurred in a dynamic reformation of identity that required not only an integration of family processes but socio-cultural processes.

Glozman and Chuang (Chap. 2) and Yoon, Johnson, Qin, and Bates (Chap. 4) both address the inadequacy of uni- or bidimensional theory to explain the complex, contextually influenced, dynamic and responsive nature of Sudanese refugee and Russian-Jewish immigrant identity formations. Yoon et al. (Chap. 4) demonstrate that emerging identity dimensions of South Sudanese refugee young adults form out of necessity where identification and misidentifications derive from contextually based social cognitions of identity. To a South Sudanese refugee, "African" as an identity is a new dimension of identity born of the lack of knowledge among others of national origins on the continent and as such it become functional to make meaning of this externally driven identity. "Sudanese American" is the term used to describe adaptation for both bicultural identity meaning and hybrid identity concepts. The chapter goes beyond segmented theories of assimilation and following tridimensional acculturation theory adopts a quad-dimensional perspective on African diaspora immigrant identity theory.

In Glozman and Chuang's (Chap. 2) study on Russian-Speaking immigrant youth, Whiteness, and the appearance of Whiteness afford choices that are typically unavailable to other immigrant groups. Identity expression and revealing identity were determined by critical factors interwoven with religion, and the desire to avoid prejudice and bias. Some decisions were influenced by factors centered around enhancing belongingness. Still, these factors are not addressed in formulations less considerate of diversity among immigrant acculturative and identity experiences. In a later segment on space and place, similar issues among Israeli and Jewish identified groups are refined even more. As Gold (Chap. 11) revealed, among immigrant subgroups some maintain identities strongly tied to national or cultural origins while taking full advantage of educational and economic opportunities available on U.S. shores.

Language, Language Role Identity, and Literacy

Among immigrant families, language, language usage, as well as the language role identities of children and youth are among key factors associated with adjustment and well-being. Yet language and literacy are intimately intertwined with evolving immigrant and ethnic-racial/cultural identity. What are the family and contextual dynamics of these identity formations? For youth with the responsibility of navigating language as a family broker across a range of contexts brokering is adultifying. The role is embedded in the family dynamic contributing to identity formation or added to identity as part of a reformation process. Simultaneously burden, family obligation and competency contributes to a sense of efficacy (Chen et al., 2020).

Investigating how youth develop and embrace language brokering as part of their identities and responsibilities, Sim, Kim, Velez, Shen, and Shang (Chap. 8) address parenting practices and parent socialization as influences on the formation of their early adolescents' language broker identities among Mexican origin youth. Parent identity reformations are also prominent in the shifts and outcomes of children's identity roles.

Two qualitative studies explore the use of language in the personal narrative of the immigrant journey. Language fuels the formation of, brings meaning to, and further shapes identity during adjustment. Quintana, Benjamin, and Turan (Chap. 7) apply a cluster analysis approach to describe the power of journey narratives among immigrant youth. Family socialization and cohesiveness are among the themes that contribute to the four typologies of the journey narratives. Quintana et al. (Chap. 7) critically point out that the narratives that immigrant identities and journey are not monolithic and do not reside in a single theoretical home. Thelamour (Chap. 9) addresses yet another aspect of language and identity among immigrants of

Caribbean and African descent who are often tasked with demonstrating alignment with African American or mainstream community. There are many drivers in the environment that impact these experiences and decisions. Adoption of the Black English (BAE) form appeared to indicate African American identity alignment.

Transnational Place and Identity: Resistant Re/Formation

Another feature of theory challenged within the volume is one of "broken ties to homelands." Traditional theory (sometimes known as "straightline" theory) would assert that the assimilation process transforms identity slowly where the immigrant person or family gradually internalizes the new culture and is accepted into the "appropriate" spaces. Gold (Chap. 11) and Mirandé (Chap. 10) both demonstrate that transnational and binational identities occur and adjustment on new shores can exist without the nullification of homeland identity and culture. In his discussion of Jewish and Israeli immigrants, Gold demonstrates that motivations for immigration to the USA are impacted by nuanced histories of these groups. These histories also change the ease of acceptance and resistance to identity reformation and the acculturative process, as in "I live here but I am not of here." Gold indicates that Israelis rarely assimilate but are successful within the society. Jewish emigrants or native-born Jewish people from other Eastern European countries tend to adopt the culture and identify as Americans. Israeli participants were shown to selectively enjoy the abundance of one nation without rejecting the values and cultural meaning at the center of their origin culture along with resolve to resist "truly becoming American."

Mirandé (Chap. 10) depicts families from a small village, Jalos, in South America where families foster and reinforce the values of and identity with the homeland community over the experiences and identities developing in the USA. These are buttressed not only by socialization strategies and parenting values but also by frequent return visits and encouraging marriage from the place of origin as youth grow into adulthood. These examples underscore the agency of immigrant groups to make clear choices about where their identities reside even while selecting the most advantageous experiences and resources of the receiving country. Their work underscores resistance interfacing with adjustment and resistance to acculturative pressures while opting for opportunities provided by homeland contexts.

Identity as a Protective Factor: Coping with Depression, Distress, and Adjustment Stress

In this section the contributors address the question of how identity may serve as protection from the psychological stressors of discrimination and misidentification. Bamaca-Colbert et al. (Chap. 13) and Banerjee and Plunkett (Chap.14) each address these stress factors as either protective or facilitative of increased mental health consequences. In the longitudinal study the increase in discrimination lowers identity affirmation and leads to increased depression in subsequent years, confirming the reject/identification hypothesis. While Banerjee and Plunkett confirm the protective function of racial-ethnic identity affirmation on psychological distress, they also provide some new domains of stress, romantic attachment distress in a Latinx sample. These studies question the short- and longer-term role of identity and whether buffering effects can weaken over time.

In their study of the acculturation gap or the disconnect between parent and youth generations Montoro, Miller-Tejada, Cross, and Rivas-Drake (Chap. 15) found no psychological consequences for parent and youth dissimilarity as youth align with the new culture. However, diminishment of family centered identities had consequences on the family system, reduced cohesion and communication among Latinx families. This finding was more pronounced among second-generation youth. Gender and identity reconstruction varied for girls of Mexican origin. Ethnic-racial identity emerges as stress-reducer but does not function alone. As alluded to earlier in our discussion of South Sudanese (Johnson et al., 2014) and Mexican origin families of Lewis' (1959) ethnography, whether the family devolution as an acceptable consequence of identity reformation is unclear.

Racial identity and cultural identities as indicated by the literature (Banks & Kohn-Wood, 2007; Mandara et al., 2009) are largely protective against mental health consequences and heightened distress. In their study of first- and second-generation immigrant youth, Clarke, Poe, Rose, and Patel (Chap. 12) describe misidentification as a microaggression leading to psychological distress and isolation. Misidentification refers to labelling as American racially without reference to cultural identity (i.e., labelling a Nigerian immigrant African American without the deeper reference to their known heritage and associated meanings). And alternatively, mislabeling cultural or national identities (i.e., mistaking an Iraqi immigrant or Puerto Rican as Mexican American). In this qualitative assessment for the former definition, they assert that integrated identities offer the most protections against psychological distress and misidentification can be incorporated as an extension of individual racial identity formation to aid in the stress reduction among African immigrants.

This group of studies suggest that identity reconstruction offers some protections but not in all areas and perhaps not for the long term. Are the tradeoffs on hard-won protections worth vulnerabilities in other systems that children and families need for the long term? Finally, these remain dynamic processes that could reform identity, yet again, in adulthood and perhaps with fewer psychological consequences.

Positive Youth and Emerging Adult Development

Positive identity development draws upon the edicts of positive youth development (PYD). PYD has its origins in positive psychology, or the study of individual, interpersonal, and environmental characteristics that enable individuals to thrive (Seligman et al., 2005), and signifies a movement towards balancing a literature that has often dwelled on pathology (Johnson, Thelamour, Sankar & Dimitrova, 2021). Instead, positive psychology emphasizes potential over suffering and seeks to understand well-being and negative outcomes (Peterson & Park, 2003). PYD is a strengths-based perspective and these chapters emphasize or in part, "turn on its head," the notion of immigrant identity development as problematic or deficit riddled. Reformation, in this instance, is associated with growth and change, as well as the positive "super-powers" that youth gain through their identity transformations that benefit and promote their development.

Strengths emerge both from the strivings and contextual supports youth discover in their environments. Attempting to harness the processes and information that make this possible for youth is the focus of all three chapters. Moreover, how PYD occurs in immigrants of African origin has not been a readily explored topic. Mpofu, Nyiransekuse, and Levers, (Chap. 18) explore youths' attempts to make meaning of their environment, the supports that exist and the tasks required to adjust successfully, often resulting in bicultural (fluid movement between two cultural identities) or what are referred to as hybrid identities (a merged or integrated identity composed of both cultures). Youth attempting to perform these developmental tasks in Australia can be more complex given smaller demographics and the racialized contexts of school and community.

Rather than criticize that segmented assimilation theory is oversimplified, particularly among multiracial multinational immigrants where their rich and complex experiences are unaccounted for including the curviness of economic history, Fajardo, Lodato, Spencer, and Graziul (Chap. 16) move away from an assimilationist argument. Instead Fajardo et al. apply phenomenological approaches that present human diversity and vulnerability as foundational. Positive development then emerges from exploration of phenomenology in PVEST theory, and immigrant identity theory. This chapter elevates an often ignored element of immigrant identity, with the consideration meaning around colorism as part of the disparity experiences and also concomitant identity transformations.

As religiously minoritized individuals, positive identity development among Muslim youth is a unique perspective of this volume, and a much-needed view of youth often demonized, "DeAmericanized," and disenfranchised socially in the USA and many Western cultures. Abo-Zena (Chap. 17) makes an interesting argument that reformation and adjustment are about accommodation of negative experiences. Identity functions "in spite of" active resistance at the societal level. Abo-Zena's critical and thoughtful exploration of Muslim youth's identity reformation processes is eye-opening and captures the essence of what this volume is all about.

Conclusion

This volume is a gateway to advanced theorizing and research on identity formation and the continual transactional process of adjustment and reformation, through the lens of developing immigration, race, and cultural experiences. Who must we become to thrive on new shores? And by what mechanisms do we arrive at a new identity destination? The chapters are windows into the many facets of this process and the incredible journeys of first- and second-generation youth and their families. Each chapter contributes to the story of innovation, new considerations, and challenges to current perspectives on immigrant identity development and change.

References

Albu, D. (2019). UNHCR Global Trends Report: Forced displacement in 2018. *Drepturile Omului*, 114

Banks, K. H., & Kohn-Wood, L. P. (2007). The influence of racial identity profiles on the relationship between racial discrimination and depressive symptoms. *Journal of Black Psychology, 33*(3), 331–354. https://doi.org/10.1177/0095798407302540

Baugh, R. (2020). *Refugees and asylees: 2019*. Annual Flow Report, September 2020. Homeland Security: Office of Immigrant Statistics. https://www.dhs.gov/sites/default/files/publications/immigration-statistics/yearbook/2019/refugee_and_asylee_2019.pdf

Berry, J. W. (1997). Immigration, acculturation, and adaptation. *Applied Psychology, 46*(1), 5–34. https://doi.org/10.1111/j.1464-0597.1997.tb01087.x

Berry, J. W. (2007). Acculturation strategies and adaptation. In J. E. Lansford, K. Deater-Deckard, & M. H. Bornstein (Eds.), *Immigrant families in contemporary society* (pp. 69–82). Guilford Press.

Berry, J. W., Phinney, J. S., Sam, D. L., & Vedder, P. (2006). Immigrant youth: Acculturation, identity, and adaptation. *Applied Psychology: An International Review, 55*, 303–332. https://doi.org/10.1111/j.1464-0597.2006.00256.x

Berry, J. W., & Sabatier, C. (2011). The acculturation and adaptation of second-generation immigrant youth in Toronto and Montreal. In S. S. Chuang & R. P. Moreno (Eds.), *Immigrant children: Change, adaptation, and cultural transformation* (pp. 125–148). Lexington Books.

Bronfenbrenner, U. (1979). *The ecology of human development: Experiments by nature and design*. Harvard University Press.

Bronfenbrenner, U. (1995). Developmental ecology through space and time: A future perspective. In P. Moen, G. H. Elder, & K. Lüscher (Eds.), *Examining lives in context: Perspectives on the ecology of human development* (pp. 599–618). American Psychological Association.

Buriel, R., & De Ment, T. (1997). Immigration and sociocultural change in Mexican, Chinese, and Vietnamese American families. In A. Booth, A. C. Crouter, & N. S. Landale (Eds.), *Immigration and the family: Research and policy on U.S. immigrants* (pp. 165–200). Lawrence Erlbaum Associates, Inc.

Buriel, R., Love, J. A., & De Ment, T. L. (2006). The relation of language brokering to depression and parent-child bonding among Latino adolescents. In M. H. Bornstein & L. Cote (Eds.), *Acculturation and parent-child relationships: Measurement and development* (pp. 249–270). Erlbaum.

Chen, H., Brenner, A., & Kim, S. Y. (2020). Discrimination, language brokering efficiency, and academic competence among adolescent language brokers. *Journal of Adolescence, 79*, 247–257. https://doi.org/10.1016/j.adolescence.2020.01.015

Chuang, S. S. (2019). The complexities of immigration and families: Theoretical perspectives and current issues. In B. H. Fiese, M. Celano, K. Deater-Deckard, E. N. Jouriles, & M. A. Whisman (Eds.), *APA handbook in psychology. APA handbook of contemporary family psychology: Applications and broad impact of family psychology* (pp. 437–455). American Psychological Association. https://doi.org/10.1037/0000100-027

Chuang, S. S., & Moreno, R. P. (Eds.). (2008). *On new shores: Understanding immigrant fathers in North America*. Lexington Books.

Chuang, S. S., & Moreno, R. P. (Eds.). (2011). *Immigrant children: Change, adaptation, and cultural transformation*. Lexington Book.

Cross, W. E., Jr. (1991). *Shades of black: Diversity in African-American identity*. Temple University Press.

Edyburn, K. L., & Meek, S. (2021). Seeking safety and humanity in the harshest immigration climate in a generation: A review of the literature on the effects of separation and detention on migrant and asylum-seeking children and families in the United States during the Trump administration. *Social Policy Report, 34*, 1–46. https://doi.org/10.1002/2379-3988.11

Ferguson, G. M., & Bornstein, M. H. (2012). Remote acculturation: The "Americanization" of Jamaican islanders. *International Journal of Behavioral Development, 36*(3), 167–177. https://doi.org/10.1177/0165025412437066

Ferguson, G. M., Bornstein, M. H., & Pottinger, A. M. (2012). Tridimensional acculturation and adaptation among Jamaican adolescent–mother dyads in the United States. *Child development, 83*(5), 1486–1493.

Ferguson, G. M., & Bornstein, M. H. (2014). Tridimensional (3D) acculturation: Culture and adaptation of Black Caribbean immigrants in the USA. In *Global perspectives on well-being in immigrant families* (pp. 31–51). Springer, New York, NY.

Fine, M. (1994). Working the hyphens: Reinventing self and other in qualitative research. In N. K. Denzin & Y. S. Lincoln (Eds.), *Handbook of qualitative research: Handbook of qualitative research* (pp. 70–82). Sage.

Gordon, M. M. (1964). *Assimilation in American life*. Oxford University Press.

Jeung, R., Yellow Horse, A., Popovic, T., & Lim, R. (2021). *Stop the AAPI hate national report. SF State Asian American Studies, San Francisco State University*. NBC News. https://secureservercdn.net/104.238.69.231/a1w.90d.myftpupload.com/wp-content/uploads/2021/03/210312-Stop-AAPI-Hate-National-Report-.pdf

Johnson, D. J., Yoon, J., Bates, L., Rana, M., & Qin, D. B. (2014, October). Fending for our families: a case study of two Sudanese refugee mothers. Paper presented at the On New Shores Conference, Ontario, Canada.

Johnson, D. J., Thelamour, B., Sankar, S., & Radosveta, D. (2021). Reframing the narrative: Revealing positive youth development in the self descriptions of Roma adolescents. In R. Dimitrova, D. L. Sam, L. & Ferrer-Wreder (Eds.). *Roma minority youths across contexts: Taking a positive approach to research, policy and practice* (pp. 109–132). Oxford University Press.

Lewis, O. (1959). *Five families*. Basic Books.

Luster, T., Johnson, D. J., & Bates, L. (2009). "Lost Boys" finding their way: Challenges, changes and small victories of resettled Sudanese refugee youth in the United States. In R. L. Dalla, J. DeFrain, J. Johnson, & D. A. Abbott (Eds.), *Strengths and challenges of new immigrant families: Implications for research, policy, education, and human services* (pp. 265–286). Lexington Books.

Mandara, J., Gaylord-Harden, N. K., Richards, M. H., & Ragsdale, B. L. (2009). The effects of changes in racial identity and self-esteem on changes in African American adolescents' mental health. *Child Development, 80*(6), 1660–1675. https://doi.org/10.1111/j.1467-8624.2009.01360.x

Massey, D. S., & Sánchez, M. (2010). *Brokered boundaries: Immigrant identity in anti-immigrant times*. Russell Sage Foundation.

Nguyen, J., & Ferguson, G. M. (2019). A global cypher: The role of hip hop in cultural identity construction and navigation for Southeast Asian American youth. *New Directions for Child and Adolescent Development, 164*, 99–115. https://doi.org/10.1002/cad.20279

Peterson, C., & Park, N. (2003). Positive psychology as the evenhanded positive psychologist views it. *Psychological Inquiry, 14*(2), 143–147.

Phinney, J. S. (1996). Understanding ethnic diversity: The role of ethnic identity. *The American Behavioral Scientist, 40*(2), 143–152. https://doi.org/10.1177/0002764296040002005

Portes, A., & Rumbaut, R. G. (1996). *Immigrant America: A portrait* (2nd ed.). University of California Press.

Portes, A., & Zhou, M. (1993). The new second generation: Segmented assimilation and its variants. *The Annals of the American Academy of Political and Social Science, 530*, 74–96. https://doi.org/10.1177/0002716293530001006

Portes, A., & Zhou, M. (2005). *The new second generation: Segmented assimilation and its variants* (pp. 221–242). Routledge.

Qin, D. B. (2008). Doing well vs. feeling well: Understanding family dynamics and the psychological adjustment of Chinese immigrant adolescents. *Journal of Youth and Adolescence, 37*(1), 22–35. https://doi.org/10.1007/s10964-007-9220-4

Quintana, S. M., & Scull, N. C. (2009). Latino ethnic identity. In F. A. Villarruel, G. Carlo, J. M. Grau, M. Azmitia, N. J. Cabrera, & T. J. Chahin (Eds.), *Handbook of US Latino psychology: Developmental and community-based perspectives* (pp. 81-98). Los Angeles, CA: Sage Publications.

Rana, M., Johnson, D. J., & Qin, D. B. (2019). Mistaken identities, discrimination, and Sikh parents' ethnoreligious socialization strategies. In H. E. Fitzgerald, D. J. Johnson, D. B. Qin, F. A. Villarruel, & J. Norder (Eds.), *Handbook of children and prejudice* (pp. 405–420). Springer.

Schwartz, S. J., Meca, A., Cano, M. A., Lorenzo-Bianco, E. I., & Unger, B. J. (2018). Identity development in immigrant youth. *European Psychologies, 23*(4), 336–349. https://doi.org/10.1027/1016-9040/a000335

Seligman, M. E., Steen, T. A., Park, N., & Peterson, C. (2005). Positive psychology progress: Empirical validation of interventions. *American Psychologist, 60*(5), 410.

Spencer, M. B., Dupree, D., & Hartmann, T. (1997). A phenomenological variant of ecological systems theory (PVEST): A self-organization perspective in context. *Development and Psychopathology, 9*(4), 817–833. https://doi.org/10.1017/S0954579497001454

Statista. (2019). *Immigration in Canada: Statistics and facts.* https://www.statista.com/topics/2917/immigration-in-canada/

Suárez-Orozco, C. (2004). Formulating identity in a globalized world. In M. Suárez-Orozco & D. B. Qin-Hilliard (Eds.), *Globalization: Culture and education in the new millennium* (pp. 173–202). University of California Press.

Tajfel, H., & Turner, J. C. (1986). The social identity theory of intergroup behavior. In S. Worchel & W. G. Austin (Eds.), *Psychology of intergroup relation* (pp. 7–24). Hall Publishers. https://www.scirp.org/(S(351jmbntvnsjt1aadkposzje))/reference/ReferencesPapers.aspx?ReferenceID=1584694

Thelamour, B., & Johnson, D. J. (2017). Exploring Black immigrants' and nonimmigrants' understanding of "acting Black" and "acting White." *Journal of Black Psychology, 43*(3), 280–304. https://doi.org/10.1177/0095798416641863

Part I
Acculturation, Multidimensional Identity, and Family Socialization Processes

Chapter 2
A Model of the Multidimensional Acculturation of Russian-Speaking Immigrant Youth in Canada

Jenny Glozman and Susan S. Chuang

International migration continues to grow worldwide, reaching 258 million in 2017, a 17.3% increase from 2010. Canada was the eighth largest host of immigrants worldwide (United Nations, 2017). In 2017–2018, immigration accounted for 79.6% of the population growth in Canada (Statistics Canada, 2018). First-generation immigrants make up 21.9% and second-generation immigrants make up 17.4% of the Canadian population (Statistics Canada, 2017a). Among youth under 15, 37.5% are first- or second-generation immigrants and it is estimated that this proportion will continue to rise to account for 39.3 to 49.1% of Canadian youth by 2036 (Statistics Canada, 2017b). Immigration is integral to the continued economic growth of Canada and, considering the aging population and low birth rates (Immigration, Refugees and Citizenship Canada, 2018), it is important to have a greater understanding of the lives of immigrants. Of particular interest are youth because of the unique connection between their immigration experiences and their development.

Research on immigration and acculturation experiences is important throughout the lifespan, although it is of particular relevance in adolescence. Definitions of adolescence vary across cultures and there is no standard age range for defining adolescence as it is a socially constructed category (Tyyskä, 2014). In fact, chronological age is only one factor to consider; other factors include physical, psychological, cognitive, and social development. Broadly, the term adolescence captures youth between the ages of 10 and 18, using the standard of the American Psychological Association (2002). During this period, individuals in the West are usually still dependent on others, and thus their experiences of acculturation are of interest.

J. Glozman (✉)
The Calm Collective Psychotherapy Clinic, Toronto, ON, Canada
e-mail: jenny@thecalmcollective.ca

S. S. Chuang
University of Guelph, Guelph, ON, Canada
e-mail: schuang@uoguelph.ca

D. J. Johnson et al. (eds.), *Re/Formation and Identity*, Advances in Immigrant Family Research, https://doi.org/10.1007/978-3-030-86426-2_2

During the adolescent years, there is significant brain and cognitive development (Steinberg, 2005). These lead to developmental adjustments in various abilities including abstract thinking, reasoning, and information processing (Bennett, 2011; Steinberg, 2005). In turn, youth have more sophisticated categorization abilities, based not only on concrete features (physical or behavioral), but on abstract and multidimensional features such as beliefs (Bennett, 2011). Therefore, youth may struggle on a deeper level with the acculturation process as they examine and try to adapt to both the concrete and abstract aspects of their environment. These cognitive advancements then allow youth to formulate (and reformulate) their sense of identity.

For youth in Western societies, they also experience the transition from middle school to high school, which exposes them to greater diversity in social contexts, relationships, roles, etc. (Tanti et al., 2011). Moreover, during the adolescent years, youth become concerned with belonging and identity (Bennett, 2011; Erikson, 1968; Newman & Newman, 1976). The increased diversity of the social context along with the search for belonging and identity may present a particular challenge for immigrant youth who also have to acculturate to new cultural dimensions of their environment that they had not been exposed to previously.

To address these issues, the chapter discusses the development of identity and sense of belonging. We specifically focus on Russian-immigrant youth because of their specific experience being invisible immigrants from a very distinct sociocultural context. However, to place youth identity development into context, it is important to first understand the acculturation processes and pathways.

Acculturation

As individuals immigrate to a new country, they experience acculturation, which has been defined as the cultural and psychological changes that groups and their individual members undergo as they come in contact with another culture (Berry, 2007; Redfield et al., 1936). At the individual level, the process of acculturation has been termed psychological acculturation and includes a long-term process of change in one's behavioral repertoire, including customs, language, values, and lifestyle (Berry, 2007; Graves, 1976).

Unidimensional Acculturation

Historically, researchers viewed acculturation as a unidimensional process whereby individuals who came in contact with a new culture either rejected it and retained their heritage culture completely, or replaced their heritage culture with the new culture (Gordon, 1964). Thus, a bipolar continuum was conceptualized with each end representing one of these two extremes, termed marginality and assimilation.

Marginality referred to the experience of individuals who maintained strong ties with their heritage cultures and rejected the receiving culture. In contrast, assimilation referred to the experience of individuals who created strong ties with the receiving culture and abandoned their heritage culture. It was assumed that assimilation was the ultimate goal. Within these assimilationist frameworks, the bicultural center of this continuum was not seen as a point of its own, but merely as a transitional state. This focus on the benefits of assimilation was rooted in a strong belief in the superiority of the mainstream host culture over the heritage cultures of the immigrants (Buriel & De Ment, 1997; Chuang & Moreno, 2012).

Bidimensional Acculturation

Over time, researchers began to acknowledge that this unidimensional approach to acculturation was not able to capture the complexities of post-migration experiences. To account for the dual cultures and their intersections, Berry (1997, 2007) developed one of the most influential and commonly used models, the Bidimensional Acculturation Framework. According to this perspective, following immigration, individuals make decisions that vary along two dimensions. The first dimension pertains to the extent to which immigrants adopt the host culture and the second dimension refers to the extent to which they preserve their heritage culture. This led to four acculturation strategies that were based on the choices individuals made along these dimensions. Individuals in the assimilation category did not maintain their heritage culture and instead embraced the host culture. Those in the integration category maintained both cultures. Individuals in the separation category maintained their heritage culture and rejected the host culture. Finally, those in the marginalization category did not maintain either culture. The individual, the heritage culture, and the host culture all influence the choice of acculturation strategy because there is an interaction between individual orientation and cultural attitudes and policies in determining which strategy will be adopted and the consequent adaptation.

However, the bidimensional framework is limited as well because of its focus on only two cultures: heritage and host. Specifically, this model does not account for the complex identities of immigrants who may have more than one culture of origin or those who settle in countries with multiple cultures of settlement. For example, the identities of Jewish immigrants from the Former Soviet Union (FSU) were described as ethnically Russian and nationally American in past research conducted in the U.S. However, in a study with adult Jewish immigrants from the FSU, Persky and Birman (2005) found that their Jewish identity was the most salient (i.e., more so than their Russian or American identities). Similarly, Birman et al. (2010) found that Jewish youth were more likely to identify as bicultural (Russian-Jewish) or tricultural (Russian-Jewish-American) than as Russian only, unlike their non-Jewish peers. This suggests that some immigrants may have additional identities that are currently unaccounted for.

Tridimensional Acculturation

To address these limitations, Ferguson and Bornstein (2014) developed a tridimensional model of acculturation, focusing on the experiences of Black Caribbean immigrants who have acculturated to more than one destination culture within the multicultural American society. In a study with 473 Jamaican mother–adolescent dyads, there were two destination cultures that Jamaican immigrants had to orient themselves towards upon their arrival in the U.S.: African American and European American. The most common acculturation option was the endorsement of three cultures (about 40%): their heritage culture and the two destination cultures. Jamaican immigrants who were typologized into the integration profile were more likely to identify themselves as tricultural than bicultural.

Overall, Ferguson and Bornstein (2014) offered a tridimensional model that added a dimension to the experiences of immigrants in their country of settlement, accounting for two destination cultures to which immigrants must acculturate. They suggest that this may be particularly relevant for immigrants who belong to a minority group, whether racialized or not. However, this model is still limited as it does not account for multiple cultures of origin. The researchers do suggest that although three dimensions seemed sufficient for their population, others may require multidimensional acculturation models. Ferguson and Bornstein (2014) also suggest that due to the added complexity of potential multidimensional models of acculturation, quantitative measures may be difficult to employ and that qualitative approaches may be better suited to capture the full complexity of these acculturation experiences.

Identity and Belonging

In order to understand the outcomes of the acculturation process on the individual, Searle and Ward (1990) defined two forms of adaptation: psychological (internal sense of personal and social identity, mental health, satisfaction) and sociocultural (external outcomes, including functioning in family, work, and school). Both types of adaptation need to be further explored with immigrant adolescents (Berry, 1997), especially considering the importance and challenge of this developmental period. Two outcomes are of particular interest in the present study. The first is the sense of identity, which is a form of psychological adaptation, and the second is sense of belonging, which represents both psychological and sociocultural adaptation.

Identity Development

Most broadly, the term identity refers to the essence that is possessed by a person, place, or group (Gleason, 1983; Wetherell, 2010). Specifically, social identity is defined as "that part of an individual's self-concept which derives from his knowledge of his membership of a social group (or groups) together with the emotional significance attached to that membership" (Tajfel, 1974, p. 69). Thus, social identity is composed of the awareness of group membership (and what it entails), as well as the emotional connection to this group and its standing in relation to others. Ethnic identity is a part of social identity that is seen as a multidimensional and dynamic construct composed of cognitive, affective, and behavioral factors (Phinney & Ong, 2007; Sinclair & Milner, 2005). It is believed to play an important role in self-concept and affect how individuals go through life, interact with, and view one another (Phinney, 1996).

Phinney (1996) developed a three-stage model of ethnic identity that considers two primary elements: ethnic identity exploration and achievement, and ethnic affirmation or belonging. This model is of interest because it acknowledges the multifaceted nature of ethnic identity both as an individual identification (personal identity) and as a group membership (social identity). The three stages of this model are: (1) unexamined ethnic identity (not salient and given little conscious thought), (2) moratorium or exploration (interest in and exploration of ethnicity), and (3) achieved ethnic identity (internalized and secure identity). Empirical findings provide partial support for stage models of ethnic identity.

In a review of the literature, Quintana (2007) found that there was consistent evidence of increase in identity exploration and identification in adolescence, though the process was gradual and not a dramatic shift or crisis as theoretically assumed. In a longitudinal study, French et al. (2006) found that although there was a significant increase in-group esteem (or belonging) over early and middle adolescence, exploration rose only in the older cohort. This suggests that ethnic identity development is not a homogeneous process and that different facets of identity may develop at different times. Moreover, it is unclear how these stage models fit for multiethnic individuals.

Research on ethnic identity is particularly relevant for immigrant youth. Phinney et al. (2001) explored the effects of a secure and strong ethnic identity and suggested that it contributed to positive psychological well-being, regardless of gender. For example, a strongly held ethnic identity has been associated with higher school achievement and psychological adjustment in Chinese immigrant adolescents (Costigan et al., 2010). In addition, low levels of ethnic identity were associated with lower than average levels of self-esteem for males only, suggesting that ethnic identity may be of particular importance to immigrant boys. Unfortunately, the researchers did not explore the reasons behind these findings. A meta-analysis examining the connection between ethnic identity and personal well-being in racialized groups found a modest positive relationship between the two constructs (Smith & Silva, 2011). However, it is unclear what role ethnic identity may play in the adjustment of white immigrant adolescents.

Over the past two decades, there has been some recognition of the importance of studying white identity in order to gain a better understanding of racism and racial inequality (McDermott & Samson, 2005). Research on whiteness has focused on the limited awareness that white individuals have of their privilege and on the unexamined nature of this racial category. For many, it was an unremarked, invisible identity (Hyde, 1995). However, research and theorizing have focused on the experiences of American-born, mainstream white European American youth. It is unclear how this work may apply to: (a) first- and second-generation immigrants, (b) immigrants coming specifically from Eastern Europe, and (c) to Canada (versus another country with significantly different immigration policies).

Sense of Belonging

Hagerty et al. (1992) developed a theoretical model of sense of belonging as part of their theory of human relatedness. According to this model, sense of belonging is an interpersonal process that is nested under the construct of relatedness (Hagerty et al., 1996). Sense of belonging was defined as "the experience of personal involvement in a system or environment so that persons feel themselves to be an integral part of that system or environment" (p. 173). Sense of belonging is composed of two dimensions (Hagerty et al., 1992). The first dimension is valued involvement, or a person's sense of being valued or needed by people, groups, or environments. The second dimension is fit, referring to a person's sense that his/her characteristics are congruent with other people, groups, or environments.

The topic of sense of belonging has been of interest in the literature for several decades (Hagerty et al., 1996) and both Maslow (1954) and Thoits (1982) described it as a basic human need. This can apply to relationships with families, peers, and community members, as well as environments such as schools, neighborhoods, countries, and cultures. Belonging is especially important to consider in immigrant research because immigrant youth not only navigate multiple possible peer groups (Sussman et al., 2007), but also two or more cultures to which they can belong (Tartakovsky, 2009).

Sense of belonging is both affected by and affects relationships. Focusing on perceptions of social support from spouses, relatives, friends, and coworkers, Hagerty et al. (1996) found that sense of belonging is negatively connected to negative social support and conflict but positively connected to positive social support actions and more perceived social support. Moreover, a high sense of belonging was also connected to psychological functioning, whereas a low sense of belonging was related to loneliness, depression, and anxiety. Peer relationships were significant for both men and women, whereas involvement in community activities was significant for women only. This suggests that sense of belonging may be more connected to social and psychological functioning for women in general, perhaps due to a tendency for women to define themselves in relation to others (Hagerty et al., 1996).

In a study with 41 countries, Chiu et al. (2012) explored immigrant and non-immigrant high school students' sense of belonging. The findings revealed immigrant status differences in youth's scores on sense of belonging; non-immigrant youth had the highest scores, then second-generation, then first-generation immigrants who scored the lowest. Although the role of peers was not explicitly explored, the researchers suggested that perhaps first-generation youth's difficulties with peer relationships may have impacted their sense of belonging.

Researchers have recently extended their investigations of sense of belonging to views on country and culture. For example, Tartakovsky's (2009) longitudinal study focused on Jewish adolescents from Russia and Ukraine who were participating in an immigration program to Israel. During the three-year post-immigration period, although youth's attitudes toward the receiving country were more favorable than towards their home country, their sense of belonging was stronger towards their home country than the receiving country. However, sense of belonging to the home and receiving countries were independent from one another, suggesting that the countries were appraised in their own context. Nevertheless, sense of belonging with one country may impact attitude towards the other, suggesting that there is some connection.

Russian-Speaking Immigrants

Canada is an important destination for immigrants from Eastern Europe (Robila, 2010), and their total number increased from 1991 to 2006 (Statistics Canada, 2016). Overall, the Russian Federation has been in the top 10 immigrant-sending countries every year since the collapse of the Union of Soviet Socialist Republics (USSR) in 1990. More recently, the Russian Federation was the third largest country of origin of international migrants (United Nations, 2017). In Canada, Russian is the tenth most common non-official (English or French) language spoken at home (Statistics Canada, 2017a). Many of those leaving the FSU were Jewish individuals and families (Anisef et al., 2002). Currently, Canada constitutes the third or fourth largest Jewish community in the world, after Israel and the USA (France's Jewish population is similar to Canada's) (Brym et al., 2018).

The current study's interest in this population is rooted in two factors. First, there are significant sociocultural differences between Eastern Europe and Western Europe/North America. Specifically, Soviet culture valued collectivism, which emphasized interdependence and cooperation among in-group members (Triandis et al., 1988). This is rooted in communist ideology which emphasizes the good of the collective over the good of the individual (Mirsky, 2001). Moreover, the socioeconomic conditions during communist times required greater material dependence between generations, resulting in closer relationships and multigenerational living arrangements (Jasinskaja-Lahti & Liebkind, 2000), as well as less autonomy for adolescents (Birman & Taylor-Ritzler, 2007; Jasinskaja-Lahti & Liebkind, 2000; Mirsky et al., 2002). Second, the invisible nature of these immigrants suggests that although they are assumed to fit in with white mainstream Canadians, due to the sociocultural differences between the groups they may feel like they are in fact caught in the middle between other racialized immigrants and the Canadian-born white population. However, their invisible nature may offer them an added layer of privilege in navigating their multidimensional immigrant and minority identities that racialized immigrants may not have access to, which is of interest in this study.

The Current Study

The current study considered the potential effects of different modes of acculturation on how Russian-Speaking immigrant youth re/define their identity as they and their families settle and adjust in Canada and the extent to which their sense of belonging is developed during this process. The present study explored the following research questions: (1) what factors play a role in the identity and sense of belonging of Russian-Speaking immigrant youth?, and (2) how do youth make meaning of the multidimensional nature of their identities and belonging experiences?

Method

Participants

The study included 24 youth who were recruited through social service and settlement agencies in the Toronto and the Greater Toronto Area. Due to the difficulties in recruiting this immigrant population and youth that may not use community programs and services, a non-probability, snowball sampling strategy was also used. The inclusion criteria were: (1) first- or second-generation immigrant, (2) self-categorized as Russian-Speaking, (3) attending high school in Grades 10–12, and (3) lived full time with their families.

The study included 24 youth, 17 females, and 7 males, ranging in age from 15 to 19 years ($M = 16.38$ years, $SD = 0.97$) (see Table 2.1 for detailed demographics). The youth's birth places included five countries, and seven for their parents (see Table 2.2). The length of residency for first-generation youth in Canada averaged 8.86 years ($SD = 5.47$ years; range = 1–15 years). Eleven youth lived in the FSU and their average length of residency was 9.30 years ($SD = 6.93$; range = 5 months to 17 years). Ten youth lived in Israel and their average length of residency was 3.85 years ($SD = 3.46$ years; range = 6 months to 12 years). Three youth experienced indirect migration via Israel, Ukraine, and the U.S Thirty of the youth's parents experienced indirect migration, including 28 through Israel and two through the U.S. One set of parents moved back to Ukraine from Israel before continuing to Canada. Three youth had fathers living outside of Canada.

Background questionnaires focused on the families' household socioeconomic status. Five youth did not provide information on five fathers who were non-residential. Generally, youth reported that their parents were highly educated, with the majority of the mothers ($n = 23$) and fathers ($n = 16$) completing a Bachelor's

Table 2.1 Demographics characteristics of participants

Characteristic	Frequency
Immigrants generation	
First	20
Second	4
Religious background	
Atheist	1
Christian/Orthodox	8
Jewish	13
Jewish/Christian	2
Parental marital status	
Married	16
Separated or divorced	6
Widowed and remarried	2
Educational context	
Jewish private school	2
Other private school	1
Semi-private school	1
Catholic public school	3
Other public school	17

Table 2.2 Birth country of participants and their parents

Country	Participants	Mothers	Fathers
Belarus		3	
Canada	4		
Israel	9		
Moldova	1	3	1
Russia	3	4	6
Siberia			1
Ukraine	7	13	14
Uzbekistan		1	2

degree or higher (one father's educational background was not completed). Most of the mothers ($n = 18$) and fathers ($n = 17$) worked full time. Nineteen youth reported that their parents owned their homes.

Procedures

Semi-structured Interviews. Semi-structured interviews were conducted with the youth. Interviews took place primarily at the youth's homes ($n = 17$, 71%), with the rest taking place in a private room at the recruitment agency, a park, coffee shop, or the library. Both the participants and their parent signed a consent form (in either

Russian or English) prior to the start of the interview and the participants also completed a background questionnaire. In the interviews, youth were asked to reflect on their own immigration experiences, the contexts within which these took place, and the impact of these experiences on their identity and sense of belonging. The interviews were conducted in English with translations of words or questions offered by the interviewer when required. Russian words and/or sentences were included in some of the responses. Interviews were transcribed verbatim (both English and Russian portions). Transcripts were then reviewed for accuracy by a trained research assistant who was fluent in both Russian and English and any discrepancies were discussed and agreed upon. The interviews lasted between 63 and 187 min ($M = 110.71$ min, $SD = 35.76$) and participants received $30 CDN in cash for their participation.

Data Analysis. The qualitative analysis was conducted using constructivist grounded theory methodology (GTM; Charmaz, 2014) because the goal was to develop a substantive theory of the role of context in identity formation and sense of belonging of Russian-Speaking immigrant youth in Canada. GTM is a good fit for the current study because there has been limited research on the experiences of white immigrant youth with multiple identities and this methodology is recommended for under-researched areas that would benefit from theory generation (Strauss & Corbin, 1990). Moreover, GTM allows researchers to not only describe data but to explain it (Glaser & Strauss, 1967), which was the goal in the current study.

The analysis using constructivist GTM (Charmaz, 2014) proceeded in three main phases: (a) initial coding using open, segment by segment coding with the goal of generating labels and concepts, (b) focused coding in order to identify the most meaningful and/or frequent initial codes and use these to analyze and categorize larger segments of data more comprehensively, and (c) theoretical coding to form a theoretical story of the data using the categories and subcategories identified in the previous phase. Overall, this process of analysis is recursive and not linear. Moreover, this study used theoretical sampling, consequently, interview questions were adapted and included in subsequent interviews with the goal of fleshing out emergent categories (Charmaz, 2014). In addition, past participants were contacted whenever possible to answer the additional questions in phone follow-up interviews. Interviews were conducted until saturation of the categories and their properties, so that the analysis of additional interviews would not produce additional concepts. Pseudonyms are used in the reporting of the results.

Results

The proposed grounded theory aims to describe how invisible immigrant youth navigate and interact with their context in ways that influence their sense of identity and belonging. This theory is based on 24 in-depth interviews with Russian-Speaking youth that took place in 2014, during the early stages of the Russo-Ukrainian war (2014 - present) and the 2014 Israel-Gaza conflict.

The majority of the youth who participated in this study stated that they were impacted by one or both of these conflicts and were confronted with their own identities and reaction to them in their households, schools, communities, and in the news. Thus, the typical process of identity exploration in adolescence was compounded for these youth, allowing for a more in depth understanding of the complexity of this process. In the current theory, this process is composed of several parts that are continuously in motion, reflecting the dynamic nature of the search for and development of identity and belonging (see Fig. 2.1). Three core elements are at the center of this process: (1) choosing identities, (2) expressing identities, and (3) fitting in. Furthermore, youth engaged in a broader process of navigating the context around them in order to facilitate their re/formation of identity.

The outcome of this overall process is individualized with certain identities being highlighted and internalized over other identities, and a sense of belonging experienced with some groups over others. Moreover, these processes are multidirectional, taking into account what the context is able and/or willing to provide, how individuals interpret and navigate this context, and what they choose to take from it.

The existence of these elements around choice and expression of identity, as well as fit, is rooted in the fact that the youth, although immigrant and often of religious minority status, were white. To some extent, the invisibility of their racial identity afforded them greater flexibility to choose their identities, when to express them, and where and when to fit in. Thus, invisibility is an underlying thread in this model, and was viewed as both an opportunity and a challenge. Finally, no gender differences were determined based on the narratives of the youth, potentially as a result of methodological issues related to the small sample size of male participants as well as self-selection bias. Thus, gender will not be addressed as part of the current model.

Choosing Identities

Choosing identities was based on the ability of youth to: (1) identify the various labels or identities that were available to them in regard to their race, ethnicity, and/or religion, (2) evaluate the components and features of the different labels or identities and make meaning of them, and (3) decide which components of each identity to embrace or not. Overall, the ability to engage in this three-step process was rooted in the youth's degree of connection to, enjoyment of, and familiarity with the particular culture or place (i.e., customs, traditions, and history).

Identifying Labels. Youth were asked to identify identity labels at different points in their lifetime: their birth country (if applicable), an intermediary country in which they lived (if applicable), and in Canada. Youth used various labels to describe their identities (see Table 2.3). Of the youth born outside of Canada, only the youth who migrated before the age of five did not report any changes to their

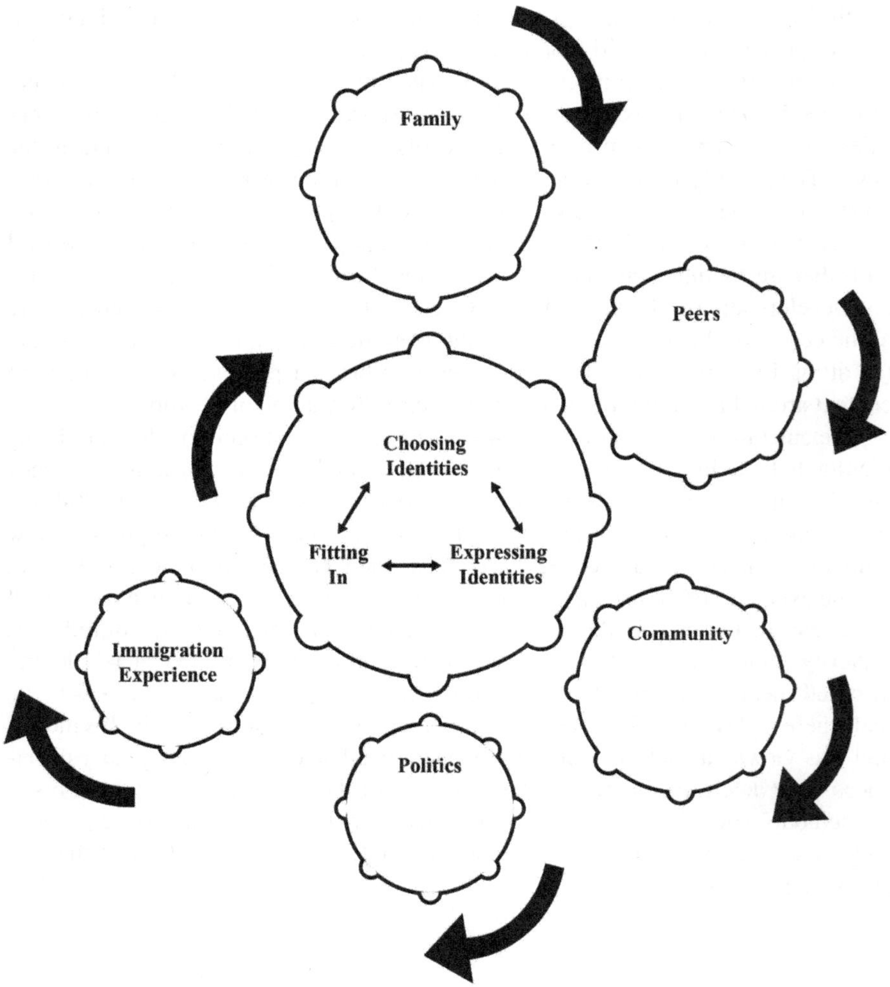

Fig. 2.1 Conceptual model for the formation of identity and belonging of Russian-speaking immigrant youth

identities over time. The other youth (18) reflected on changes in the relative importance of each identity from one context or time period to the next. For seven youth, identities were also additive and by the time they were in Canada, 15 of the 24 youth listed between three and four identities. For five of the youth, their religious identity was replaced with an atheist/agnostic identity.

Importantly, the youth did not necessarily list identities based on their place of birth and/or current place of residence. For example, five of the nine youth born in Israel did not list their birthplace as an identity. However, all of these youth listed Russian and Jewish as identities, reflecting their families' heritage and religious faith.

Table 2.3 Demographics and identity rankings by country

Gender	Place of Birth	Time in Canada (years)	Age	Identity Rankings: Birth Country	Intermediate Country	Host Country (Canada)
M	Israel	5	17	Russian		Jewish > Russian
F	Israel	12	17	Russian > Jewish		Jewish > Russian
F	Israel	12	16	Russian > Jewish > Israeli		Russian > Jewish > Israeli
M	Israel	12	15	Jewish > Russian		Jewish > Russian
F	Israel	13	15	Jewish > Israeli > Russian		Jewish > Israeli > Russian
M	Israel	14	16	Russian		Russian = Israeli
F	Israel	14	16	Jewish > Israeli > Russian		Israeli/Jewish > Canadian > Russian
F	Israel	15	16	Russian > Israeli > Jewish > White		Russian > Jewish/ Atheist > Israeli > White
F	Israel	15	16	Russian > Orthodox > White	Russian > Orthodox > White[a]	White > Russian > Atheist
M	Ukraine	1	17	Ukrainian > Christian		Ukrainian > Christian
M	Ukraine	1.5	18	Atheist > Ukrainian > White		Ukrainian > Atheist > White
F	Ukraine	4	16	White > Russian > Christian		Russian- Ukrainian > White > Christian
F	Ukraine	4	16	Ukrainian > Christian		Ukrainian > Agnostic
F	Ukraine	9	16	Ukrainian or Russian	Jewish > Russian[b]	Russian > Atheist
M	Ukraine	13	17	Ukrainian		Canadian > Soviet > Ukrainian > Orthodox
F	Ukraine	15	17	Ukrainian = Orthodox		Ukrainian > Canadian > White > Orthodox
F	Russia	1	17	Orthodox > Russian > White		Orthodox > Russian > White
F	Russia	1.5	19	Russian > European > Christian		Russian > European > Christian

(continued)

Table 2.3 (continued)

Gender	Place of Birth	Time in Canada (years)	Age	Identity Rankings Birth Country	Intermediate Country	Host Country (Canada)
F	Russia	12	15	Russian > Orthodox > White	Russian > Orthodox > White[c]	Atheist > Russian > Canadian > White
F	Moldova	3	17	Romanian > Orthodox		Romanian > Orthodox
M	Canada	–	17	Jewish > Israeli > Canadian > Russian		
F	Canada	–	16	Jewish > Canadian > Russian		
F	Canada	–	15	Jewish/Israeli > Canadian > Russian		
F	Canada	–	16	Russian > Jewish		

Note: M: Male; F: Female
[a]Former Soviet Union
[b]Israel
[c]United States

Evaluating Components. In their process of evaluating the identity labels available to them, youth did not provide a single definition of any of their identities. Instead, they offered a multidimensional web of components that were based on various factors, including individual, familial, social, political, and historical. At times, these were presented as separate elements, but often their interwoven and complex nature was reflected. As Ella explained:

> I have Russian roots so I guess you can say I am Russian. My parents have Russian roots. They have parents from Russia, all of them are from Russia. So, I guess I could say I'm Russian. I do usually say that...if people ask you 'What are you?' I usually say 'Russian-Jewish' but really I just speak Russian fluently at home and it's always been that way so I say I'm Russian-Speaking. Sometimes I'll say I'm Russian I guess if in that day I'll think about the fact that my parents are from Russia and they have Russian roots. (Female, 16, born in Israel)

The complex definitions of these identities often required youth to reflect on them in relation to others, comparing and contrasting the various elements. In particular, they made distinctions between internal (e.g., psychological) and external (e.g., physical appearance) elements of identity. As Sabrina said:

> Russian would be first because it's more a part of me physically. It's more a part of me because I look Russian. I was born in Russia but I grew into Canadian customs so I just like

> putting Russia first because it's more the physical part of me. Canada is more a mental part of me. (Female, 15, born in Russia)

In effect, it was the youth's experience of living within multiple cultures that allowed them to become more insightful, reflective, and able to critique their identities. Through their process of comparison between countries and cultures, youth were able to recognize differences in mentality, cultural values, and lifestyle.

Choosing Components. The third element, the process of evaluating different aspects from each culture, provided youth with the information they needed to decide which components to incorporate into their cultural practices and identities, and which to replace or discard. For cultural practices, this process was bidirectional. Specifically, the more youth learned about the Canadian culture, the more they viewed their heritage culture differently and sought different aspects from it. For example, some youth made choices around issues of values and mentality, such as holding on to the value around closeness with family, but letting go of the judgmental mentality (e.g., about appearance, behavior) they reported was common in the FSU. This process would become more complex the more cultures an individual was navigating. Thus, youth created an internal mosaic of these cultures, taking what they considered to be the best elements from each.

For identities, their ability to think critically enabled youth to not simply incorporate the components they gathered from external or traditional narratives. Instead, they interacted with and critically evaluated these views and produced their own multidimensional definitions of their identities. This was particularly common in evaluating the religious components of identity, highlighting the importance of separating faith from culture/tradition and the connection these provide to history and community. Jewish youth, in particular, discussed being Jewish as important to them, but also described themselves as not very religious. In this way, youth created a place for their identity both within and outside of the traditional narrative and held on to components of their identities that were personally meaningful to them. As Stella described:

> I think being Jewish is more than just praying and I know people who are Jewish and they don't pray or go to synagogue every day. Same thing with people who are Christian. They don't necessarily go to church ever even. They just don't feel a connection to God, they don't feel like He could exist, but according to them, it's very unrealistic, but you can still...There's more culture to it than that, and it's more of tradition I guess is a better word, to carry on the traditions that you grew up with. It's comfortable to, for example, talking about the origin of Islam and Jewish culture from the holidays and I've learned about it so much and the events that happened. I don't disregard them. I feel like you can still celebrate a culture and agree with it and you're allowed to remember it and celebrate it, but you don't necessarily have to connect to God. You can still recognize the events that happened before you, and the traditions, what other people believe, but you don't really have to connect to the spirit or a higher power. (female, 16, born in Ukraine)

Expressing Identities

The Russian-Speaking immigrant youth had the ability not only to choose which identities to claim and internalize, but also to decide when and how to disclose them. This decision was linked to the avoidance of bias and relied on the ability to reflect on factors in their environment.

Invisibility played an important part in the decisions and experiences around identity expression, both in positive and negative ways. The benefit of the invisibility was based on the youth's ability to control external markers of identity (e.g., attire, language, etc.). In addition, they were able to choose which label to provide when asked the questions: "where are you from?" or "what are you?" This was particularly the case around the Jewish and Israeli identities. One youth even expressed relief in being able to hide that identity behind the more visible white identity:

> I think that I'm more privileged by the fact that people would think that I'm European because I know there are people in my school that don't like Jewish people. They hate Jewish people. Especially with everything that's going on right now. I think that I'm lucky to look white, I guess so I'm not discriminated against. (Liron, male, 17, born in Israel)

The challenge of the invisible identity arose when youth heard negative comments about their identities without the speakers knowing that they were members of the group. Youth then had the choice to speak up or remain silent and allow the negative comments to continue. Often, they chose to stand up for themselves and their group, but at other times, for reasons of safety, discrimination, or simply a desire to avoid a fight, they remained silent. Thus, their internal struggle was unresolved.

Youth spoke about their ability to discern when it was or was not a good idea to express or disclose an identity. Specifically, they trusted their own ability to judge the safety of a situation. This was a particular concern for the Jewish and Israeli-born youth. The Ukrainian- and Russian-born youth did not feel that there was a danger in disclosing their identities and generally felt that their identities were accepted in Canada (though not always understood), despite the conflict that was in the media at the time. For the Jewish and Israeli identities, fear of antiseemitism and avoidance of bias and conflict around these identities played a role in decisions around identity expression.

Youth who had more than one ethnic/cultural identity to choose from often highlighted and expressed the identity they viewed as more accepted in Canada, even though they did not choose the Canadian label itself. For some youth, this was the Russian identity whereas for others, it was the Jewish identity. As Anna-Lee explained: "Here I emphasize the fact that I'm more Jewish because here Jewish people are more accepted than Russian people are" (female, 17, born in Israel).

Despite the youth's decisions to conceal their identities in certain situations, expression of identity was still important in the creation of social connection and sense of community. Thus, identities were expressed when they allowed youth to form a connection with others, but were hidden whenever the youth sensed that they may interfere with connection. Identities were also disclosed as part of the youth's desire to highlight the ways in which they were different and stood out in Canada. Particularly, in situations when they felt that this difference would be handled positively.

Decisions around identity expression were not only made in the context of strangers and concerns about safety, but also with friends and family. With friends, in certain situations, youth highlighted one identity over another because they believed that identity was more accepted within that group. With family, youth who moved away from religion made decisions around the expression of their atheist identity to their parents. Their considerations revolved around the potential reactions of their parents and, in some cases, youth chose to keep this new invisible identity hidden in order to avoid conflict with their parents.

Fitting in the Social World

The term "fitting in" was chosen instead of the term "belonging" because it captured the conscious and active engagement of the youth with their social world and was more grounded in their narratives. Fitting in consisted of a sense of comfort and connection within oneself and with others. Specifically, youth could have an internal sense of fit and be comfortable with who they were. Youth could also fit within their families, peer groups, schools, broader community, as well as the country at large. The two facets of fitting in are interconnected, as Bella described: "I'm with these girls and I just feel like I found who I am... I feel wanted... I actually have someone to be there for me and I don't have to be worried that I'll be alone" (female, 15, born in Israel).

Overall, the process of fitting in was closely connected to identity and could be viewed as the culmination of choosing and expressing identities. Moreover, a dynamic, bidirectional relationship existed between the processes around identity and belonging. Specifically, if youth felt that they fit in, they were more likely to choose and express certain identities. In turn, their choices and expressions of certain features of identities led to a greater sense of fitting in, or belonging. Thus, once identities were chosen, youth were able to navigate the context around them, assess their level of fit within it, and alter the expression of identity if more or less fit was required or desired. As Jackie explained:

> If people around me didn't speak Russian then I would just say 'oh I'm Jewish.' You kind of more adapt to not really what other people think... but you kind of just go with what makes sense wherever you are. It depends, you don't always want to stand out but you don't always want to blend in either, kind of depends on your situation. (Female, 16, born in Canada)

Fitting in within their social world was a central issue for these youth, and the challenges they described experiencing were often more connected to belonging than survival. Youth expressed feeling different from individuals from their birth countries once they adjusted to Canada, and were not able to see their old friends regularly or spend time with them: "I kind of changed a little bit, they also changed, and probably we don't have as much in common anymore" (Andrey, male, 18, born in Ukraine). Others discussed being different from people in Canada because they grew up elsewhere: "The social norms and the way people talk here and the things they talk about, the way they act. It's completely different than it was in Israel so it was kind of hard for me to integrate here" (Liron, male, 17, born in Israel).

Their description of lack of fit was twofold: they had an internal feeling of being different and they perceived themselves as externally or visibly different, based on factors such as appearance or spoken language. Specifically, experiences of language loss or inability to master a language could hinder their sense of fit within certain communities, both Canadian and heritage. As Maria explained: "This English-speaking group, I was not really comfortable. Because… I couldn't talk in my language and I couldn't explain what I think because it's English. I couldn't explain my emotions so that was hard" (female, 17, born in Moldova).

As with identity choice and expression, fitting in was not simply the result of an internal decision-making process. There was an external perception of the youth's identities and how well they fit in and these perceptions were not always in line with their own sense of identity or fit. Thus, the processes of identity denial and assumption may impact internal sense of identity as well as fit and belonging.

Navigating the Context

This process refers to the youth's experiences in interacting with the individuals and institutions around them in multidirectional ways. Overall, in the current study, five components appeared to be significant and salient: (1) family (immediate and extended), (2) peers (in-groups and out-groups), (3) community (neighborhoods, schools, and organizations), (4) political context (within the country and international), and (5) and immigration (family history and personal experience). Each youth conceptualized these components and made meaning of them in varying ways. For example, a component such as family for one youth may only include a mother, but for another, it included both parents, two siblings, and their maternal grandparents.

Thus, the process of navigating the context differed among individuals. The variations of the process were based on: (1) the makeup of the context the individual must navigate (number and type of components), (2) the importance and meaning of each component to the individual, (3) how the different components interact among each other, and (4) how they interact with the individual (the multidirectional nature of the interaction and the level of individual control over it). Moreover, all of these factors may change over the course of time and circumstances.

Much of the time, the process of navigating the context may be automatic and require little conscious thought. However, when a shift occurred, such as immigration, it caused a disruption in the automatic movement of this process and it became more purposeful. As a result, youth may become more alert and aware of the context, think about how it impacts them more, and may notice when something is not working well or changing. As Olena described:

> Something changed in understanding of how life is going, because here it's different. So, sometimes you should understand that you should calm down, or no, you should smile! When we just came here I didn't believe that people can just smile when they're going across the street they smile and I'm like, 'what's going on? Is it ok?' (female, 19, born in Russia)

Thus, youth may become more aware of processes around identity and belonging, and how these connect to the context around them, upon their move to a new country. This process of interaction between the youth and the context can be both positive and negative, and either experience can impact one's identity and belonging. For example, one youth explained how purposefully seeking friendships with Canadian-born youth can be beneficial: "They definitely brought in more culture. They taught me more North American culture. I became more accustomed to what my friends were teaching me, and it's made me more Canadian" (Sabina, female, 15, born in Russia). However, not all the experiences the youth had with their environment were positive. In order to avoid some of these challenges, youth adjusted their identity expression (e.g., by highlighting one identity over another, or avoiding discussions of religion or politics) in order to increase their level of fit and comfort within a group. As Anna-Lee explained:

> I don't feel like it's uncomfortable for me that with every group I identify myself as a different person. It's just that I try to make it, to get closer with them, and maybe make it seem like I'm more this or more that so there won't be any miscommunication or misunderstandings or some bad feelings from their side. I don't really care personally, I just want to feel more comfortable. (female, 17, born in Israel)

War and political conflict are another example of a shift that can disrupt the process of identity development and bring it into focus Overall, the political conflicts in both Russia-Ukraine and Gaza-Israel appeared to strengthen the youth's ethnic, cultural, and/or religious identity and most of them discussed their pride in their people and/or countries of origin when reflecting on the conflicts. As Adam contended:

> I think as I was always raised to be proud to be Jewish and never be ashamed of it and when this situation came up, the way that I was taught showed more by being more proud now. What's going on in Gaza now and Israel. (male, 15, born in Israel)

Similarly, war and political conflict can also have an impact on sensing of belonging. Specifically, the youth reported that the political conflict had a unifying effect on their communities, bringing Israelis, Jews, Russians, and Ukrainians together. In this way, the political conflicts touched on both identity and belonging for these youth. The conflicts even had the ability to unite groups that previously experienced tension, such as Russian-Speaking and Ukrainian-speaking Ukrainians:

> I feel more accepted as a Russian-Speaking Ukrainian because before there was a divide in Ukraine between Russian-Speaking Ukrainians and Ukrainian-speaking Ukrainians. However, now when Russia intervenes there's more of a difference between Ukrainians and Russians… Ukrainians don't really care that much anymore about whether you're Russian-

speaking or Ukrainian-speaking, as long as you're just on the same side. So if anything, I feel more comfortable with my identity. (John, male, 17, born in Ukraine)

However, in some cases political conflicts also created strife within friendships, peer groups, as well as within and between communities. The youth had to navigate this, which then impacted their sense of identity, belonging, and identity expression.

Overall, the process of navigating the context is based on individual agency and the ability to directly interact with the environment in complex ways. These interactions are multidirectional, not the environment simply acting on the individual, but the individual making conscious choices about what to take in and also what to express. In addition to the direct and indirect effects of each component on the individual, there is continued and active decision making and exploration on the part of the individual. Thus, the sociocultural context (e.g., family, school, community, culture, religion, ideology, etc.) and the demands and challenges posed by the environment will influence the extent of agency the youth would use to navigate the environment around them. Moreover, this process impacts the multidimensional nature of the youth's identities, to the extent that it influences not only the various identities, but also their content and the relationships among all of them. Thus, all of these components, and the youth's interactions with them, contribute to both socialization and acculturation to varying degree.

Discussion

The present study extends our current understanding of the multidimensional acculturation experiences of Russian-Speaking immigrant youth in Canada. Specifically, the primary goal of this study was to generate a substantive theory of the identity and sense of belonging of these youth, using a constructivist grounded theory methodology (Charmaz, 2014). The model (see Fig. 2.1) was developed around two core aims: (1) to understand how youth make meaning of the multidimensionality of their identities and belonging experiences, and (2) to identify the contextual factors that play a role in the identity and sense of belonging of this population and how youth navigate these.

The model focused on two main features. First, the youth approached the multidimensionality of their identities and belonging through the core processes of choosing identities, expressing identities, and fitting in. Second, the youth used input and resources from their social contexts, by engaging in a broader process of navigating the context. In addition, five contextual factors were identified as being of particular importance to the youth in the current study: (1) family (immediate and extended), (2) peers (in- and out-groups), (3) community (schools, neighborhoods, and agencies), (4) politics (national and international), and (5) immigration experiences (timing in particular). The processes that youth engaged in were dynamic and reciprocal and thus interacted with and affected one another. Choosing and expressing identities, as well as fitting in, influenced the youth's navigation of their context.

In turn, as youth navigated their context, their identities and belonging were further altered.

The model created also demonstrated that identity is fluid and identity choice and expression may shift according to the situation and through the process of navigating the context around them. These results are consistent with past research and theorizing on the phenomenon of cultural frame switching (Hong et al., 2000) that suggests that bicultural individuals are able to switch between cultural lenses in response to environmental cues. The frame switching can occur with or without conscious intent, but it does require a certain degree of familiarity with all of the cultures one is switching between (Schwartz et al., 2017). Similarly, past research on multiracial individuals found that some youth held a situational identity whereby different elements of their identity became more salient, depending on the context (Renn, 2000, 2008). Thus, when considering acculturation, not only individual choices and experiences, but also the context within which these take place, must be examined (Schwartz et al., 2014).

Moreover, these results fit outside of the bidimensional and tridimensional models of acculturation (Berry, 2007; Ferguson & Bornstein, 2014). In reflecting on their current identities, 15 of the 24 youth listed between three and four labels. These identities accounted for factors outside of their places of birth, residence, or citizenship (the factors commonly considered in bidimensional acculturation research). This finding is consistent with research conducted with undergraduate, first-generation immigrant students in Montreal, Canada, which found that the participants used nearly three distinct identities to describe themselves during the course of a single day (Doucerain et al., 2013). This study focused on identity shifts over the course of a day, whereas the current study looked at youth's reflections on changes in their identity over time.

It is important to note that the largely invisible and, in some cases, majority (e.g., white, Christian) nature of the youth's identities allowed the youth to experience less struggle around their identities overall. The youth did make choices around these identities in ways that reflected their degree of belonging and comfort in various situations, but the choice to hide certain identities to increase these experiences was available to them in a way that it is not often available to visible minority youth. In addition, although their invisible identity did lead to some indirect experiences of victimization, it also allowed them to avoid some direct experiences by hiding their identity in different situations. Thus, the largely invisible nature of their identities allowed the youth an added layer of control and flexibility in the active process of identity management around choice, expression, and fitting in.

The current study has a few limitations. First, single informants (the youth) were used in this study, who reported not only on their own experiences but also on their perceptions of their environment. Future research should interview other key developmental figures, such as parents and peers. This would allow researchers to more deeply explore the active role of such figures in the development of the youth. Second, this study was retrospective requiring youth to report about their experiences over time. Future research should be longitudinal and examine pre- and post-migration experiences in real-time. Finally, the sample size was small and although the primary goal of theoretical saturation was achieved, there was a lack of statistical equivalence in the numbers of recent and non-recent immigrants, Jews and Christians, boys and girls, etc. Future research should compare the experiences of these youth in order to further examine their experiences of immigration, identity, and belonging.

Conclusions

Future research should examine multidimensional acculturation among other immigrant groups, including those who were ethnic minorities in their countries of origin, those who experienced indirect migration, and those who are ethnic minorities within their host countries. It would be of interest to explore the different contextual components that impact acculturation, identity, and belonging in various populations. Moreover, these results indicate the importance of research accounting for multidimensionality and moving away from using quantitative measure to assess bidimensional acculturation, using birth and host countries as proxies. Individuals should be allowed to not only select their own identity labels, but to define these and express their meaning-making around them. Finally, future research should further explore the role of war and political conflict in identity and sense of belonging across diverse populations.

References

American Psychological Association. (2002). *Developing adolescents: A reference for professionals*. Author. https://www.apa.org/pi/families/resources/develop.pdf

Anisef, P., Baichman-Anisef, E., & Siemiatycki, M. (2002). *Multiple identities and marginal ties: The experience of Russian Jewish immigrant youth in Toronto* (Working Paper No. 19) (Toronto, Joint Centre of Excellence for Research on Immigration and Settlement). http://www.ceris.metropolis.net/wp-content/uploads/pdf/research_publication/working_papers/wp19.pdf

Bennett, M. (2011). Children's social identities. *Infant and Child Development, 20*(4), 353–363. https://doi.org/10.1002/icd.741

Berry, J. W. (1997). Immigration, acculturation, and adaptation. *Applied Psychology: An International Review, 46*(1), 5–68. https://doi.org/10.1111/j.1464-0597.1997.tb01087.x

Berry, J. W. (2007). Acculturation. In J. E. Grusec & P. Hastings (Eds.), *Handbook of socialization* (pp. 543–558). Guilford.

Birman, D., Persky, I., & Chan, W. Y. (2010). Multiple identities of Jewish immigrant adolescents from the former Soviet Union: An exploration of salience and impact of ethnic identity. *International Journal of Behavioral Development, 34*(3), 193–205. https://doi.org/10.1177/0165025409350948

Birman, D., & Taylor-Ritzler, T. (2007). Acculturation and psychological distress among adolescent immigrants from the Former Soviet Union: Exploring the mediating effect of family relationships. *Cultural Diversity and Ethnic Minority Psychology, 13*, 337–346. https://doi.org/10.1037/1099-9809.13.4.337

Brym, R., Neuman, K., & Lenton, R. (2018). *2018 survey of Jews in Canada: Final report*. https://www.environicsinstitute.org/docs/default-source/project-documents/2018-survey-of-jews-in-canada/2018-survey-of-jews-in-canada%2D%2D-final-report.pdf?sfvrsn=2994ef6_2

Buriel, R., & De Ment, T. (1997). Immigration and sociocultural change in Mexican, Chinese, and Vietnamese American families. In A. Booth, A. C. Crauters, & N. Landale (Eds.), *Immigration and the family: Research and policy on U.S. immigrants* (pp. 165–200). Erlbaum.

Charmaz, K. (2014). *Constructing grounded theory* (2nd ed.). Sage.

Chiu, M. M., Pong, S., Mori, I., & Chow, B. W. (2012). Immigrant students' emotional and cognitive engagement at school: A multilevel analysis of students in 41 countries. *Journal of Youth and Adolescence, 41*(11), 1409–1425. https://doi.org/10.1007/s10964-012-9763-x

Chuang, S. S., & Moreno, R. P. (2012). Theoretical perspectives on immigrant families. In M. A. Fine & F. D. Finchman (Eds.), *Handbook of family theories: A content-based approach* (pp. 316–337). Taylor and Francis/Routledge.

Costigan, C. L., Koryzma, C. M., Hua, J. M., & Chance, L. J. (2010). Ethnic identity, achievement, and psychological adjustment: Examining risk and resilience among youth from immigrant Chinese families in Canada. *Cultural Diversity & Ethnic Minority Psychology, 16*(2), 264–273. https://doi.org/10.1037/a0017275

Doucerain, M., Dere, J., & Ryder, A. G. (2013). Travels in hyper-diversity: Multiculturalism and the contextual assessment of acculturation. *International Journal of Intercultural Relations, 37*(6), 686–699. https://doi.org/10.1016/j.ijintrel.2013.09.007

Erikson, E. H. (1968). *Identity: Youth and crisis*. Norton.

Ferguson, G. M., & Bornstein, M. H. (2014). Tridimensional (3D) acculturation: Culture and adaptation of Black Caribbean immigrants in the United States. In R. Dimitrova, M. Bender, & F. van de Vijver (Eds.), *Global perspectives on well-being in immigrant families* (pp. 31–51). Springer.

French, S. E., Seidman, E., Allen, L., & Aber, J. L. (2006). The development of ethnic identity during adolescence. *Developmental Psychology, 42*(1), 1–10. https://doi.org/10.1037/0012-1649.42.1.1

Glaser, B. G., & Strauss, A. L. (1967). *The discovery of grounded theory: Strategies for qualitative research*. Aldine de Gruyter.

Gleason, P. (1983). Identifying identity: A semantic history. *The Journal of American History, 69*(4), 910–931. https://doi.org/10.2307/1901196

Gordon, M. M. (1964). *Assimilation in American life: The role of race, religion, and national origins*. Oxford University.

Graves, T. D. (1976). Acculturation, access, alcohol in a tri-ethnic country. *American Anthropologist, 59*, 306–321. http://www.jstor.org.subzero.lib.uoguelph.ca/stable/3629450

Hagerty, B. M., Lynch-Sauer, J., Patusky, K. L., Bouwsema, M., & Collier, P. (1992). Sense of belonging: A vital mental health concept. *Archives of Psychiatric Nursing, 6*(3), 172–177. https://doi.org/10.1016/0883-9417(92)90028-H

Hagerty, B. M., Williams, R. A., Coyne, J. C., & Early, M. R. (1996). Sense of belonging and indicators of social and psychological functioning. *Archives of Psychiatric Nursing, 10*(4), 235–244. https://doi.org/10.1016/S0883-9417(96)80029-X

Hong, Y.-Y., Morris, M. W., Chiu, C.-Y., & Benet-Martínez, V. (2000). Multicultural minds: A dynamic constructivist approach to culture and cognition. *American Psychologist, 55*(7), 709–720. https://doi.org/10.1037/0003-066X.55.7.709

Hyde, C. (1995). The meanings of whiteness. *Qualitative Sociology, 18*, 87–95. https://doi.org/10.1007/BF02393197

Immigration, Refugees and Citizenship Canada. (2018). *2018 annual report to parliament on immigration.* https://www.canada.ca/content/dam/ircc/migration/ircc/english/pdf/pub/annual-report-2018.pdf

Jasinskaja-Lahti, I., & Liebkind, K. (2000). Predictors of the actual degree of acculturation of Russian-speaking immigrant adolescents in Finland. *International Journal of Intercultural Relations, 24*(4), 503–518. https://doi.org/10.1016/S0147-1767(00)00010-9

Maslow, A. (1954). *Motivation and personality*. Harper.
McDermott, M., & Samson, F. L. (2005). White racial and ethnic identity in the United States. *Annual Review of Sociology, 31*, 245–261. https://doi.org/10.1146/annurev.soc.31.041304.122322
Mirsky, J. (2001). Psychological independence among immigrant adolescents from the Former Soviet Union. *Transcultural Psychiatry, 38*(3), 363–373. https://doi.org/10.1177/136346150103800306
Mirsky, J., Baron-Draiman, Y., & Kedem, P. (2002). Social support and psychological distress among young immigrants from the former Soviet Union in Israel. *International Social Work, 45*(1), 83–97. https://doi.org/10.1177/0020872802045001321
Newman, P. R., & Newman, B. M. (1976). Early adolescence and its conflict: Group identity versus alienation. *Adolescence, 11*, 261–274.
Persky, I., & Birman, D. (2005). Ethnic identity in acculturation research: A study of multiple identities of Jewish refugees from the former Soviet Union. *Journal of Cross-Cultural Psychology, 36*, 557–572. https://doi.org/10.1177/0022022105278542
Phinney, J. S. (1996). Understanding ethnic diversity: The role of ethnic identity. *The American Behavioral Scientist, 40*(2), 143–152. https://doi.org/10.1177/0002764296040002005
Phinney, J. S., Horenczyk, G., Liebkind, K., & Vedder, P. (2001). Ethnic identity, immigration, and well-being: An interactional perspective. *The Journal of Social Issues, 57*(3), 493–510. https://doi.org/10.1111/0022-4537.00225
Phinney, J. S., & Ong, A. D. (2007). Conceptualization and measurement of ethnic identity: Current status and future directions. *Journal of Counseling Psychology, 54*(3), 271–281. https://doi.org/10.1037/0022-0167.54.3.271
Quintana, S. M. (2007). Racial and ethnic identity: Developmental perspectives and research. *Journal of Counseling Psychology, 54*(3), 259–270. https://doi.org/10.1037/0022-0167.54.3.259
Redfield, R., Linton, R., & Herskovits, M. J. (1936). Memorandum for the study of acculturation. *American Anthropologist, 38*(1), 149–152. https://doi.org/10.1525/aa.1936.38.1.02a00330
Renn, K. A. (2000). Patterns of situational identity among biracial and multiracial college students. *The Review of Higher Education, 23*(4), 399–420. https://doi.org/10.1353/rhe.2000.0019
Renn, K. A. (2008). Research on biracial and multiracial identity development: Overview and synthesis. *New Directions for Student Services, 123*, 13–21. https://doi.org/10.1002/ss.282
Robila, M. (2010). *Eastern European immigrant families*. Routledge.
Schwartz, S. J., Birman, D., Benet-Martínez, V., & Unger, J. B. (2017). Biculturalism: Negotiating multiple cultural streams. In S. J. Schwartz & J. B. Unger (Eds.), *The Oxford handbook of acculturation and health* (pp. 29–47). Oxford University Press.
Schwartz, S. J., Vignoles, V. L., Brown, R., & Zagefka, H. (2014). The identity dynamics of acculturation and multiculturalism: Situating acculturation in context. In V. Bennet-Martinez & Y. Hong (Eds.), *The Oxford handbook of multicultural identity* (pp. 57–93). Oxford University.
Searle, W., & Ward, C. (1990). The prediction of psychological and socio-cultural adjustment during cross-cultural transitions. *International Journal of Intercultural Relations, 14*, 449–464. https://doi.org/10.1016/0147-1767(90)90030-Z
Sinclair, J., & Milner, D. (2005). On being Jewish: A qualitative study of identity among British Jews in emerging adulthood. *Journal of Adolescent Research, 20*(1), 91–117. https://doi.org/10.1177/0743558404271132
Smith, T. B., & Silva, L. (2011). Ethnic identity and personal well-being of people of color: A meta-analysis. *Journal of Counseling Psychology, 58*(1), 42–60. https://doi.org/10.1037/a0021528
Statistics Canada. (2016). *150 years of immigration in Canada*. Author. https://www150.statcan.gc.ca/n1/pub/11-630-x/11-630-x2016006-eng.htm
Statistics Canada. (2017a). *Focus on geography series, 2016 Census*. Author. https://www12.statcan.gc.ca/census-recensement/2016/as-sa/fogs-spg/Facts-can-eng.cfm?Lang=Eng&GK=CAN&GC=01&TOPIC=1
Statistics Canada. (2017b, October 25). Immigration and ethnocultural diversity: Key results from the 2016 Census. *The Daily*. https://www150.statcan.gc.ca/n1/en/daily-quotidien/171025/dq171025b-eng.pdf?st=W%2D%2D5cv9S

Statistics Canada. (2018). Canada's population estimates: Total population. *The Daily*. https://www150.statcan.gc.ca/n1/en/dailyquotidien/180927/dq180927c-eng.pdf?st=yJP4F2jR

Steinberg, L. (2005). Cognitive and affective development in adolescence. *Trends in Cognitive Sciences, 9*(2), 69–74. https://doi.org/10.1016/j.tics.2004.12.005

Strauss, A., & Corbin, J. (1990). *Basics of qualitative research: Grounded theory procedures and techniques*. Sage.

Sussman, S., Pokhrel, P., Ashmore, R. D., & Brown, B. B. (2007). Adolescent peer group identification and characteristics: A review of the literature. *Addictive Behaviors, 32*(8), 1602–1627. https://doi.org/10.1016/j.addbeh.2006.11.018

Tajfel, H. (1974). Social identity and intergroup behaviour. *Social Science Information/sur Les Sciences Sociales, 13*(2), 65–93. https://doi.org/10.1177/053901847401300204

Tanti, C., Stukas, A. A., Halloran, M. J., & Foddy, M. (2011). Social identity change: Shifts in social identity during adolescence. *Journal of Adolescence, 34*(3), 555–567. https://doi.org/10.1016/j.adolescence.2010.05.012

Tartakovsky, E. (2009). Cultural identities of adolescent immigrants: A three-year longitudinal study including the pre-migration period. *Journal of Youth and Adolescence, 38*(5), 654–671. https://doi.org/10.1007/s10964-008-9370-z

Thoits, E. A. (1982). Conceptual, methodological and theoretical problems in studying social support as a buffer against life stress. *Journal of Health and Social Behavior, 23*, 145–159. https://doi.org/10.2307/2136511

Triandis, H. C., Bontempo, R., Villareal, M. J., Asai, M., & Lucca, N. (1988). Individualism and collectivism: Cross-cultural perspectives on self-ingroup relationships. *Journal of Personality and Social Psychology, 54*, 323–338. https://doi.org/10.1037/0022-3514.54.2.323

Tyyskä, V. (2014). *Youth and society: The long and winding road* (3rd ed.). Canadian Scholars.

United Nations. (2017). *International Migration Report 2017: Highlights*. Author. https://www.un.org/en/development/desa/population/migration/publications/migrationreport/docs/MigrationReport2017_Highlights.pdf

Wetherell, M. (2010). The field of identity studies. In M. Wetherell & C. T. Mohanty (Eds.), *The SAGE handbook of identities* (pp. 3–26). Sage.

Chapter 3
"We Keep the Traditions Going": Intergenerational Transmission of Cultural Identities Among Asian Indian Multigenerational Households

Sailaja Subramaniam and Marsha T. Carolan

"Because of the children only. My children and grandchildren are here, why would I want to return?"

Throughout much of the history of the United States (U.S.), immigration has always been a source of population growth and cultural change. The recent changing demographics of the U.S. reflects the increasing number of older immigrants moving to the U.S., to reunite with their children as a result of the opportunities afforded by the Family Reunification Act of 1990 (Wilmoth, 2010). This act offered many naturalized and legal immigrants the opportunity to encourage their parents to relocate to the U.S.

Extended family living among immigrants tends to be higher compared with the U.S. born population. However, this varies based on country of origin and length of residence in the U.S. (Glick & Hook, 2002). There has been a steady increase in multigenerational families in the U.S. over the past decade. According to Pew Research Center, the analysis of U.S. Census Bureau data in 2016 revealed that the number of multigenerational households increased to 20% of the total U.S. population (Cohn & Passel, 2018). Specifically, about 29% of Asians lived in multigenerational households, compared to about 27% of Hispanics, 26% of Blacks, and 16% of Whites. These families typically consist of three or more generations co-residing, including grandparents and grandchildren younger than 25 years of age (Wilmoth, 2001). As Dykstra (2010) observed, "The new demographic circumstances in which family members of multiple generations share several decades together compel us to recognize that individuals are embedded in a complex web of vertical and horizontal ties" (p.1).

S. Subramaniam (✉) · M. T. Carolan
Michigan State University, East Lansing, MI, USA
e-mail: subram37@msu.edu; carolan@msu.edu

D. J. Johnson et al. (eds.), *Re/Formation and Identity*, Advances in Immigrant Family Research, https://doi.org/10.1007/978-3-030-86426-2_3

Grandparents in ethnic minority families are conduits for their grandchildren's ethnic socialization (Ishizawa, 2004). They serve as family historians and play a primary role in transmitting ethnic heritage, family traditions, and norms, all of which contribute to the saliency of their role (Nakazawa et al., 2018; Silverstein et al., 2012). The purpose of this study was to examine late-life Asian Indian immigrants' unique contributions in multigenerational families. We provide an overview of the pattern of immigration among Asian Indians followed by the contextual framework for our study with Asian Indian multigenerational families. We then present the voices of the late-life immigrants, the adult children, and spouses to capture the significant contribution of the grandparents in shaping the cultural identities of their grandchildren.

Pattern of Immigration among Asian Indians in the United States

Asian Indians are the third largest Asian group in terms of foreign-born immigrants after China and the Philippines. Currently, there are more than two million Indian-born immigrants residing in the United States, accounting for 4.7 million of the 41.3 million foreign-born population (US Census, 2017a, b). Asian Indians are the second largest Asian group in terms of foreign-born immigrants after China (U.S. Census, 2017b).

Asian Indian immigration to the U.S. can be traced back to the 1800s. However, it is only after the mid-twentieth century did a major influx of immigration take place (Dasgupta, 1998). They immigrated to the U.S. in three distinct waves. The first wave immigrated during the mid-1960s when restrictions on Asian immigrants were lifted. Most of the Asian Indians were young adults in pursuit of the "American Dream" as they attended institutions of higher learning or sought better professional opportunities (Rangaswamy, 2000).

The U.S. Immigration and Naturalization Act 1965 vastly increased the opportunities for immigrants from Asia, with Asian Indians being one of the fastest growing immigrant groups in the U.S. (Lamb, 2002). During the economic boom of the 1970s, most of the second wave immigrants immigrated with their immediate families to meet the demand for highly skilled professionals (e.g., engineers, doctors, and scientists) (Mogelonsky, 1995).

The enactment of the Family Reunification Act of 1990 sparked the third wave of arrivals as it offered many naturalized and legal immigrants the opportunity to sponsor the immediate blood relatives of immigrants already settled in the U.S. The percentage of Asian Indians who came under the family reunification scheme increased from 27% to 86% between 1969 and 1985, and a considerable portion of this third wave were aging parents of Asian Indians who came to live with their children in the U.S. (Diwan et al., 2011; Hing, 1993; Nandan, 2007).

There is significant diversity among Asian Indians in terms of educational levels, socio-economic status (SES), and within-group differences in language, religious beliefs, English proficiency, and immigration status (e.g., first, second generation), as compared with other minority groups. However, Asian Indians who immigrated to the U.S. brought with them the rich traditions of India and many have sought to retain their Indian identities and their family ties as they adjusted and adapted to living in the U.S. (Bhattacharya & Shibusawa, 2009; Sharma & Kemp, 2012).

Conceptual Framework

Life Course Perspective

A life course perspective examines the biological, mental, and social aspects of changes throughout the life course of an individual (Hendricks, 2012). It provides a framework for understanding development over the life span and the role and implication for families in society.

One of the great achievements of the twentieth century is the increase in life expectancy, as advances in medicine and technology began to insulate people from the ravages of diseases and other hazards. People who would have died early of preventable causes began to live long enough to experience aging-related diseases such as cancer, cardiovascular diseases, osteoporosis, and sensory impairment (Olshanky, 2011). The global population age 65 and older is increasing rapidly and projected to reach 1.6 billion by 2050, with the fastest growing age group consisting of people in their eighties and older (He et al., 2016).

Aging also occurs in context, as people age within their families, homes, and communities (Pruchno, 2018). Developmental life course change emphasizes the embeddedness of grandparents within the larger context of multigenerational family systems and this includes how older adults negotiate their roles as grandparents (Hayslip & White, 2008). Kivnick (1982), was of the view that Erickson's psychosocial stages indicate that the grandparent identity is encapsulated within meanings of immortality through descendants, continuity of family, and being a valued elder (as cited in Reitzes & Mutran, 2004, p. S213).

Hayslip and Page (2012) identified the four dimensions that define the role of grandparents: (1) symbolic (the meaning assigned to the role), (2) emotional (feelings about being grandparents), (3) attitudinal (belief's about activities that are central to the role), and (4) behavioral (based on the quality of interactions with grandchildren). As individuals traverse through their life course, they acquire new roles and exit others and grandparenting provides a sense of identity and purpose especially during middle and later-life when generative developmental needs are important (Reitzes & Mutran, 2004; Thiele & Whelan, 2008).

Scholars exploring the grandparent role have also highlighted the variability within these relationships as it depends on the age and mobility of the grandparents, the distance between grandparents and grandchildren, the age and gender of the grandparent-grandchild, the family dynamics, and socio-economic variables (Connidis & Barnett, 2019; Stelle et al., 2010). The level of involvement will likely vary when grandchildren are young and the grandparents are in good health versus when grandchildren are adults and the grandparents are older or frail (Boon et al., 2008). However, the contributions of grandparents in terms of assistance, guidance, and socialization of their grandchildren cannot be underestimated. Research findings highlight the positive impact that grandparents have in the lives of their grandchildren through transmission of values (Pratt et al., 2008), serving as a support system in times of crisis (Ruiz, 2004), or being attachment figures by providing stability and security (Seibert & Kerns, 2009).

Scholars have also noted that among ethnic and racial minority groups in the U.S., grandparents play a central role because grandchildren are growing up amidst dual cultures (mainstream U.S. culture and their own ethnic culture) (Greenfield, 2011; Hunter & Taylor, 1998). Among East Asians and South Asians, grandparents are traditionally highly respected, revered, honored, and placed at the pinnacle of the traditional family hierarchy. They are perceived as being the custodians of family customs, traditions, and values (Babu et al., 2018; Nakazawa et al., 2018). Saxena and Sanders (2009) found that among Asian Indian immigrant grandchildren, having more contact and assigning more importance to their grandparents' influence contributed to better relationship quality with their grandparents.

Racial and Ethnic Identity Development

A number of scholars have highlighted that first-generation Asian Indian immigrants tended to selectively acculturate (Ibrahim et al., 1997; Inman et al., 2007; Wakil et al., 1981). Selective acculturation is a process whereby immigrant groups adopt some aspects of the dominant culture while maintaining elements of their own ethnic culture (Portes & Rumbaut, 2014). For first-generation Asian Indians, bicultural functioning based on selective acculturation patterns can be attributed to their initial reason for immigrating, namely educational, and economic opportunities (Inman et al., 2007). Despite the physical distance from the homeland, psychological closeness is maintained by "reinventing Indian culture" on foreign soil (Dasgupta, 1998, p. 954) which in turn is also significant in contributing toward ethnic group identity (Phinney, 2003).

Among first-generation immigrants, this is central to the concept of ethnic identity, a component of self-concept, that consciously anchors an individual to a particular ethnic group (Ong et al., 2006; Phinney, 2000). Ethnic identity includes affective components such as a sense of belonging, pride, and affirmation; cognitive components such as knowledge of heritage and traditions; and value orientations such as individualism or collectivism and other components such as age and context

(Phinney, 2000; Phinney et al., 2000). Portes and Rumbaut (2014) noted that communities with selective acculturation patterns and in-group support tend to have higher ethnic identity retention than communities without such patterns and support. It denotes a sense of belonging, as well as a commitment to the group's values, beliefs, conventions, and customs.

Second-generation children of immigrants were likely to acquire their ethnic identities through their parents and relatives or through their ethnocultural network (Chen et al., 2014; Iwamoto et al., 2013; Knight et al., 2011; Umaña-Taylor et al., 2014). Iwamoto et al. (2013) found that there were unique features and multidimensional processes of racial and ethnic identity development among second-generation Asian Indians based on the strength of parental and reference group influences and experiences with discrimination that continued throughout the life span.

Asian Indian first-generation immigrant parents were disposed to either bicultural or traditional socialization goals (John & Montgomery, 2012). Parents with bicultural socialization goals believed that it was important for their children to learn autonomy but at the same time encouraged some observation of their ethnic culture such as maintaining ties to their extended families. Whereas parents with more traditional socialization goals were more likely to endorse family-centered and ethnic behaviors like stressing the importance of speaking in the mother tongue, cultural or religious observances, or implicit obedience to parental authority. Regardless of the socialization goals, both categories of parents gave high priority to their children forming close relationships with their grandparents.

Although the transmission of cultural socialization from parents to children has been documented, few studies have examined the impact of multiple cultural orientations, especially in multigenerational families as a result of the grandparents' role in the cultural orientation of their grandchildren. For example, late-life immigrant grandparents identify themselves as being Indians with more traditional socialization beliefs. Grandchildren who have had their grandparents living with them in multigenerational homes from a younger age were more likely to have a multidimensional cultural orientation (Ferguson et al., 2014; Kim & Hou, 2016; Treas & Mazumdar, 2004), as these grandparents have played a significant role in shaping their cultural orientation. These intergenerational relationships play a key role in the social and behavioral development of the second-generation children of immigrants.

Method

Participants

There were 19 participants from eight families who were interviewed in this study. They resided in five states in the U.S., namely Colorado, Michigan, New Mexico, North Carolina, and Texas. Individual interviews were conducted with late-life immigrant parents, adult children, and their respective spouses where available. All the participants identified as Asian Indians and Hindus.

Late-Life Asian Indian Parents. There was a total of nine participants. Eight participants self-identified as females with one male participant. They ranged in age from 71 to 89 years old (*M* age = 77.88 years; *SD* = 6). They described their ethnicities as Malayali, Tamil, Telugu, and Punjabi. They were all born in India and identified their mother tongue (Malayalam, Tamil, Telugu, and Punjabi) as being their dominant language, followed by other regional Indian dialects and English. They were all sponsored by their adult children and relocated to the U.S. between 5 and 20 years ago (*M* = 14.44 years; *SD* = 5.91). Six participants had obtained U.S. citizenship and the remaining three participants were green card holders waiting to transition to U.S. citizenship. Six of the participants were widows and only two couples remained together. The husband of one participant was unable to participate due to health reasons.

Adult Children and Spouses. Eight adult children (four sons, four daughters) participated in this study together with two of their spouses. One couple had both their mothers living with them and they were both interviewed separately with respect to their role as an adult child and as a spouse living with a mother-in-law. The participants were either naturalized U.S. citizens or green card holders. Three of the adult children came to the U.S. to pursue higher education at the graduate or professional level after obtaining their undergraduate degrees in India, another three came as undergraduate students and two of the daughters came after they married U.S. based spouses and established their own careers. Seven of the participants were still married, two participants were divorced, and one was widowed. The participants were between 44 and 60 years old (*M* age = 53.25 years; *SD* = 5.28), and they each had between one to three children and between 15 and 25 years of age (*M* age = 20.50 years; *SD* = 5.37).

The spouses were between 50 and 55 years old (*M* age = 54 years; *SD* = 1). They had established careers. One participant had recently retired. The spouses had their mothers-in-law living with them for several years and were involved in the caregiving aspects of the older participants. Tables 3.1, 3.2 and 3.3 provide an overview of the participants' demographic information. Pseudonyms have been assigned to the participants.

Interview Structure

The interviews were in-depth face to face interviews which lasted between 60 and 75 minutes. The participants' informed consent was obtained prior to the commencement of the interviews. Two instruments were used in gathering information: (1) a demographic questionnaire that sought information relating to the participants' age, gender, ethnicity, religion, marital status, income, and educational level, and (2) a semi-structured interview questionnaire.

Table 3.1 Demographic characteristics of participants as a percentage of the sample

Characteristics	Parents ($n = 9$)	Adult children ($n = 8$)	Spouses ($n = 2$)
Gender			
F	88.8	50.0	100.0
M	11.1	50.0	
Age			
40–50		25.0	
51–60		75.0	100.0
71–80	33.3		
81–90	66.6		
Ethnicity			
Malayali	33.3	37.5	50.0
Tamil	22.2	25.0	
Telugu	33.3	25.0	50.0
Punjabi	11.1	12.5	
Education			
Graduate	22.2	100.0	100.0
Bachelors			
High School	77.7		
Relationship Status			
Married	33.3	62.5	100.0
Divorced		25.0	
Widowed	66.6	12.5	
Employment Status			
Employed		100.0	50.0
Retired	33.3		50.0
Homemaker	66.6		
Annual Income (per annum)			
None	66.6		
Less than $10000	33.3		
$30000–59,988		12.5	50.0
>$60000		87.5	50.0

Data Analysis

Data were analyzed using the recursive phases of thematic analysis. Thematic analysis is a method for identifying, analyzing, and reporting themes (patterns) within data (Boyatzis, 1998; Braun et al., 2014), that involves discovering and identifying common threads that extend across the entire interview or sets of interviews (Braun et al., 2014; Braun & Clarke, 2006). Trustworthiness measures included collecting data from multiple sources in the same household as part of triangulation, maintaining an audit trail, memoing, providing clear and accurate descriptions and member checks.

Table 3.2 Demographic details of late-life participants

Name	Age	Gender	Marital status	Education	Children	Years in US	Living with	Reasons for immigration
Seema	81	F	Widow	H/School	4	20	Daughter	Childcare
Usha	76	F	Widow	H/School	3	15	Son	Death of Spouse
Indra	89	F	Widow	H/School	3	18	Son	Childcare
Leela	85	F	Widow	H/School	3	20	Son	Reunite with children
Gita	74	F	Married	H/School	2	10	Daughter	Childcare
Vasu	73	M	Married	Graduate	2	18	Daughter	Childcare
Radha	71	F	Married	Graduate	2	18	Daughter	Childcare
Padmini	78	F	Widow	H/School	2	6	Daughter	Death of Spouse
Dhana	74	F	Widow	H/School	1	5	Son	Childcare

Table 3.3 Demographic details of adult children and spouses

Name	Age	Gender	Ethnicity	Marital status	Education	Children	Relationship
Rajiv	53	M	Malayali	Married	College	2	Son/Son-in-law
Maya	53	F	Punjabi	Married	College	2	Daughter/ Daughter-in-law
Ashok	60	M	Tamil	Married	College	2	Son
Sheela	54	F	Malayali	Married	College	2	Daughter-in-law
Ram	60	M	Telugu	Married	College	2	Son
Dipa	55	F	Telugu	Married	College	2	Daughter-in-law
Devi	54	F	Malayali	Married	College	2	Daughter
Asha	44	F	Telugu	Divorced	College	3	Daughter
Priti	49	F	Malayali	Divorced	College	1	Daughter
Siva	53	M	Tamil	Widower	College	2	Son

Results

The analysis of the interviews with the older parents, adult children, and spouses focused on the identification of themes that were a part of each of their experiences. The manifestation of the themes was minimally varied among the families. The primary themes which were identified revolved around transmission of cultural identity through intergenerational caregiving and the role of grandparents in inculcating cultural identity.

Transmission of Cultural Identity through Intergenerational Caregiving

All the adult children spoke about fulfilling their filial obligations to their parents by taking care of them. They had a sense of duty which was a cultural expectation that was ingrained as part of their early socialization (de Valk & Schans, 2008; Diwan et al., 2011). Devi, age 54, described her decision to sponsor her parents as being inevitable after the demise of her brother: "I think there's no choice I have to. I mean this is what I am supposed to do, I can't just leave them somewhere. They're my parents I love them." Priti, age 44, and a divorcee with one child decided to sponsor her mother after the demise of her father. She summed up the sense of filial obligation as follows:

> It was somehow not said to you, maybe it was not even expected of you but there was that responsibility that as your parents grow older you would be the one to step in and take care of them.

Dipa, age 55, has been providing care to her mother-in-law, Seema since her health deteriorated. Her mother-in-law, who is in her eighties has multiple health problems, mobility issues and requires assistance. She felt that her mother-in-law's presence helped her reinforce family values and compassion in her two sons.

> I tried to instill compassion and family values in them and having her here has reinforced those. She always says things are so much better when they come. When I'm going to get water, they'll go run and bring it for me. If she is struggling to get somewhere they'll hold her hand and take her. They're not here a lot but when they are they're good with her and they get along very well. She feels very happy that she can spend time with them, and they try to spend time with her.

The close bonds shared with the grandchildren have been a tremendous comfort for the grandparents, many of whom have been caregivers to their grandchildren from infancy. Even though many of their grandchildren were much older and have their own lives, they described how happy they were when they visited. Indra, age 89 relocated to the U.S., 20 years ago just after her husband's retirement to assist her daughter. She was the primary caregiver to her daughter's children in the early years of her daughter's and son-in-law's medical careers and later to her son's children as well. She described her relationship with her grandchildren as follows:

> I find all my grandchildren have so much affection for me. They come for me, whenever they come here, even (granddaughter's name) also, whenever she comes she comes first here, and then goes.

All the adult children and spouses were unanimous in their appreciation for their parents' and/or in-laws' presence and influence in their children's lives and the difference it made in terms of the upbringing for some of the children who have had their grandparents present since birth. Rajiv and Maya, both age 53, came to the U.S. as young university students and stayed on to pursue their careers. They were in the unique position of having both their mothers live with them for almost 20 years. Maya shared that her two boys have definitely grown to be caring individuals as a result of their grandparents' influence in their lives and the close bond they shared:

> They love their grandparents. They're very caring individuals. For boys, they're very caring. Even like vacations sometimes when we don't take them they're like 'Oh *Nana* and *Nani, Achemma* why didn't we bring them.' Even though they are 18 and 20-year olds, the grandmothers will drive them crazy like 'when are you going to come and eat this, eat that' but they put up with it all and they're such good kids that way.

Rajiv felt that the benefits of living with the older parents far outweighed any drawbacks and served as an important learning experience for their two sons.

I think at the same time they got to see how we all lived together right? How we all interact, how we all care for each other which I think is very good for the kids to learn. And I think it is benefitted all of us.

Asha, age 44, has had her parents with her since the birth of her first child. She reflected that her three boys were more wholesome as a result of her parents' involvement in their lives. She shared that her older boys would sometimes mediate when there was an argument between her and her mother and that meant that they were able to appreciate the dynamics at play in family functioning:

> I feel they are more wholesome. They know when mum and I have differences, they actually talk to my mum – '*Ammama,* mum is trying to say this or say that or it's okay you know *Ammama.*' They are not growing up in a sterile environment of just mum and dad, they are used to this whole dynamic of a regular normal healthy family, where it's okay to have differences, but love prevails all the time.

The Role of Grandparents in Inculcating Cultural Identity

All the older participants regarded themselves as being the cultural conservators of their heritage and traditions by ensuring that they continued to observe many of the Hindu celebrations, prayers, and rituals, especially for the benefit of their grandchildren. Although the tenor of the traditional celebrations has been modified to fit with the environment they were in, they still held steadfast to many of the traditions of their communities. Usha, age 76, has been living with her son and daughter-in law for almost 15 years. She continued to observe all the traditions, customs, and practices of her heritage.

> We are very much immersed in the culture. We just follow our own culture we don't need to follow other cultures. *Diwali, Navaratri pooja, Onam,* we usually invite people over for a meal. We observe everything just like back home.

They also took it upon themselves to promote their culture and traditions by speaking in their mother tongue to their adult children and grandchildren, teaching grandchildren values through storytelling and songs, trying to infuse spiritual lessons with teachings from the Hindu Vedic sacred texts, and encouraging the grandchildren to learn Indian classical dance or music.

Vasu and Radha, a couple in their early 70s, relocated to the U.S. after the birth of their first grandchild. Their daughter sponsored them as she needed their assistance with childcare. They were both active in their respective careers in India but decided to relocate after retirement. They both took an active part in the education of their three grandsons especially in ensuring that they were able to appreciate their culture and heritage. Radha described how they fostered their traditions and culture as follows:

> Our grandchildren, they have the values, they have knowledge and respect. We speak Telugu in the house, so they don't forget it, recite Telugu poems full of values to them and slokas from the *Bhagavadgita*.

Vasu and Radha recalled spending a lot of time with their three grandchildren in terms of telling them stories from Hindu mythology, teaching them how to speak in their mother tongue, Telugu and inculcating spiritual values in them. As Vasu explained:

> I teach them all the stories, the *Bhagavatam, Bhagavadgita, Telugu.* I used to bring books also for our grandchildren. It's called *Amar Chitra Katha.* I used to bring it. You know the volumes are bound. These two kids whenever they take a break for lunch they will be reading one book. We don't keep quiet when they are around here. We chant and sing with them.

For the adult children and their families, strengthening ties to their culture and heritage was a significant outcome of having the parents live with them. Their presence ensured that traditional food prepared in the household in addition to the rituals, celebrations, values, beliefs, and customs would be incorporated into the family lifestyle. Many of the adult children and spouses revealed that without their parents' presence in their household, it was more than likely that they would not be as steadfast in observing many of the traditions and cultural values. Siva, age 53, is an engineer, educated in the U.S. He is a first-generation Asian Indian who has lived in the U.S. for more than 20 years. He sponsored his mother to live with him in the U.S. after the demise of his wife a few years ago. He still observes the Indian customs and traditions with the help of his mother, which he feels is his way of setting an example for his two teenaged daughters on appreciating their culture and heritage:

> I try to pretty much do the same things that my parents did. Growing up you watch and then you do the same things and hopefully the children will do the same you know; they'll keep the traditions going whether it is the religious rites that you perform or celebrating the Hindu festivals.

For Rajiv and Maya both their parents served as the main caregivers for their two sons. Since Rajiv is South Indian and Maya is North Indian, they conveyed how having their parents with them has helped them appreciate the rich traditions from North and South India. As Rajiv explained:

> There are things with them around we'd probably do more, whether it be culturally or festivities, because if it was just me and Maya, we wouldn't have been getting up in the morning and observing all that. Getting something from South India and North India, getting to see all the different things they do, the customs, the behaviors, what they do. I think it's great.

Maya described how both her sons participated and enjoyed the various festivities that they celebrated in their homes. Although they identified as Hindus, they also celebrated the Christmas and Thanksgiving as it was their way of feeling that they were a part of the fabric of this country:

> We do have our South Indian festivals, *Vishu* and *Onam* that we do together, and the kids are all into it and they love the food and they wear their *kurtas* and eat on the banana leaf and it's lovely. Thanksgiving and Christmas are the main festivals for us as well. My favorite is Thanksgiving, which is an American holiday, but I love it because I feel like this country has given me everything that I have today so I'm just very grateful.

A traditional family-based value system was also important for Dipa, age 54, who described herself as having a close relationship with her mother-in-law. Her own experience of living in an extended family in India during her childhood years helped to pave the way for her own experience of having her mother-in-law living in her home in the U.S., ensuring that they observed all the rituals and festivities of their culture:

> I grew up in a joint family, it was a family where we had more than just the husband and wife. I come from a family that has a lot of traditional values and we do things traditionally. Having her around has definitely helped to enhance that. We celebrate pretty much everything. Birthdays and *poojas,* like we did *Ganesh Chaturthi* together, we had *Lakshmi Pooja.*

Ram, age 60 and his siblings sponsored both their parents to relocate to the U.S. The parents used to rotate their stay between him and his two siblings in the U.S. and the parents played an active role in assisting with childcare and other activities. Ram felt that having his parents with them since his sons were young children has given them an appreciation for multigenerational living and intergenerational relationships. He was grateful for having his parents around as it allowed his sons to gain an understanding of their heritage and a sense of connection with their past:

> They also get to hear stories about how their grandparents grew up and what they did. They were always fascinated with that, so they used to talk to my father about that, and talk about me, how I grew up. They like to hear all those stories. There's a lot of things you can learn from grandparents and so we actually feel lucky that they had the opportunity.

Discussion

The qualitative in-depth interviews with the late-life parents and adult children revealed two major themes: the transmission of cultural identity through intergenerational caregiving and the role of grandparents in inculcating cultural identity. The study demonstrates that families provide the primary setting through which culture is defined and interpreted (Hayslip et.al., 2019; Wiscott & Kopera-Frye, 2000). Positive relationships with adult children and grandchildren support socialization processes and provide a route to adaptation for older parents in multigenerational families that also contribute to family well-being.

The older participants considered their role as grandparents as the most important family role they had. They not only identified closely with their role and identity as grandparents, but they also derived a great sense of accomplishment and meaning in terms of their contributions to their grandchildren's development, in addition to shaping the cultural orientation of the grandchildren (Silverstein et al., 2012; Silverstein & Marenco, 2001). The grandparents in these families were instrumental in helping to facilitate the development of cultural identity in their grandchildren. Grandchildren with greater exposure to grandparents earlier in life have also been found to have more positive attitudes toward older people in general and stronger ties to their cultural heritage (Hayslip et al., 2019; McGuinn & Mosher-Ashley, 2002). Some studies have also highlighted how a strong emotional bond with a grandparent effectively models a healthy relationship and serves to mitigate the negative effects of parenting (Silverstein & Ruiz, 2006).

Treas and Mazumdar (2004, p.106) described the grandparents' contribution to shaping the cultural orientation of their grandchildren as "symbolic kin keeping" which refers to activities that go beyond the functional requirements to keep families together, specifically extending to activities that preserve the meaning of kin relationships and the ethnic culture. Our findings suggest that the reciprocity inherent in intergenerational caregiving allows grandparents to have a direct and indirect influence on their grandchildren, as in teaching them values and social norms, in addition to being important conduits through which language, customs, and traditions of the culture were communicated to children and grandchildren (Silverstein et al., 2012; Usita & Shakya, 2012). These functions take on special importance for grandparents because they depend on the younger generation to keep the culture and traditions alive. Therefore, being cultural conservators to the younger generation was an instrumental method of creating continuity and provided older parents with ongoing status and identity (Becker et al., 2003).

Implications for Research and Policy

The narratives of the participants of this study highlight the type of family relationships that multigenerational living can generate. The implication of these stories is that children of immigrants stand to benefit by learning more about their cultural heritage, traditions, and identity. The role of grandparents and the processes within multigenerational families should also be considered as important mechanisms that shape the cultural orientations and identities of the younger generations of Asian Indians in the U.S. Clinicians and service providers working with multigenerational immigrant families must be prepared to understand the nuances of ethnic socialization within these families and how each generation would navigate their identities within these family systems.

There are a number of limitations of this study related to a small and homogeneous sample size. A second limitation is data collection, which was not specifically geared toward identity development in these families. More empirical studies are required in this area with data from the grandchildren in these families as well. Notwithstanding the limitations, this study provides some initial evidence that indicates that children growing up in multigenerational Asian Indian minority families are likely to have multiple frames of cultural references which points to a more multidimensional socialization process.

The changes being proposed to the immigration provisions in the U.S., on ending chain migration policies will be problematic for many families as it will impact the long-standing American policy of supporting family reunification (Qiu, 2018). Domestic immigration policies must continue to support the entry of older immigrants to join their children especially in the light of the benefits of multigenerational living which far outweighs the challenges.

In conclusion, the findings of this study are consistent with a number of studies which have found that late-life immigrants to the U.S. were able to navigate the challenges of aging in a new environment as a result of the reciprocity of care within multigenerational households. The giving and receiving of care and support played an important function in bridging generations together within the immediate and extended family, fostering a sense of generativity. The late-life parents were able to actively construct their identities in their role as grandparents by providing practical and emotional support and were conduits for their grandchildren's ethnic socialization in terms of serving as family historians and transmitting ethnic heritage, family traditions, and norms. The cultural socialization that children of immigrants are exposed to from their traditional grandparents and bicultural parents in multicultural homes may result in their ability to synthesize aspects of both cultures into a unique blend of Indian and Asian Indian and American.

References

Babu, N., Hossain, Z., Morales, J. E., & Vij, S. (2018). Grandparents in Bangladesh, India, and Pakistan: A way forward with traditions and changes in south Asia. In D. Shwalb & Z. Hossain (Eds.), *Grandparents in cultural context* (pp. 159–186). Routledge. https://doi.org/10.4324/9781315642284

Becker, G., Beyene, Y., Newsom, E., & Mayen, N. (2003). Creating continuity through mutual assistance: Intergenerational reciprocity in four ethnic groups. *Journal of Gerontology: Social Sciences, 58B*, S151–S159. https://doi.org/10.1093/geronb/58.3.S151

Bhattacharya, G., & Shibusawa, T. (2009). Experiences of aging among immigrants from India to the United States: Social work practice in a global context. *Journal of Gerontological Social Work, 52*(5), 445–462. https://doi.org/10.1080/01634370902983112

Boon, S. D., Shaw, M. J., & MacKinnon, S. L. (2008). Grandparent health and young adults' judgments of their grandparent-grandchild relationships. *Journal of Intergenerational Relationships, 6*(2), 155–173. https://doi.org/10.1080/15350770801955115

Boyatzis, R. E. (1998). *Transforming qualitative information: Thematic analysis and code development*. Sage.

Braun, V., & Clarke, V. (2006). Using thematic analysis in psychology. *Qualitative Research in Psychology, 3*, 77–101. https://doi.org/10.1191/1478088706qp063oa

Braun, V., Clarke, V., & Terry, G. (2014). Thematic analysis. In P. Rohleder & A. C. Lyons (Eds.), *Qualitative research in clinical and health psychology* (pp. 95–114). Macmillan. https://doi.org/10.1007/978-1-137-29105-9

Chen, S. H., Hua, M., Zhou, Q., Tao, A., Lee, E. H., Ly, J., & Main, A. (2014). Parent–child cultural orientations and child adjustment in Chinese American immigrant families. *Developmental Psychology, 50*, 189–201. https://doi.org/10.1037/a0032473

Cohn, D. V, & Passel, J. S. (2018). *A record 64 million Americans live in multigenerational household*. https://www.pewresearch.org/fact-tank/2018/04/05/a-record-64-million-americans-live-in-multigenerational-households/

Connidis, I. A., & Barnett, A. E. (2019). *Family ties and aging*. Sage.

Dasgupta, S. D. (1998). Gender roles and cultural continuity in the Asian Indian immigrant community in the US. https://psycnet.apa.org/doi/10.1023/A:1018822525427

de Valk, H. A. G., & Schans, D. (2008). "They ought to do this for their parents": Perceptions of filial obligations among immigrant and Dutch older people. *Ageing & Society, 27*, 49–66. https://doi.org/10.1017/s0144686x07006307

Diwan, S., Lee, S. E., & Sen, S. (2011). Expectations of filial obligation and their impact on preferences for future living arrangements of middle-aged and older Asian Indian immigrants. *Journal of Cross-Cultural Gerontology, 26*(1), 55–69. https://doi.org/10.1007/s10823-010-9134-6

Dykstra, P. (2010). *Intergenerational family relationships in ageing societies*. United Nations Economic Commission for Europe, United Nations. https://www.unece.org/index.php?id=10581

Ferguson, G. M., Iturbide, M. I., & Gordon, B. P. (2014). Tridimensional (3D) acculturation: Ethnic identity and psychological functioning of tricultural Jamaican immigrants. *International Perspectives in Psychology: Research, Practice, Consultation, 3*(4), 238–251. https://doi.org/10.1037/ipp0000019

Glick, J. E., & Hook, J. (2002). Parents' co-residence with adult children: Can immigration explain racial and ethnic variation? *Journal of Marriage and Family, 64*(1), 240–253. https://doi.org/10.1111/j.1741-3737.2002.00240.x

Greenfield, E. A. (2011). Grandparent involvement and parenting stress among nonmarried mothers of young children. *Social Service Review, 85*(1), 135–157. https://doi.org/10.1086/658395

Hayslip, B., Jr., Toledo, R. M., Henderson, C. E., Rodriguez, R. M., & Caballero Vela, D. M. (2019). Cross-cultural differences in the experience of grandparent–grandchild relationships and related psychosocial outcomes. *The International Journal of Aging and Human Development, 89*(2), 151–171. https://doi.org/10.1177/0091415018769468

Hayslip, B., & Page, K. S. (2012). Grandparenthood: Grandchild and great-grandchild relationships. In R. Bleiszner & V. Hilkevitch-Bedford (Eds.), *Handbook of families and aging* (pp. 183–212). Praeger.

Hayslip, B., & White, D. (2008). Grandparents as grievers. In M. S. Stroebe, R. O. Hansson, W. Stroebe, & H. Schut (Eds.), *Handbook of bereavement research (pp. 441–460)*. American Psychological Association.

He, W., Goodkind, D., & Kowal, P. (2016). *US Census Bureau, international population reports*. P95/ 16-1. *An aging world: 2015*. US government printing office.

Hendricks, J. (2012). Considering life course concepts. *Journals of Gerontology Series B: Psychological Sciences and Social Sciences, 67*(2), 226–231. https://doi.org/10.1093/-geronb/gbr147

Hing, B. O. (1993). Making and remaking Asian America through immigration policy: 1850–1990. Stanford University Press. https://doi.org/10.1086/ahr/100.5.1701-a

Hunter, A. G., & Taylor, R. J. (1998). Grandparenthood in African American families. In M. E. Szinovacz M. E. (Ed.), *Handbook on grandparenthood (pp.70–86)*. Greenwood.

Ibrahim, F., Ohnishi, H., & Sandhu, D. S. (1997). Asian American identity development: A culture specific model for south Asian Americans. *Journal of Multicultural Counseling and Development, 25*(1), 34–50. https://doi.org/10.1002/j.2161-1912.1997.tb00314.x

Inman, A. G., Howard, E. E., Beaumont, R. L., & Walker, J. A. (2007). Cultural transmission: Influence of contextual factors in Asian Indian immigrant parents' experiences. *Journal of Counseling Psychology, 54*(1), 93–100. https://doi.org/10.1037/0022-0167.54.1.93

Ishizawa, H. (2004). Minority language use among grandchildren in multigenerational households. *Sociological Perspectives, 47*(4), 465–483. https://doi.org/10.1525/sop.2004.47.4.465
Iwamoto, D. K., Negi, N. J., Partiali, R. N., & Creswell, J. W. (2013). The racial and ethnic identity formation process of second-generation Asian Indian Americans: A phenomenological study. *Journal of Multicultural Counseling and Development, 41*(4), 224–239. https://doi.org/10.1002/j.2161-1912.2013.00038.x
John, A., & Montgomery, D. (2012). Socialization goals of first-generation immigrant Indian parents: A Q-methodological study. *Asian American Journal of Psychology, 3*(4), 299–312. https://doi.org/10.1037/a0025556
Kim, S. Y., & Hou, Y. (2016). Intergenerational transmission of tridimensional cultural orientations in Chinese American families: The role of bicultural socialization. *Journal of Youth and Adolescence, 45*(7), 1452–1465. https://doi.org/10.1007/s10964-016-0423-4
Kivnick, H. Q. (1982). *The meaning of grandparenthood*. UMI Research Press.
Knight, G. P., Berkel, C., Umaña-Taylor, A. J., Gonzales, N. A., Ettekal, I., Jaconis, M., & Boyd, B. M. (2011). The familial socialization of culturally related values in Mexican American families. *Journal of Marriage and Family, 73*(5), 913–925. https://doi.org/10.1111/j.1741-3737.2011.00856.x
Lamb, S. (2002). Intimacy in a transnational era: The remaking of aging among Indian Americans. *Diaspora: A Journal of Transnational Studies, 11*(3), 299–330. https://doi.org/10.1353/dsp.2011.0003
McGuinn, K. K., & Mosher-Ashley, P. M. (2002). Children's fears about personal aging. *Educational Gerontology, 28*(7), 561–575. https://doi.org/10.1080/03601270290099769
Mogelonsky, M. (1995). Asian-Indian Americans. *American Demographics, 17*(8), 32–39.
Nakazawa, J., Hyun, J.-H., Ko, P.-C., & Shwalb, D. W. (2018). Grandparents in Japan, Korea and China: From filial piety to grandparenthood. In D. Shwalb & Z. Hossain (Eds), *Grandparents in cultural context* (pp. 18–219). Routledge.
Nandan, M. (2007). "Waves" of Asian Indian elderly immigrants: What can practitioners learn? *Journal of Cross-Cultural Gerontology, 22*, 389–404. https://doi.org/10.1007/s10823-007-9042-6
Olshanky, S. J. (2011). Trends in longevity and prospects for the future. In R. H. Binstock & L. K. George (Eds.), *Handbook of aging and the social sciences* (pp. 47–56). Elsevier.
Ong, A. D., Phinney, J. S., & Dennis, J. (2006). Competence under challenge: Exploring the protective influence of parental support and ethnic identity in Latino college students. *Journal of Adolescence, 29*(6), 961–979. https://doi.org/10.1016/j.adolescence.2006.04.010
Phinney, J. S. (2000). Ethnic and racial identity: Ethnic identity. In A. E. Kazdin (Ed.), *Encyclopedia of psychology* (Vol. 3, pp. 254–259). American Psychological Association. https://doi.org/10.1037/10518-122
Phinney, J. S. (2003). Ethnic identity and acculturation. In K. M. Chun, P. Balls Organista, & G. Marín (Eds.), *Acculturation: Advances in theory, measurement, and applied research* (pp. 63–81). American Psychological Association. https://doi.org/10.1037/10472-006
Phinney, J. S., Ong, A., & Madden, T. (2000). Cultural values and intergenerational value discrepancies in immigrant and non-immigrant families. *Child Development, 71*, 528. https://doi.org/10.1111/1467-8624.00162
Portes, A., & Rumbaut, R. G. (2014). *Immigrant America: A portrait*. University of California Press. https://doi.org/10.1525/9780520959156
Pratt, M. W., Norris, J. E., Hebblethwaite, S., & Arnold, M. L. (2008). Intergenerational transmission of values: family generativity and adolescents' narratives of parent and grandparent value teaching. *Journal of Personality, 76*(2), 171–198. https://doi.org/10.1111/j.1467-6494.2007.00483.x
Pruchno, R. (2018). Aging in context. *The Gerontologist, 58*(1), 1–3. https://doi.org/10.1093/geront/gnx189
Qiu, L. (2018). What is chain migration? *Here is the controversy behind it*. https://www.nytimes.com/2018/01/26/us/politics/the-facts-behind-the-weaponized-phrase-chain-migration.html
Rangaswamy, P. (2000). *Namaste America: Indian immigrants in an American metropolis*. State Press. https://doi.org/10.2307/2693070

Reitzes, D. C., & Mutran, E. J. (2004). Grandparent identity, intergenerational family identity, and well-being. *The Journals of Gerontology Series B: Psychological Sciences and Social Sciences, 59*(4), S213–S219. https://doi.org/10.1093/geronb/59.4.s213

Ruiz, D. S. (2004). *Amazing grace: African American grandmothers as caregivers and conveyers of traditional values*. Praeger. https://doi.org/10.5860/choice.42-4338

Saxena, D., & Sanders, G. F. (2009). Quality of grandparent-grandchild relationship in Asian-Indian immigrant families. *The International Journal of Aging and Human Development, 68*(4), 321–337. https://doi.org/10.2190/ag.68.4.c

Seibert, A. C., & Kerns, K. A. (2009). Attachment figures in middle childhood. *International Journal of Behavioral Development, 33*(4), 347–355. https://doi.org/10.1177/0165025409103872

Sharma, K., & Kemp, C. L. (2012). "One should follow the wind": Individualized filial piety and support exchanges in Indian immigrant families in the United States. *Journal of Aging Studies, 26*(2), 129–139. https://doi.org/10.1016/j.jaging.2011.10.003

Silverstein, M., Lendon, J., & Giarrusso, R. (2012). Ethnic and cultural diversity in aging families: Implications for resource allocation and well-being across generations. In R. Blieszner & V. H. Bedford (Eds.), *Handbook of families and aging* (pp. 287–308). Praeger.

Silverstein, M., & Marenco, A. (2001). How Americans enact the grandparent role across the family life course. *Journal of Family Issues, 22*(4), 493–522. https://doi.org/10.1177/019251301022004006

Silverstein, M., & Ruiz, S. (2006). Breaking the chain: How grandparents moderate the transmission of maternal depression to their grandchildren. *Family Relations, 55*(5), 601–612. https://doi.org/10.1111/j.1741-3729.2006.00429.x

Stelle, C., Fruhauf, C. A., Orel, N., & Landry-Meyer, L. (2010). Grandparenting in the 21st century: Issues of diversity in grandparent–grandchild relationships. *Journal of Gerontological Social Work, 53*(8), 682–701. https://doi.org/10.1080/01634372.2010.516804

Thiele, D. M., & Whelan, T. A. (2008). The relationship between grandparent satisfaction, meaning, and generativity. *The International Journal of Aging and Human Development, 66*(1), 21–48. https://doi.org/10.2190/ag.66.1.b

Treas, J., & Mazumdar, S. (2004). Caregiving and kin keeping. Contributions of older people to America's immigrant families. *Journal of Comparative Family Studies, 35*, 105–122. https://doi.org/10.1016/s0890-4065(02)00048-8

U.S Census Bureau. (2017a). Place of birth for foreign born populations. https://www.census.gov/topics/population/foreign-born.html

U.S Census Bureau. (2017b). *American Community Survey*. https://www.census.gov/acs/www/data/data-tables-and-tools/data-profiles/2017/

Umaña-Taylor, A. J., Quintana, S. M., Lee, R. M., Cross, W. E., Rivas-Drake, D., Schwartz, S. J., Syed, M., Yip, T., & Seaton, E. (2014). Ethnic and racial identity during adolescence and into young adulthood: An integrated conceptualization. *Child Development, 85*(1), 21–39. https://doi.org/10.1111/cdev.12196

Usita, P. M., & Shakya, H. B. (2012). Aging families and immigration. In R. Bleiszner & V. Hilkevitch-Bedford (Eds.), *Handbook of families and aging* (pp. 337–359). Praeger.

Wakil, S. P., Siddique, C. M., & Wakil, F. A. (1981). Between two cultures: A study in socialization of children of immigrants. *Journal of Marriage and the Family*, 929–940. https://doi.org/10.2307/351349

Wilmoth, J. M. (2001). Living arrangements among older immigrants in the United States. *The Gerontologist, 41*, 228–238. https://doi.org/10.1093/geront/41.2.228

Wilmoth, J. M. (2010). Health trajectories among older movers. *Journal of Aging and Health, 22*(7), 862–881. https://doi.org/10.1177/0898264310375985

Wiscott, R., & Kopera-Frye, K. (2000). Sharing of culture: Adult grandchildren's perceptions of intergenerational relations. *International Journal of Aging and Human Development, 51*, 199–215. https://doi.org/10.2190/0uy5-mxxp-w81k-vxcu

Chapter 4
Multidimensional Acculturation and Identity Re-Formation Processes Among South Sudanese Refugee as Emerging Adults in the United States

Junghee Yoon, Deborah J. Johnson, Desiree B. Qin, and Laura Bates

Between 1999 and 2001, approximately 3800 unaccompanied South Sudanese refugee youth arrived in the United States (U.S.) after years of trauma and hardship from the violent war and political oppression in their home country (Bates et al., 2005; Luster et al., 2008; United States Conference of Catholic Bishops [USCCB], 2013). They were separated from their parents as children and displaced at various times, driven from Ethiopia at gunpoint or via sweeping internally set fires; their long and dangerous desert treks led many of them to Kenya's refugee camps with other children and youth. The loss of parents during the war aggravated their traumatic childhood experience (Berman, 2001; Garmezy & Rutter, 1983). At the time of their arrival in the U.S., approximately 500 were minors (under the age of 18), and they were placed under the Unaccompanied Refugee Minors Program (URMP) across the country (Geltman et al., 2005; USCCB, 2013). Some of the minors were raised by European American foster parents, but in general, most were resettled in the U.S. and have been educated in this country (Bates et al., 2005).

South Sudanese unaccompanied refugee minors (URMs) have transitioned from adolescence to young adulthood without the care and guidance from their biological parents. Over time, their integration into the American society has deepened across educational, social, cultural, and economic domains (Luster et al., 2010; Qin et al., 2015; Rana et al., 2011). However, little is known about how they have developed, modified, and strengthened their self-identities while acculturating into American society as URMs for more than a decade. To fill this gap, the current study focuses on the processes of ethnic, racial, and immigrant identity re-formation and acculturation among South Sudanese URMs as emerging adults in the U.S.

J. Yoon (✉) · D. J. Johnson · D. B. Qin · L. Bates
Michigan State University, East Lansing, MI, USA
e-mail: yoonjun8@msu.edu; john1442@msu.edu; dqin@msu.edu; bateslau@msu.edu

D. J. Johnson et al. (eds.), *Re/Formation and Identity*, Advances in Immigrant Family Research, https://doi.org/10.1007/978-3-030-86426-2_4

Given the distinct differences between traditional culture and American culture, acculturation is an inevitable and challenging task for URMs from South Sudan across a range of developmental domains, including identity. In the psychology literature, researchers have discussed important aspects of ethnic identity development in the process of acculturation such as exploration and commitment, among immigrant adolescents and emerging adults (Phinney, 1992, 2003). A recent literature review of 110 studies on the relationships among acculturation, ethnic identity, and mental health concludes that these areas are closely related, affecting individuals, as well as communities (Balidemaj & Small, 2019).

In addition, it is well known that the family is a critical context for the acculturation of immigrant adolescents (Phinney & Vedder, 2006). For example, the immigrant parents play a significant role in transmitting traditional cultural values and fostering their children's development of self and identity in the multicultural context (Kim & Hou, 2016). However, less is known about how the experience of acculturation during adolescence is associated with the identity formation in emerging adulthood when the individuals were significantly disadvantaged by the separation from their parents in early childhood.

Adolescence and emerging adulthood are critical periods for identity exploration and maturation (Arnett, 2000, 2004; Erikson, 1968). Compared to adolescence, emerging adulthood is especially tasked with sociocultural identity development in terms of race and ethnicity (Blake, 2019). Exploring ethnic and racial identity in this population is particularly compelling because race and ethnicity may be more salient for emerging adults, as their social interactions are broadened by making important developmental transitions from high school to college or to employment (Phinney & Chavira, 1992). Through these transitions, emerging adults engage in developing a more complex self-concept as they attempt to fit in and to understand how others view them in the larger sociocultural and political contexts. Therefore, considering the macro-level contextual factors that influence self-construction at the microlevel seems developmentally appropriate when examining identity changes and retention during emerging adulthood.

Acculturation and identity re-formation processes are likely to be more complex and influential in a variety of contexts for immigrants of color and even more so for violently and suddenly displaced refugees. Refugees from Africa often experience complex acculturation processes in terms of their race/ethnicity and religion, though their experiences are understudied (Ferguson & Birman, 2016). It is likely that processes of acculturation and identity development studied with convenience samples of college students from immigrant backgrounds may be insufficient to explain the unique experience of the young refugee populations. For this reason, the multicultural psychology perspective suggests purposeful sampling to understand the developmental experiences of ethnoculturally diverse populations who are highly underrepresented in research (Hall et al., 2016).

In a Charles Smith's (2017) play, Objects in the Mirror, a young refugee, Shedrick, charts a treacherous path out of Africa to Australia, fueled by war. However, the passage to Australia requires an identity trade-off by the young travelers, first, a loss of homeland, then the tenuous safety in a new land, marred by their crafted lives and identities. Shedrick must give up his identity (name and country) in order to survive. The youth at the center of the play struggles desperately with the inability to be who he truly is because the accommodations required and assimilation demands prevent him from using his real name, which is his last tie to his homeland and family. For the young man, these are bitter sacrifices of identity, survival, and family.

Similarly, diverse African immigrant populations find themselves quickly subsumed under the U.S. racial categories of Blacks or African Americans and hidden within these contexts. However, the experiences of this group of South Sudanese refugee youth are quite different, not only from immigrant communities of color, but also from African American communities. Additionally, their experiences as URMs cannot be completely aligned with the refugee children who arrived with their parents. Although a few studies have explained different developmental aspects between African immigrants and nonimmigrant African Americans based on diverse ethnicities and cultures (e.g., Thelamour & Johnson, 2017), these distinctions do not prevail within the developmental literature.

For more than a decade, Berry's (1997) bicultural perspective that merely focuses on mainstream American culture and immigrants' culture of origin has dominated previous research on acculturation across many countries. However, Ferguson et al. (2012) point out that diverse cultures exist in the U.S., arguing that the traditional bidimensional framework of acculturation does not fully capture the experience of Black immigrants, who are likely to be exposed to more than two cultures. The researchers propose a tridimensional (3D) model of acculturation tested by their study involving Jamaican immigrant youth and adults. Their findings show that African immigrants are likely to incorporate Jamaican, European American, and African American cultures in their acculturation process (Ferguson et al., 2012; Ferguson & Bornstein, 2014). Moreover, the integration of these three cultures (i.e., tricultural) is prominent and associated with more positive behavioral adaptation among the youth compared to those who assimilate (Ferguson et al., 2012).

Interestingly, however, tricultural Jamaican immigrant adults do not necessarily express themselves with the multicultural self-label but report diverse self-labels based on their country of origin, race, and ethnicity, depending on salience (Ferguson et al., 2014). This finding shows how Black and Brown immigrants and refugees internalize and express their identities can vary by sociocultural situations, especially where they have to align themselves with -and fit into the American racial hierarchy. Although this seems to be an important part of their integration process, why they identify themselves as they do in certain times and situations remains unclear (Glozman & Chuang, 2018). It is also vague how acculturation in public domains (e.g., behavior, language use) is interconnected to acculturation in private domains (e.g., values, ethnic, and racial identity) and which processes, managed in what ways, lead to healthy outcomes.

Sociologists have also emphasized race and ethnicity as important factors that significantly influence the experiences of immigrants of color in the U.S. For example, Portes and Zhou (1993) propose the segmented assimilation theory to demonstrate how the racial minority categorization affects the immigrants' opportunities in life and adjustment outcomes in education and economic status. In particular, the researchers claim "downward assimilation" as a problematic path that second-generation immigrants are likely to experience as a consequence of assimilation toward the minority culture and behaviors in the U.S. (Portes & Rumbaut, 2001; Portes & Zhou, 1993). Although the perspective of segmented assimilation provides a framework to describe a different typology of acculturation outcomes (i.e., upward, downward), the concept has evolved based on a constrained Eurocentric hypothesis (Rudmin et al., 2017; Stepick & Stepick, 2010). Thus, the theory's assertion about the negative influence of adopting U.S. minority cultures and orientations on the children of immigrants is controversial (Stepick & Stepick, 2010; Waters et al., 2010) and contradictory to the recent findings that show the prevalence and advantage of incorporating African American culture and identification in the lives of Black African immigrant youth in the U.S. (Ferguson et al., 2012; Habecker, 2017).

Based on this current literature, this study attempts to provide an understanding of the experiences of African refugees in a Western society, with a special focus on South Sudanese URMs in the U.S. The current study describes important psychosocial and contextual factors that have influenced their multidimensional acculturation and identity re-formation during their long-term processes of American resettlement. Specifically, the current study is the first set of studies to investigate how the multidimensional acculturation processes, intertwined with identity re-formation in the African URM population in America at different points of emerging adulthood.

Psychosocial Context of South Sudanese Refugees in the U.S.

Centering on ecological systems of development (Bronfenbrenner & Morris, 1998, 2006) as a framework, we provide a discussion of the unique psychosocial context of South Sudanese refugees to better understand the results and developmental implications presented in the current study. For this group of South Sudanese refugees, identity development has been under the influences of sociocultural, political, and historical contexts across two different countries: South Sudan and the U.S. Additionally, race and ethnicity (personal characteristics), acculturation experiences (representing both processes and contextual factors in this study), and American sociocultural aspects (also representing both processes and contextual factors) all appear to contribute to South Sudanese refugees' identity re-formation. In turn, identity reconstruction seems to be an important component of a general adjustment outcome in this population. Thus, knowing their dynamic situations and statuses across the different systems of development is necessary to understand and interpret the participants' narratives (see Fig. 4.1).

Past experiences of war, parent separation, and American resettlement are the important contexts for the developing person to understand their formation and re-formation of identities (e.g., South Sudanese, refugees) over time. During

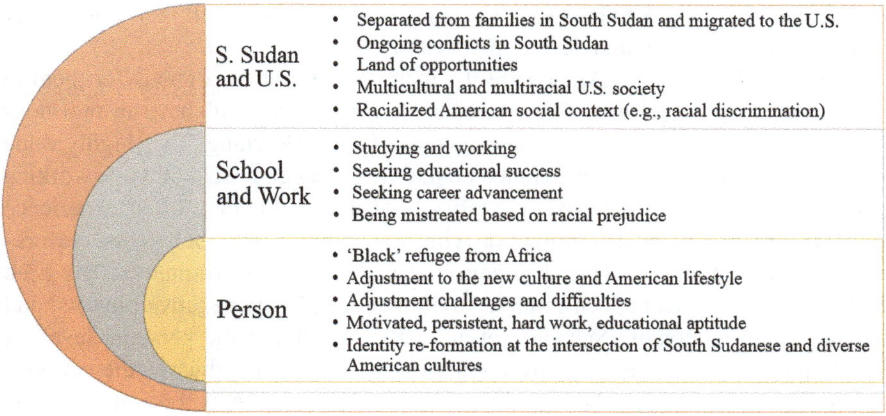

Fig. 4.1 Key psychosocial context of South Sudanese refugees in the U.S.

resettlement, these young refugees from South Sudan have strived not only to overcome the trauma of war but also to thrive as new residents of America. In general, these refugee youth, despite being critically disadvantaged by separation from their parents as children, have maintained a positive outlook for the future through the current resettlement process (Luster et al., 2010; Qin et al., 2015). Many of these refugees have sought better educational opportunities and good jobs in the hope of helping those who were left behind in their country of origin and helping themselves, too, to make it easier for them to acculturate in American society (Lawrence et al., 2016; Lese & Robbins, 1994; Rana et al., 2011; Wallin & Ahlström, 2005).

At the macro-level of human development, two nations have influenced their lives from early childhood to young adulthood. First, they were born and raised in South Sudan, forced to migrate as children, and have resettled as adolescents in the U.S. Despite having lived in the U.S. for more than 15 years, many of them still perceive themselves as South Sudanese based on their early socialization and memories of their birthplace. There have been ongoing political conflicts in South Sudan, and many of South Sudanese in the U.S. lost their family members due to violent incidents. Many refugees have been supporting their families transnationally by sending money, and some of them have formed nongovernmental organizations (NGOs) to contribute to rebuilding their home country. They have founded schools and a hospital for those who remain in the country, suffering from hunger, diseases, lack of education, poor infrastructure, and lack of access to the resources essential for survival. This transnational bond plays an important role in influencing their lives in the U.S.

Further, the host country—the U.S.—has exposed them to a completely different social environment. During their early period of resettlement, they encountered cultural and environmental differences as they experienced new ways of living, thinking, behaving, and interacting (Luster et al., 2010). However, during the latter course of their resettlement, overcoming cultural differences and being able to function in an "American" way are considered crucial among these refugees in terms of managing their daily activities to the extent that they could fulfill their purpose by acquiring an education, and developing a career.

The school and the workplace are the major acculturative contexts for them to adjust (i.e., adaptive cultures; García Coll et al., 1996). The youth have an enormous drive to be successful and to complete degrees of higher learning. They highly value education and hope for career advancement. Often, they go to school while working part-time for financial reasons. At both school and work, they often experience social mistreatment and discrimination when they interact with their peers, coworkers, and even strangers. This experience of racism and discrimination has been reported as a significant acculturative stress associated with negative physical and mental health outcomes (Schwartz et al., 2010). Lastly, at the personal level, as Black immigrants and refugees from Africa, they struggle to adjust to the new culture and be accustomed to the American lifestyle. They are highly motivated, persistent, enduring, and work hard to achieve their goals. They pursue education not just to have a better life but also to make the best of their purpose of contributing to their country and helping their people gain economic stability. Psychologically, they have also undergone the process of reconstructing their identity (e.g., bicultural identity) and negotiating different cultures as a way of adapting (e.g., biculturalism) (Qin et al., 2015).

Quasi-longitudinal Qualitative Study

The present study is part of a larger research project on South Sudanese refugee youth, known as the "Lost Boys," that began in 2001 and focused on their risks, resilience, and adaptation to a new culture (Luster et al., 2006). The data for this study have been collected in two phases over time. A former group of researchers conducted the first phase of the study in 2007 by interviewing 19 South Sudanese URMs to explore their experiences of adjustment to living in a new culture. Questions were asked about their goals when they came to the U.S., the progress that they made in achieving their goals, and the changes and maintenance of their identity and culture (Luster et al., 2010; Rana et al., 2011).

In previous studies on South Sudanese URMs, many researchers found a strong sense of purpose as a resilience factor for the participants (Bates et al., 2005, 2013; Duncan, 2001; Luster et al., 2010; Rana et al., 2011). Motivated and inspired by these reports, the research team conducted Phase 2 of the study from 2014 to 2015 to follow up with the participants in Phase 1. Although the South Sudanese URMs comprise a unique ethnocultural group that is difficult to access for a follow-up study due to high levels of geographic mobility, the long-term engagements and trusting relationships between the research team members and the South Sudanese URM community allowed the second group of researchers to conduct 11 follow-up interviews 7 years after the first interview. Based on these qualitative data drawn from in-depth individual interviews in 2007 and 2014–2015, the research question for the present study is focused on how South Sudanese URMs reconstructed their identities after resettling in the U.S.

Participants

All participants were born in South Sudan and were separated from their biological parents as children. They lived in refugee camps in Kenya for several years before they resettled in the U.S. At the time of resettlement, the mean age of these 19 South Sudanese URMs was 15 years old, and the youngest was 11 years old.

In Phase 1, the sample comprised 17 male and 2 female South Sudanese emerging adults aged 18 to 26 years, with the mean age of 22.42 years old ($SD = 2.24$ years). In Phase 2, follow-up interviews were conducted with 11 South Sudanese URMs who completed their first interviews in Phase 1, which comprised 57.9% of the total number of participants from Phase 1 ($N = 19$). During this second phase of data collection, their ages ranged from 25 to 32 years with a mean age of 29.27 ($SD = 2.28$ years). Two female and nine male participants were included.

Among the 11 participants (Phase 2), four were married, five were in relationships, and five became parents. All participants completed high school; many finished a range of college degree and advanced studies. Four participants reported obtaining a master's degree as their highest-level education, four reported four-year college degrees, and the remaining participants reported either a two-year college degree or some college education without a degree.

In terms of employment status, 10 participants (91%) reported being currently employed, and 8 were employed full-time (Table 4.1); however, of the employed, only four secured skilled professional jobs, such as a medical technician or an accountant. Six of the employed held semi-skilled manual labor jobs, such as a factory line worker, despite having completed their education (Table 4.2). Nine participants indicated that they were financially supporting their families back in South Sudan.

Table 4.1 Occupational achievement of South Sudanese URMs

Occupational achievement	Phase 1[a] (%)	Phase 2[b] (%)
Currently working		
Yes	12 (63.2)	10 (91.0)
No	7 (36.8)	1 (9.0)
Employment status		
Not employed	7 (36.8)	1 (9.1)
Part-time	11(57.9)	2 (18.2)
Full-time	1 (5.3)	8 (72.7)

[a] $n = 19$
[b] $n = 11$

Table 4.2 Types of occupations among the employed South Sudanese URMs

Types of occupations	Phase 1[a] (%)	Phase 2[b] (%)
Semi-skilled, manual labor jobs (e.g., factory line worker, restaurant server, cashier, etc.)	12 (100)	6 (60.0)
Skilled, professional jobs (e.g., medical technician, accountant, etc.)		4 (40.0)

[a] $n = 12$
[b] $n = 10$

Interviews

For both Phases 1 and 2, the study conducted open-ended semi-structured interviews in which the participants were comfortable with speaking about their experiences, and the researchers were able to follow the interview protocol. The interviews were audio-recorded with the participants' consent and later transcribed. Each interview began with a short demographic questionnaire and took approximately 1 to 1.5 hours to complete.

In Phase 1, the questions centered on the participants' goals when they came to the U.S. and the progress they made in achieving their goals during the first seven years of resettlement. They were also asked about the challenges they experienced and the support they received. They discussed their views on why some of the youth were relatively successful since resettlement, whereas others struggled. Parts of the interview focused on their experiences in the American social context (e.g., experiences of harassment) and identity issues.

In Phase 2, follow-up interviews were conducted with 11 participants among the 19 from Phase 1. The follow-up interviews focused on the successes, changes, and challenges in achieving their goals and adapting to life in the U.S. since the first interview. Other sections focused on their identity and acculturation experiences, especially their views on race and sociocultural identity during the 14 years of resettlement.

Modified Grounded Theory Approach

For qualitative data analysis, the group of researchers used the modified grounded theory approach (Creswell, 2013; Cutcliffe, 2005; Strauss & Corbin, 1990) and followed the three-step coding procedure—open, axial, and selective coding (Strauss & Corbin, 1990). The original grounded theory approach was modified to employ existing theories and augment or challenge its assumptions when generating a grounded theory (Creswell & Creswell, 2017). In this study, the data analysis was guided by the theoretical frameworks of 3D acculturation (Ferguson et al., 2012) and segmented assimilation (Portes & Zhou, 1993). However, this process is a hybrid, in which the study remains open to the emergence of new theoretical elements or elaboration arising from the data, which shift, change, and advance the existing perspectives.

Bronfenbrenner and Morris (2006) refer to this mode of research as the "discovery mode" rather than the mode of verification. Specifically, the modified grounded theory used in the current study takes a multicultural psychology perspective to develop an "inside-out" model that emphasizes "the perspectives of those in ethnocultural communities that are underrepresented in research" (Hall et al., 2016, p. 40) and deemphasizes generalizability. In this way, this study attempts to avoid merely fitting the voices to the theoretical structure. Rather, it allows the voices to teach the researchers what sits beyond the structure that they need to recognize and incorporate.

Data Analysis

For data analysis, a three-step coding procedure was used: open, axial, and selective coding (Creswell, 2013; Strauss & Corbin, 1990). In the open coding phase, the first author reviewed the 30 interview transcripts (19 interviews from Phase 1, and 11 interviews from Phase 2), defined salient categories of information supported by the participants' narratives, and described the data into codes. Next, the researchers conducted axial coding to identify the central phenomena based on the codes and categories that were defined previously. As a result, the data were classified into main themes such as acculturation processes and dimensions, and acculturative experience associated with identities. In the last step of analyses, the researchers conducted selective coding and interpreted the data to generate a theory in which the categories were interrelated to represent the dynamic processes of identity re-formation and quad-dimensional acculturation throughout the two phases. As a result of this analysis, a grounded theory has been formulated to describe both reactionary and integrative identity re-formation, along with quad-dimensional acculturation processes in the lives of South Sudanese URMs during their emerging adulthood.

Results

This section describes the detailed processes of acculturation that have influenced the identity formation and re-formation of South Sudanese refugees as emerging adults. The two key themes that emerged included: (1) quad-dimensional acculturation processes (subthemes: exploration, commitment, internalization, retainment, and coping associated with multiple ethnic and racial identities—South Sudanese, mainstream American, African, and African American), and (2) the changes in these identities between Phase 1 and Phase 2 from bicultural to quad-cultural.

Acculturation Processes: Exploration, Commitment, Internalization, Retainment, and Coping

Exploration: Facing Cultural Differences, Exploring Diverse Cultures, and Gaining a Multicultural Perspective

Early in the resettlement period, the emerging adult refugees faced starkly different ways of living and were exposed to a more diverse society and multiple cultures in the U.S. These perhaps constituted a broader set of lived experiences than in the camps or previous in-continent placements. The participants discussed mastering basic life skills (e.g., driving, using the toilet, cooking using a stove), adjusting to American ways of being (e.g., language, dress, music, food, housing, banking) and its cold weather during winter, and learning social skills and cultural knowledge from interpersonal relationships across a range of unique environments (e.g., in school with peers and teachers, at home with foster families, at work with coworkers).

> The first one was, the first one was to understand the culture of United States which I do not know. And that was the big challenge. And then the second was the cold. It was daily too cold here. And the third one was like food…The problem was the cultural differences cause what I was seeing, it was I had never seen in my life. What even through the way they sit, living, eating, and drinking, I have never seen that—Diu, 24 years old.

> It was just very hard and different… whether, food, culture, socializing, what I say how to speak out to interact, but I have sort of learned a lot of that—David, 18 years old.

The participants were highly resilient, adaptive, and fast learners regarding what counted as necessities in American life during the early years of their adjustment (Luster et al., 2006). Being selected for the U.S. resettlement program and arriving in America were life-saving experiences that fueled their hopes and dreams. Wek, 24 years old, recalled the day he first arrived in the U.S.:

> It was a dream come true. It was nice. We were lucky. They waited for us at the airport, after we came, a lot of people came and greeting us…It's like we were getting a new life, we still have feelings about it.

As the South Sudanese learned more about the culture and early challenges were resolved, school and work challenges replaced the challenges of early adaptation and adjustment, taking the forefront in their lives and requiring them to rethink their identities, reasons for being here, and finally, to recast their strategies for coping with the situations and the very core of their self-concepts. Adut and Garang described,

> Whenever I do something I think about my background, how did I come here and why am I here. So, all these questions helped me to formulate what I will do. They motivated me to do my stuff—Adut, 25 years old.

> It's so typical for us, we put ourselves saying every single of us has to go to college because we need to go back and help—Garang, 22 years old.

Commitment: Seeking the American Dream to Help Others

These young South Sudanese refugees not only devoted themselves to learn complex daily management and new ways of functioning in the U.S. but also strived to thrive in education. Completing degrees of higher learning was an indispensable part of their lives because of their strong drive toward educational attainment. They described school success as the bridge between the past and the future. Thus, the participants showed an enormous drive to be successful in school for the betterment of their future careers, to have financial security, and to help those left behind in South Sudan.

> My first priority was to get as much education as possible. Here, if you decided to work instead of going to school, I think in the long run, they're not getting a lot. It'd be hard for you to help because it comes from the knowledge you gain from here. So, my main objective was to focus on school—James, 24 years old.

> I have to go to school here and finish what I was coming for and get a better job so that when I go back, [I] have to help people [family] out too. Those are the goals I think I have to accomplish for stay in the U.S.—John, 23 years old.

> One of my goals is to become a CPA type of accountant and after that it's probably to work for a CPA firm and then down the road is to start my own business and then I also see myself someday, starting a school somewhere in Africa. So, I would say, being a CPA, starting a school and working in a public accounting firm. Part of the reason I want to run my own business someday is because I go back to South Sudan and be able to employ people, so they can have their own job so now I'm not only paying them, but they are able to help their own. That's how I look at it—Ater, 28 years old.

However, achieving the American Dream was full of unexpected challenges and struggles. More than half of the participants (11) in Phase 1 and nine of the 11 participants in Phase 2 stated that they were supporting their family members in South Sudan by sending them money on a regular basis. Most frequently, the money was used for the medical, educational, and basic expenses of family members in Africa. To meet this transnational financial demand, many had been economically active since they were adolescents.

Despite their diligent work and perseverance, financial constraints sometimes thwarted their plans for pursuing their education. With overwhelming financial demands and limited resources, a number of participants (10) experienced course failure requiring them to repeat classes. Sometimes the result was to end their programs in order to earn enough money to meet immediate financial needs. Majak explained:

> I've tried to, but it's not easy. I'm doing that, but sometimes this semester I failed my classes, because I was working, because my uncle was sick and dying...I tried to work many hours this semester, and I failed my class. (Interviewer: You wanted to make enough money to send back.) So, he can get treatment. That keeps me away from going to school—Majak, 23 years old.

Besides, they faced many social contexts where they were mistreated and devalued because of their race. They often had to deal with stressful racial prejudice from their peers, coworkers, and even from strangers. Dau, 24 years old, said:

> I've been stopped here by the police, maybe more than six times… for some reason they just stop me and they would say something like, 'oh your driver's license plate seemed like expired' and it's not, it's not expired, but they just stop me. They will, they will stop sometime and say, 'is that your car, is this your car?' Those things happen. I will get a ticket that I don't deserve...I got treated differently, not because I'm a Sudanese, but because I'm Black … Because there are people who are like me here, [although] I have [a] different ethnic background, maybe we are Black together, but that doesn't mean we do the same thing. But I cannot blame the police; police do not know who I am, maybe will not.

Despite the racism they experienced, they stayed focused, maintaining an extraordinary level of commitment and a positive outlook for the future. They reasoned that there was a definitive reason for their survival and existence in the U.S. as South Sudanese. In this sense, the participants' psychological process of transforming their experience of war as children to encourage meaningful commitments, promote optimism, and successful adjustment is noteworthy.

> You have to think hard of what you have been through and then to make a good decision—Garang, 22 years old.

> I think my past experience kinda influence me. We went through the toughest situations to where anything that comes. It's like, 'okay it's tough but I can make it through'… From Sudan to Kenya and from Kenya to here, I was basically a war child. We were always getting shot at and I grew up in a refugee camp where we barely had food or water, and we were sleeping in hutches and stuff. So, I just look at the big picture like 'I'm here, I can do so much better'. Like when something faces me, it's like, 'okay, this is difficult but, I can make it through'. I try to be positive about it—Nya, 25 years old.

> Staying in Kenya without, sometime would go days without food and all those. Those kinds of struggle make me stronger. I know how to survive well and, when I came here, it's kind of easy—Adut, 32 years old.

Internalization of Ethnicity and Race: Evaluating African Ethnicity and Integrating African American Misidentification Based on Skin Color

Based on their experiences, they internalized the racialized social understanding that contributed to their new perspective of the broader society. In the case of the South Sudanese refugees, skin color and discriminatory experiences were the triggers that caused the participants to explore multiple layers of identities, distinguishing and binding African and African American. They learned about the meanings attached to their skin color that had been stigmatized, devalued, and mistreated by the larger society, as well as in the same racial group, according to the darkness of their skin. Alek described:

> I'm African, no doubt. African Americans know that. But within a certain case I'd be African American, like I can defend African Americans because I fall under that class.

> Internally, I call myself African, but I can still relate myself to African Americans, too … If you say something in the class toward African Americans and I'm there, I would guarantee I would say a word, I would defend African Americans against anything. But, if it is something that is just normal, I'll consider myself African. But at the same time, if you say something and I'm the only Black person there, I'm definitely African American.

Thus, the participants were able to align themselves with and be sympathetic to U.S. born African Americans regarding racial discrimination, yet they also differentiated themselves from the latter. While the South Sudanese understood the others' frequent categorization of them as African Americans and the pervasive racism against Blacks in America, the South Sudanese refugees did not think of themselves as "true" African Americans.

Rather, they identified themselves as Africans but primarily as South Sudanese. This adaptive social understanding allowed them to define themselves accurately between the two cultures (i.e., African and African American), often overlooked as the same by the mainstream culture in the U.S. Their internalization of the complex nature of race and ethnicity in American society promoted identity re-evaluation and re-formation, adding an additional identity as African. Perhaps they would not have had to develop this identity if they had resettled in an African country.

> To be politically correct, that's what the documents say. I don't see myself as African American but, it's easier to say I'm African American—David, 25 years old.

> I just get along with it and yeah sure. Well usually most the time when somebody calls me African American, I usually say no I'm African-African. So, there is time when people will say why? I always tell them I was born in Africa I was not born here, so—Bor, 28 years old.

Retainment: Maintaining the South Sudanese National Identity and Culture

Negotiating the American culture(s) while maintaining their South Sudanese cultural identity was the major acculturative task of the participants. Garang commented, "If you forget Sudan, forget those cultures, and cut the whole thing [out] just America, you might make a bad choice." In essence, they had adopted external markers of culture, such as styles of dressing and the music common of their young American peers, particularly, African American, yet had not changed the core values of their South Sudanese culture. As a result, they had established balanced ways of conducting themselves, remembering, celebrating, and participating in both cultures (i.e., American and South Sudanese).

For example, all participants in Phase 2 discussed the meaning of dating, establishing a romantic relationship, and recreating a family in the U.S. and how to accomplish these relational milestones. The participants also talked about how marriage traditions would be honored or modified in the new context. These processes of acculturation also created an important psychological mechanism that supported their South Sudanese identity. Most participants maintained a strong South Sudanese identity even though they acquired other cultural aspects of identity, such as American. Adut said:

> While I'm in America now, I'm using American culture, I approach everything using the American way and keeping my culture there. That's how I, and my culture help me in making good decision. When I am making a good decision about something I thought about my culture and my culture helped me make a better decision.

In everyday life, they kept abreast of the current events in South Sudan through the news, radio, the Internet, and gatherings. Their nationalism encouraged them to organize social and political action for South Sudan by raising money, launching, and joining NGOs, participating in U.S. based South Sudanese networks, and maintaining strong bonds with fellow South Sudanese refugees.

Coping with Racism

As African refugeesAfrican refugees, they often faced racial mistreatment based on pervasive prejudices in American society. The participants frequently discussed experiences with racism, which were mainly reported based on their interpersonal relationships with peers in school, individuals in neighborhoods, coworkers, and customers in their workplaces. Diu and Garang described:

> From work a lot of people did, you know like I don't know when I was in keeping for washing dishes for 3 years (laughing), not moving to next level, then I think that there is another reason something like that. So, I don't know, cause may be someone treated me like I don't know anything—Diu, 25 years old.

> There were lot of problems, such as when kids are teasing you, how do you deal with it. Or when some kids are making fun of you, how do you deal with so, honestly that's the fair question that you asked but it was very very challenging—Garang, 21 years old.

Nevertheless, the participants emphasized coping with racial discrimination and harassment experiences in American society to maximize their time and efforts toward success and minimize their goals being affected or obstructed by the incidents. In other words, to cope with racism, the participants tended to focus their attention on what was considered more important—their goals and purposes—so they continued to work diligently. By doing so, they attempted to avoid unnecessary conflicts and setbacks along the path of success to achieve their objectives. They did not internalize such racism in ways that led to enhanced vulnerability and destructive outcomes. Rather, they committed themselves to American education and tried to make a sound transition into the labor force. John and Adut commented:

> We kind of like if you face those things, if you smart, you don't want to worry about it because that's how thing goes. This is my color. I can't get rid of it. If somebody laugh at you or wrong understand you, that's the culture differences. You gotta stay focused and work it out—John, 23 years old.

> Because it doesn't make me leave the focus of what I was doing, and I focus on what I'm here for… Here's what we were talking about when hear the topic –I'm not gonna talk

> about it and what you trying to say, but this is what we're talking about. Let's keep focus here—Adut, 32 years old.

Thus, the experience of racism seemed to support their evaluation of South Sudanese ethnicity and refugee experiences as well as provide a deeper understanding of the Black experience in American society.

Identity Re-Formation Associated with Multidimensional Acculturation: Bicultural to Quad-cultural

Through the processes of acculturation, the coded data suggest four relevant cultures in discussing their identity: South Sudanese, mainstream American, African, and African American. However, these four acculturation dimensions emerged over time as the participants demonstrated changes in their self-descriptions over the two phases. First, integrating South Sudanese and mainstream American dimensions in their identity was prominent in Phase 1. For example, David described:

> I am one hundred percent Sudanese. I know I have to live in America, I have to adapt to American way, but I am one hundred percent Sudanese... I miss living in Africa. Before, I didn't value it, but now I value it so much like my home was there. I might be very Americanish but very traditional at heart like, because my culture is very away, but I value it so important now than before like Africa is so beautiful than before I even thought of it was before—David, 18 years old.

As a result, in Phase 1, of the 19 participants, 12 expressed a bicultural identity. However, in Phase 2, six participants extensively discussed how the four cultures constituted different parts of their identity, depending on the situation, and expressed a quad-dimensional identity (see Table 4.3). For instance, in the follow-up interview, David described his identity:

> I'm American to the way I feel like I can represent American culturally. I'm very attached to my South Sudanese community at the same time. I feel [I] belong to America; I feel

Table 4.3 Self-descriptions associated with acculturation dimensions in both phases

	South Sudanese and American (bicultural)	South Sudanese only	American only	South Sudanese, African, and African American	Dinka[a] only
Phase 1 (*N* = 19)	12 (63.1%)	3(15.8%)	2 (10.5%)	1 (5.3%)	1 (5.3%)
	South Sudanese, American, African, and African American (quad-cultural)	South Sudanese and American (bicultural)	South Sudanese only		
Phase 2 (*N* = 11)	6 (54.5%)	2 (18.2%)	3 (27.3%)		

[a]Dinka is the largest tribal group and a language in South Sudan

> wanted here. I have family, I have friends, so America is home. I came here when I was younger, so it is pretty much through my progress and development. Everything that I know that's been very productive is through America (……) I cannot just say I'm African American. I feel like I'm not equipped enough to respond to any demand for or any need for an African American. It's not my experience. I don't have the tools. I might know what to say, but at the same time, it's just not in my body. I feel comfortable saying that I'm not African American, I'm African—David, 25 years old.

The current findings from Phase 2 show a similar proportion of quad-cultural individuals among the South Sudanese URMs compared to that of tricultural individuals in Ferguson et al.' (2014) study focusing on Jamaican immigrant adults. Compatible proportions of bicultural individuals (31% and 18.1%) and monocultural individuals (21% and 27.2%) were also documented (see Table 4.3).

In terms of changes in identity, six of the bicultural participants (54.5%) in Phase 1 became quad-cultural in Phase 2. This indicated an important characteristic of increased cultural dimensions and incorporation, especially when the participants more frequently mentioned African and African American cultural dimensions in Phase 2 in comparison with their first interviews in Phase 1. Additionally, two participants appeared to maintain a bidimensional identity, whereas three participants consistently concentrated on their South Sudanese identity, reluctant to embrace other cultural dimensions when identifying themselves from a diverse perspective (see Table 4.4).

These findings not only suggest multiple cultural dimensions that the participants assimilated, adopted, and adapted but also indicate propose dynamic processes of identity reconstruction: integrating the multiple cultures over time to develop a strong, flexible sense of self in the complex psychosocial context of South Sudanese refugees as emerging adults. The successful integration of the four cultures was linked to the emerging identities and cultural orientations that seemed to be healthy for them to obtain fluidity in their sense of self, adjustable in various situations where the priority culture and orientation vary, without losing a core sense of self. Garang discussed the benefits of integrating multiple cultures for a successful adjustment to the U.S. He commented:

> People who are actually winning now in my three cultures or four cultures, so I try to take best out of it and leave the junk out of it and then to have both. I think it is very useful rather than having Dinka, but it's having both cultures.

Table 4.4 Characteristics of identity re-formation over time

Changes in identity from Phase 1 to Phase 2	Number of participants (%)
Increased cultural dimensions of identity transformed from bicultural to quad-cultural	6 (54.5)
Maintained bicultural identity	2 (18.1)
Maintained and insisted on strong south Sudanese identity without incorporating other cultural dimensions	3 (27.2)

$N = 11$

Discussion

Acculturation processes have shaped how South Sudanese refugees redefine themselves in the American sociocultural context where race and ethnicity play a significant role. Over time, the participants have provided extensive and meaningful self-descriptions, reflecting the four cultural dimensions experienced in their acculturation processes. In this situation, identity re-formation, particularly the presentation of multiple identity perspectives, is beneficial for societal adjustment and promotive relationships. Importantly, having these identities emerge as new self-constructions without feeling a significant loss of their South Sudanese identity is a strength of the acculturative process. Thus, maintaining their South Sudanese identity is critical for them to have a strong sense of self while they embrace different cultures and social identities arising from their American resettlement experience.

Of great significance to the discussion about identity re-formation is the flexibility and responsivity of the acculturative process for the protection of the self and cultural adjustment. The lived experiences of the South Sudanese refugee participants involved responding to a novel set of coping challenges including redefinition of the self in a system where their ethnic, national, or language identities carry little meaning for the others in their daily transactions. The formation of an "African" identity serves at least several critical purposes; it ties them to their origins, retains their pride in those origins, asserts their immigrant and displacement statuses, helps distinguish them from African Americans in managing misidentification, and imposes identity placement (societally located social cognitions). According to Medina et al. (2019), for Black Canadian immigrants, Black racial group membership and their African heritage culture are both important positive linkages between White culture and life satisfaction.

The qualitative methods used in this study have enabled the research group not only to formulate research questions conceptually derived from the existing approaches (i.e., 3D acculturation and segmented assimilation) but also to expand and re-form these frameworks with attendant elaborations to capture the development of the understudied ethnocultural group of African URMs. First, the 3D or multidimensional acculturation approach (Ferguson et al., 2012; Ferguson & Bornstein, 2014) is especially preferred for its perspective that multiple identities, especially minoritized alignments, can be beneficial and indicate a healthy pathway to adjustment or acculturative outcomes for the individual. The tridimensional or multidimensional acculturation theory provides the important central framework, allowing the current researchers to discover four different cultural dimensions influencing the South Sudanese URM participants' sense of self, transformed from bicultural to quad-cultural between the two phases.

Nguyen and Ferguson (2019) suggest that the identity re-formations and cultural variability are meaningful psychosocial mechanisms for immigrant youth that promote their adjustment in multiple transactional contexts and support strong self-systems. Among the study participants, multiple identity systems have grown across time as a necessary set of developments for operating and responding to diverse contexts while not necessarily losing the participants' sense of self. Consistent with this study and other ethnic/racial identity scholars' views (Douglass & Umaña-Taylor, 2016; Phinney, 2003), the present findings illustrate that identity is transformative and subject to change at varying points (early and late) during emerging adulthood in the South Sudanese URMs' settlement in the U.S. Besides, it demonstrates fluidity in identity according to the situation as an adaptive psychosocial mechanism for South Sudanese URMs transitioning to adulthood as they encounter more racial incidents in college or in workplaces.

The segmented assimilation theory (Portes & Zhou, 1993) provides a highly contested view of immigrant identity development and acculturative process, often ending in the promulgation of the mainstream or the majority identity development as the best outcome of the adjustment process. In contrast to this segmented assimilation perspective, this study's findings suggest that South Sudanese URMs have evolved with a multifaceted understanding of ethnicity and race through the active learning, evaluations, and transactions among the four cultures relevant to their life in the U.S.

In sum, maintaining their strong South Sudanese identity while incorporating African and African American dimensions of identity has helped them reconcile the sociocultural conflicts that they have experienced. In this study, it is evident that this integration is adaptive rather than destructive, as the segmented assimilation theory assumes. The discrimination and racism that they have experienced have not become reasons to align with the African American identity. Rather, these are reasons to understand and sometimes take up the perspective against racism together and move forward. This work helps in understanding that misidentification can be met with not only reproach or rejection, but a constructive identity response and, an incorporation of a new meaningful identity to address that less meaningful misidentification and subjugated experience of racism. Importantly, immigrant and personal identity are bolstered by the new formulation.

Limitations

A few limitations of this study should be noted. The current findings are largely based on sets of data collected from South Sudanese male URMs. However, a small but notable number of female URMs will likely develop a different pathway of identity re/formation at the intersection of race and gender. Another analysis (Johnson et al., 2015) suggests the high possibility that female refugees experience a different series of psychosocial challenges and cultural contexts in both South Sudanese and American communities. A full exploration of the women's experience would perhaps yield new insights that are not accessible here. It should also be noted that this study's participants comprised a highly selected group of URMs and might not be representative of the whole URM population in the U.S. Furthermore, not finding the full complement of the sample in the follow-up segment of the study and therefore the complete range of the identity process, the data rest on the strength of the depth of the data in multiple time points.

Implications

The current study's findings elevate multiple identity perspectives and minoritized alignments as healthy acculturation and adjustment by challenging the models that rest on duality in their formulations, which is inadequate for explaining the lived experience and flexibility of identities in a multicultural setting. As Bornstein (2017) emphasizes the specificity principle in acculturation science to gain a clearer understanding of how "specific setting conditions of specific people at specific times" impact the "specific domains in acculturation by specific processes" (p. 3), the current study argues for a more precise explanation of acculturation and identity re-formation by building a theory and elaboration within the specific ecology of South Sudanese URMs. By displaying the unique processes of acculturation and identity reconstruction among South Sudanese URMs, the current study identifies the contextual and sociocultural features and their processes that underlie how race and ethnicity shape this group's experiences, behavior, and adjustment.

The present study has implications for mental health services targeting URMs. Many refugees come to the U.S., and particularly, children are placed through social services of one kind or another. Due to war-related and displacement traumas, social services and mental health services are provided by local communities and organizations, such as Lutheran Social Services. Support from these programs was valuable in the experiences of South Sudanese refugee youths' resettlement (Luster et al., 2010). In these services, it may be useful to support the development and retention of a strong national identity while making a place for new and necessary cultural identities to be formed and nurtured. South Sudanese URMs demonstrate their resilience through these adaptation and acculturation processes and the identities that evolve over time.

Conclusions

This chapter has demonstrated that ethnic and racial identity re-formation is an adaptive process among immigrant youth and occurs in a transactional, transformational, and contextually meaningful relationship. Among the refugee youth of African origins, the re-formation process has special utility for understanding their acculturation and adaptive identity development in the context of their displacement and alignment experiences over time. Illuminated also is how they use new information from their environment to form novel and enhanced identity features that are necessary for successful resettlement. African immigrant and refugee youth and emerging adults have the additional dynamic of societal alignment with a highly racialized group, further complicating their identity re-formation process. Still, there is little existing work on these increasingly important and globally distributed populations. Expanding the research on these unique groups, particularly the youth, is essential for deepening the response to and support of their long-term adjustment and success. Future work requires a better understanding of the influence of not only identity stress and trauma but also resilience as factors contributing to the re-formation process.

Appendix

Identity re-formation, educational achievement, and types of occupation among South Sudanese URM transitioning to adulthood

Participants (Pseudonym)	Identity Re-formation		Highest level of education (Phase 2)	Types of occupation (Phase 2)
	Phase 1	Phase 2		
1. Diu	South Sudanese	South Sudanese	4-year college degree	Semi-skilled, manual labor job
2. Dau	South Sudanese	South Sudanese	4-year college degree	Skilled, professional job
3. Bor	South Sudanese	South Sudanese	Some college but no degree	Semi-skilled, manual labor job
4. Gabriel	Bicultural	Quad-cultural	Associate's 2-year college	Not employed
5. David	Bicultural	Quad-cultural	4-year college degree	Semi-skilled, manual labor job
6. Bul	Bicultural	Quad-cultural	Some college but no degree	Semi-skilled, manual labor job
7. Nya	Bicultural	Quad-cultural	4-year college degree	Semi-skilled, manual labor job
8. Garang	Bicultural	Quad-cultural	Master's degree	Skilled, professional job
9. James	Bicultural	Quad-cultural	Master's degree	Skilled, professional job
10. Ater	Bicultural	Bicultural	Master's degree	Skilled, professional job
11. Adut	Bicultural	Bicultural	4-year college	Semi-skilled, manual labor job

References

Arnett, J. J. (2000). Emerging adulthood: A theory of development from the late teens through the twenties. *American Psychologist, 55*(5), 469–480. https://doi.org/10.1037/0003-066X.55.5.469

Arnett, J. J. (2004). *Emerging adulthood: The winding road from the late teens through the twenties*. Oxford University Press.

Balidemaj, A., & Small, M. (2019). The effects of ethnic identity and acculturation in mental health of immigrants: A literature review. *International Journal of Social Psychiatry, 65*(7–8), 643–655. https://doi.org/10.1177/0020764019867994

Bates, L., Baird, D., Johnson, D. J., Lee, R. E., Luster, T., & Rehagen, C. (2005). Sudanese refugee youth in foster care: The "lost boys" in America. *Child Welfare, 84*(5), 631–648.

Bates, L., Luster, T., Johnson, D. J., Qin, D. B., & Rana, M. (2013). Sudanese refugee youth: Resilience among undefended children. In D. Johnson, D. Abényiga, & R. Hitchcock (Eds.), *Vulnerable children: Global challenges in education, health, well-being, and child rights* (pp. 167–183). Springer. https://doi.org/10.1007/978-1-4614-6780-9_12

Berman, H. (2001). Children and war: Current understandings and future directions. *Public Health Nursing, 18*(4), 243–252.

Berry, W. B. (1997). Immigration, acculturation, and adaptation. *Applied Psychology, 46*(1), 5–34. https://doi.org/10.1111/j.1464-0597.1997.tb01087.x

Blake, M. K. (2019). Self and group racial/ethnic identification among emerging adults. *Emerging Adulthood, 7*(2), 138–149. https://doi.org/10.1177/2167696818799013

Bornstein, M. H. (2017). The specificity principle in acculturation science. *Perspectives on Psychological Science, 12*(1), 3–45. https://doi.org/10.1177/1745691616655997

Bronfenbrenner, U., & Morris, P. A. (1998). The ecology of developmental processes. In W. Damon & R. M. Lerner (Eds.), *Handbook of child psychology: Theoretical models of human development* (pp. 993–1028). Wiley.

Bronfenbrenner, U., & Morris, P. A. (2006). The bioecological model of human development. In R. M. Lerner & W. Damon (Eds.), *Handbook of child psychology* (Theoretical models of human development) (Vol. 1, 6th ed., pp. 793–828). Wiley.

Creswell, J. W. (2013). *Qualitative inquiry and research design: Choosing among five approaches*. Sage.

Creswell, J. W., & Creswell, J. D. (2017). Chapter 3. The use of theory. In J. W. Creswell & J. D. Creswell (Eds.), *Research design: Qualitative, quantitative, and mixed methods approaches* (pp. 49–74). Sage.

Cutcliffe, J. R. (2005). Adapt or adopt: Developing and transgressing the methodological boundaries of grounded theory. *Journal of Advanced Nursing, 51*(4), 421–428. https://doi.org/10.1111/j.1365-2648.2005.03514.x

Douglass, S., & Umaña-Taylor, A. J. (2016). Time-varying effects of family ethnic socialization on ethnic-racial identity development among Latino adolescents. *Developmental Psychology, 52*(11), 1904–1912. https://doi.org/10.1037/dev0000141

Duncan, J. (2001). Sudanese "lost boys" in the United States: Adjustment after six months. United States Catholic Conference, Washington, DC. http://www.brycs.org/documents/upload/lostboys.pdf

Erikson, E. H. (1968). *Identity, youth, and crisis*. Norton & Co.

Ferguson, G. M., & Birman, D. (2016). Acculturation in the United States of America. In D. Sam & J. Berry (Eds.), *Cambridge handbook of acculturation psychology* (2nd ed., pp. 396–416). Cambridge University Press. https://doi.org/10.1017/CBO9781316219218.023

Ferguson, G. M., & Bornstein, M. H. (2014). Tridimensional (3D) acculturation: Culture and adaptation of black Caribbean immigrants in the USA. In R. Dimitrova, M. Bender, & F. van de Vijver (Eds.), *Global perspectives on well-being in immigrant families* (pp. 31–51). Springer. https://doi.org/10.1007/978-1-4614-9129-3_3

Ferguson, G. M., Bornstein, M. H., & Pottinger, A. M. (2012). Tridimensional acculturation and adaptation among Jamaican adolescent–mother dyads in the United States. *Child Development, 83*(5), 1486–1493. https://doi.org/10.1111/j.1467-8624.2012.01787.x

Ferguson, G. M., Iturbide, M. I., & Gordon, B. P. (2014). Tridimensional (3D) acculturation: Ethnic identity and psychological functioning of tricultural Jamaican immigrants. *International*

Perspectives in Psychology: Research, Practice, Consultation, 3(4), 238–251. https://doi.org/10.1037/ipp0000019
García Coll, C., Lamberty, G., Jenkins, R., McAdoo, H. P., Crnic, K., Wasik, B. H., & Vázquez García, H. (1996). An integrative model for the study of developmental competencies in minority children. *Child Development, 67*(5), 1891–1914.
Garmezy, N. E., & Rutter, M. E. (1983). *Stress, coping, and development in children*. Johns Hopkins University Press.
Geltman, P. L., Grant-Knight, W., Mehta, S. D., Lloyd-Travaglini, C., Lustig, S., Landgraf, J. M., & Wise, P. H. (2005). The "lost boys of Sudan": Functional and behavioral health of unaccompanied refugee minors resettled in the United States. *Archives of Pediatrics & Adolescent Medicine, 159*(6), 585–591. https://doi.org/10.1001/archpedi.159.6.585
Glozman, J., & Chuang, S. S. (2018). Multidimensional acculturation and identity of Russian-speaking youth in Canada: The role of parents. *Journal of Adolescent Research, 34*(4), 464–488. https://doi.org/10.1177/0743558418767624
Habecker, S. (2017). Becoming African Americans: African immigrant youth in the United States and hybrid assimilation. *Journal of Pan African Studies, 10*(1), 55–76.
Hall, G. C. N., Yip, T., & Zárate, M. A. (2016). On becoming multicultural in a monocultural research world: A conceptual approach to studying ethnocultural diversity. *American Psychologist, 71*(1), 40–51. https://doi.org/10.1037/a0039734
Johnson, D. J., Carolan, M., & Bates, L. (2015). A woman is always to blame: Community-sanctioned violence in a Sudanese refugee woman's experience. *International Psychology Bulletin, 19*(3), 24–29.
Kim, S. Y., & Hou, Y. (2016). Intergenerational transmission of tridimensional cultural orientations in Chinese American families: The role of bicultural socialization. *Journal of Youth and Adolescence, 45*(7), 1452–1465. https://doi.org/10.1007/s10964-016-0423-4
Lawrence, J. A., Kaplan, I., & Collard, A. H. (2016). Understanding the perspectives of refugee unaccompanied minors using a computer-assisted interview. *Forum: Qualitative Social Research, 17*(2). https://doi.org/10.17169/fqs-17.2.2483
Lese, K. P., & Robbins, S. B. (1994). Relationship between goal attributes and the academic achievement of Southeast Asian adolescent refugees. *Journal of Counseling Psychology, 41*(1), 45–52. https://doi.org/10.1037/0022-0167.41.1.45
Luster, T., Bates, L., & Johnson, D. J. (2006). Risk and resilience. In F. Villaruel & T. Luster (Eds.), *Crisis in youth mental health* (Vol. 2, pp. 27–52). Praeger Press.
Luster, T., Qin, D. B., Bates, L., Johnson, D. J., & Rana, M. (2008). The lost boys of Sudan: Ambiguous loss, search for family, and reestablishing relationships with family members. *Family Relations: An Interdisciplinary Journal of Applied Family Studies, 57*(4), 444–456. https://doi.org/10.1111/j.1741-3729.2008.00513.x
Luster, T., Qin, D. B., Bates, L., Rana, M., & Lee, J. A. (2010). Successful adaptation among Sudanese unaccompanied minors: Perspectives of youth and foster parents. *Childhood, 17*(2), 197–211. https://doi.org/10.1177/0907568210365664
Medina, A. C., Rowley, S. J., & Towson, S. (2019). Race matters for black Canadians: The intersection of acculturation and racial identity in emerging adulthood. *Emerging Adulthood, 7*(2), 97–108. https://doi.org/10.1177/2167696818785099
Nguyen, J., & Ferguson, G. M. (2019). A global cypher: The role of hip hop in cultural identity construction and navigation for southeast Asian American youth. *New Directions for Child and Adolescent Development, 2019*(164), 99–115. https://doi.org/10.1002/cad.20279
Phinney, J. S. (1992). The multigroup ethnic identity measure: A new scale for use with diverse groups. *Journal of Adolescent Research, 7*(2), 156–176. https://doi.org/10.1177/074355489272003
Phinney, J. S. (2003). Ethnic identity and acculturation. In K. M. Chun, P. Balls Organista, & G. Marin (Eds.), *Acculturation: Advances in theory, measurement, and applied research* (pp. 63–81). American Psychological Association. https://doi.org/10.1037/10472-006

Phinney, J. S., & Chavira, V. (1992). Ethnic identity and self-esteem: An exploratory longitudinal study. *Journal of Adolescence, 15*(3), 271–281. https://doi.org/10.1016/0140-1971(92)90030-9
Phinney, J. S., & Vedder, P. (2006). Family relationship values of adolescents and parents: Intergenerational discrepancies and adaptation. In J. W. Berry, D. L. Phinney, D. L. Sam, & P. Vedder (Eds.), *Immigrant youth in cultural transition: Acculturation, identity, and adaptation across national contexts* (pp. 167–184). Erlbaum.
Portes, A., & Rumbaut, R. G. (2001). *Legacies: The story of the immigrant second generation.* University of California Press.
Portes, A., & Zhou, M. (1993). The new second generation: Segmented assimilation and its variants. *The Annals of the American Academy of Political and Social Science, 530*(1), 74–96. https://doi.org/10.1177/0002716293530001006
Qin, D. B., Saltarelli, A., Rana, M., Bates, L., Lee, J. A., & Johnson, D. J. (2015). "My culture helps me make good decisions": Cultural adaptation of Sudanese refugee emerging adults. *Journal of Adolescent Research, 30*(2), 213–243. https://doi.org/10.1177/0743558414547097
Rana, M., Qin, D., Bates, L., & Saltarelli, A. J. (2011). Factors related to educational resilience among Sudanese unaccompanied minors. *Teachers College Record, 113*(9), 2080–2114.
Rudmin, F., Wang, B., & de Castro, J. (2017). Acculturation research critiques and alternative research designs. In S. J. Schwartz & J. B. Unger (Eds.), *Oxford library of psychology. The Oxford handbook of acculturation and health* (pp. 75–95). Oxford University Press.
Schwartz, S. J., Unger, J. B., Zamboanga, B. L., & Szapocznik, J. (2010). Rethinking the concept of acculturation: Implications for theory and research. *American Psychologist, 65*(4), 237–251. https://doi.org/10.1037/a0019330
Smith, C.. (2017, April 29—Jun 4). Objects in the mirror, Goodman's Albert Theatre, Chicago. (Director).
Stepick, A., & Stepick, C. D. (2010). The complexities and confusions of segmented assimilation. *Ethnic and Racial Studies, 33*(7), 1149–1167. https://doi.org/10.1080/01419871003599518
Strauss, A., & Corbin, J. M. (1990). *Basics of qualitative research: Grounded theory procedures and techniques.* Sage.
Thelamour, B., & Johnson, D. J. (2017). Exploring Black immigrants' and nonimmigrants' understanding of "acting Black" and "acting White." *Journal of Black Psychology, 43*(3), 280–304. https://doi.org/10.1177/0095798416641863.
United States Conference of Catholic Bishops [USCCB]. (2013*). The United States unaccompanied refugee minor program: Guiding principles and promising practices.* http://www.usccb.org/about/children-and-migration/unaccompanied-refugee-minor-program/upload/united-states-unaccompaniedrefugee-minor-program-guiding-principles-and-promising-practices.pdf
Wallin, A.-M. M., & Ahlström, G. I. (2005). Unaccompanied young adult refugees in Sweden, experiences of their life situation and well-being: A qualitative follow-up study. *Ethnicity & Health, 10*(2), 129–144. https://doi.org/10.1080/13557850500071145
Waters, M. C., Tran, V. C., Kasinitz, P., & Mollenkopf, J. H. (2010). Segmented assimilation revisited: Types of acculturation and socioeconomic mobility in young adulthood. *Ethnic and Racial Studies, 33*(7), 1168–1193. https://doi.org/10.1080/01419871003624076

Chapter 5
Family Engagement and Identity Development Among Immigrant-Origin Latinx Emerging Adults

Dalal Katsiaficas and Haeyoon Chung

In the United States, emerging adults are increasingly from immigrant backgrounds—either immigrating to the U.S. themselves with parents (referred to as first-generation) or are born in the U.S. as the children of immigrant parents (referred to as second-generation). Currently, the number of immigrant-origin emerging adults (age 18–34) has grown to nearly 20 million (Rumbaut & Komaie, 2010), with one in four 16 to 24-year-olds in the U.S. being either first- or second-generation immigrants (Batalova & Fix, 2011; U.S. Census Bureau, 2013). Families play critical roles in immigrant-origin emerging adults' lives (Fuligni & Pedersen, 2002), yet are often understudied in how they shape the identity development process, particularly during emerging adulthood. In this chapter we explore the ways in which family supports and obligations among Latinx immigrant-origin emerging adults shape their identities as they come into adulthood.

We begin with a brief overview of emerging into adulthood for Latinx immigrant-origin students before weaving together the extant literature on the role of family with the work on multiple identities and Hyphenated Selves (Fine & Sirin, 2007). We then present our methodological approach to understanding the role of families in shaping the identity development process and provide mixed-methods findings from our sample of Latinx immigrant-origin emerging adults. Lastly, we draw conclusions and discuss future directions about where to take this burgeoning field of study.

D. Katsiaficas (✉) · H. Chung
University of Illinois at Chicago, Chicago, IL, USA
e-mail: dalalkat@uic.edu; hchung58@uic.edu

D. J. Johnson et al. (eds.), *Re/Formation and Identity*, Advances in Immigrant Family Research, https://doi.org/10.1007/978-3-030-86426-2_5

Emerging into Adulthood for Latinx Immigrant-Origin Undergraduate Adults

Emerging adulthood is an important developmental period for identity development (Arnett, 2004), particularly for those from immigrant-origin families. Emerging adulthood is characterized by continued identity explorations (Arnett, 2004) when significant identity development processes take place (Côté, 2006). Specifically, ethnic identity exploration increases dramatically during the college-going years (Syed & Azmitia, 2009) as this is often a time when young people find themselves in shifting contexts that promote examination and negotiation of their ethnic identities between the contexts of home, school, country of origin, and the society they live within (Phinney, 2008). Further, identity explorations often take place in the domains of love and work as young people try out new roles prior to settling into enduring commitments (Arnett, 2008). In addition, emerging adults are exploring their identities in the domains of sexuality, religion, spirituality, and personal values during this phase of life (Schwartz et al., 2013; Schwartz, Pantin, et al., 2005).

Multiple Identities of Immigrant-Origin Emerging Adults

Identity processes are somewhat unique for immigrant-origin emerging adults. Immigrant-origin emerging adults occupy multiple spaces—they often maintain transnational ties to their country of origin while "learning a new land" (Suárez-Orozco et al., 2008). Living at the intersection of these multiple contexts, immigrant-origin emerging adults embody multiple selves (Cooper et al., 1998; Deaux & Perkins, 2001; Fine & Sirin, 2007; Suárez-Orozco, 2004). Multiple identities are both achieved by and ascribed to immigrant-origin emerging adults by the social spheres they are embedded within (Suárez-Orozco, 2004). During this phase of life, immigrant-origin emerging adults cognitively perceive input from their social surroundings. They become aware of the "social mirror" or reflections of themselves through the eyes of those in their social surroundings (Spencer, 1999; Suárez-Orozco, 2004).

Immigrant-origin populations, especially visible and racial minorities, contend with the current anti-immigrant socio-political climate in the U.S. marked by xenophobia, discrimination, and racism (Suárez-Orozco et al., 2018). For Latinx emerging adults specifically, this specific socio-political context has been marked by distinct and potent messages with the rise in popularity of Donald Trump, who won the 2016 presidential election with promises to "build a wall" between the U.S. and Mexico as a form of "protection" from Mexican immigrants who he referred to as "rapists" and "drug smugglers" (Hooghe & Dassonneville, 2018). What unfolded was a presidency marked by disregard for the law, reinforcement of extraordinary prejudice, cronyism, misogyny, and violence towards many marginalized groups. Such xenophobia and racism manifest in both overt and subtle forms of discrimination like openly targeting immigrant populations with discriminatory laws and practices (López et al., 2010), the perpetuation of negative stereotypes, as well as more veiled forms of discrimination such as microaggressions (Sue et al., 2007). Given this socio-political environment and shifting cultural norms, values, and attitudes broadly (Galliher et al., 2017), this context has potential to powerfully shape Latinx immigrant-origin emerging adults' identity development.

Hyphenated Selves Framework

Hyphenated Selves provides a theoretical framework that incorporates understanding how socio-political contexts shape young people's multiple identities. The Hyphenated Selves framework encourages researchers to theorize the social psychological processes that connect individuals to the socio-political contexts in which they live without exoticizing them (Fine & Sirin, 2007; Sirin & Fine, 2008). By hyphen, Fine (1994) refers to the embodiment of the intersection of multiple politicized identities that are at once joined and fragmented, by the current socio-political climate, history, geography, and belonging. This framework is particularly helpful in studying the identities of youths living on the borderlines of moral exclusion (Opotow, 1990) such as Muslim-American youth in the U.S. following 9/11 and youth of color and immigrant-origin youth in the U.S. (Sirin et al., 2008).

Immigrant-origin emerging adults are navigating multiple, often conflicting, contexts in their everyday lives as they make the daily migrations between contexts (e.g., home and school), across cultures (e.g., host and home cultures), through space (e.g., here and there), and within relationships (Katsiaficas et al., 2011). These "everyday hyphens" may act as resources for immigrant-origin emerging adults to draw on socially, cognitively, and even economically (Katsiaficas et al., 2011; Suárez-Orozco et al., 2008), rather than markers of socially fracturing forces. The study of Hyphenated Selves is inextricably linked to context. Taken together, the Hyphenated Selves framework allows researchers to understand how identities are formed and informed by one another, providing a theoretical and methodological framework that allows for youth to narrate the stories of how pieces and wholes may sit together side by side, innovate to create new selves and sometimes hold together conflict.

Family Engagement Among Immigrant-Origin Emerging Adults

Immigrant-origin emerging adults from Latinx backgrounds are comprised of diverse populations, yet are regarded as sharing characteristics, experiences, and beliefs that structure their development. *Familism* is one representative cultural term that emphasizes family closeness and loyalty among this population (Vega, 1990).

In contrast to emphasizing the values of independence and individuality in the dominant U.S. culture, familism reinforces family members to remain in family boundaries not only for parents but also for their children to place family values first. This family-oriented cultural value can strongly influence life experiences throughout development. In particular, this cultural value manifests as family engagement in their behaviors, particularly during emerging adulthood. Family engagement consists of family obligations as well as family support. Family obligations and responsibilities refer to how youth and emerging adults contribute to their families. According to Fuligni and colleagues (1999), family obligations are comprised of three aspects: attitudes toward providing family assistance, respect for the family, and the importance of providing support in the future when they become adults. Family support, in contrast, consists of emotional and financial supports from youth and emerging adults' family members.

From a young age, many Latinx young people have family responsibilities for parents and extended family members (Baer & Schmitz, 2007; Phinney et al., 2000). In comparison to their non-immigrant-origin peers, children from immigrant families have a strong sense of familial duty beginning in early developmental stages and continuing throughout emerging adulthood (Schwartz, et al., 2012). As an example, this can manifest in terms of language brokering, where immigrant-origin youth translate and interpret school forms, parent–teacher conferences, and report cards; household credit card bills, bank statements, and insurance forms; and social interactions in post offices, stores, doctor offices, and government agencies (Morales & Hanson, 2005; Schwartz et al., 2012). Taking on these language brokering tasks leads to a renegotiation of their roles in their families. For the first generation, the difficulties that often arise during migration require family members to renegotiate their roles, positions, attitudes, and expectations in relation to one another in the family and behave according to this shift in their relationships subsequent to migration (Lanuza, 2017). Furthermore, these obligations and responsibilities often take on gendered patterns. For example, Latinx young women experience expectations to be caretakers within families (Muñoz & Maldonado, 2012), and young men experience expectations as financial contributors to the family (Saenz & Ponjuan, 2011).

Research suggests that this tendency increases throughout development until emerging adulthood when emerging adults are often most engaged in family obligations, both financially and emotionally (Fuligni & Pedersen, 2002). Emerging adults from Latinx families reported the strongest sense of familial duty during emerging adulthood than their East Asian and European peers, and those Latinx emerging adults have a tendency to live longer with their parents and financially support their families than did their European American peers (Fuligni & Pedersen, 2002; Goldscheider & Goldscheider, 1999). Given that the developmental period of emerging adulthood is characterized by self-focus (Arnett, 2000), these findings suggest that Latinx emerging adults retain their familial values for assistance and respect to parents and siblings rather than increased autonomy and exploration of adult roles. Latinx emerging adults may experience greater expectations to provide assistance to their families as they have more access to economic and social resources. The elevated financial status by working part-time or full-time reinforces responsibility to repay their families in ways not available to them as children or adolescents (Fuligni & Pedersen, 2002).

These obligations are not one-sided, however. Instead, recent work regarding Latinx emerging adults' development characterizes this period of life as being marked by "collective contribution," whereby immigrant parents and their immigrant-origin children share a sense of mutually constitutive contribution for their survival (Katsiaficas et al., 2016). Specifically, Latinx immigrant parents provide a number of family supports to their children financially, emotionally, and academically, however, it is important to note that often these academic supports may be constrained by parents' own level of education, particularly as they move forward in their education (Sanchez, Reyes, & Singh, 2005).

Many immigrant-origin Latinx parents are considered to be role models of hard work and sacrifice for their children who observe the physically demanding jobs and in the case of undocumented status, risk of deportation incurred by their parents (Cervantes et al., 2015; Louie, 2012; Smith, 2006). This in turn produces a sense of responsibility amongst Latinx emerging adults, as well as a desire for academic achievement as a means to improve the standard of living for themselves and their family members (Louie, 2012; Smith, 2006). For example, Guan and Fuligni (2016) examined how the longitudinal trajectory of parent, sibling, and peer support changed during the transition to emerging adulthood among Asian, European, and Latinx communities. Exploring the association of social support with expectations to assist the family within individuals, they found that individuals who reported higher mean levels of family support also reported higher expectations to assist their families, such as financially supporting parents, living near them, or taking care of siblings. Therefore, the interactional nature of social support among immigrant families is supported, not only by cultural background, but also, the nature of immigration and the acculturation process. In these ways, the relation between family obligations and family supports is a dynamic process of social exchange and reciprocity (Parrott & Bengtson, 1999). Given that these familial interdependent cultural values play a critical role in Latinx IOE adults' emotional well-being (Fuligni & Pedersen, 2002; Tseng, 2004), college life (Gandara, 1995; Hurtado et al., 1996), and academic performance (Fuligni et al., 2005; Tseng, 2004), it is necessary to explore how these obligations and support shape their identities as they come of age.

Latinx Emerging Adults' Identity Development Within the Family Context

Significant identity development processes take place during emerging adulthood (Côté, 2006) as key identity domains remain unresolved for young people well beyond their adolescence (Arnett, 2008; Schwartz et al., 2005). Ethnic/racial identity exploration takes on new importance during emerging adulthood. Findings from longitudinal research with an ethnically diverse sample of emerging adult college students suggests that ethnic identity exploration increased dramatically during the 4 years of college (Syed & Azmitia, 2009). For Latinx emerging adults specifically, Syed et al. (2007) found that within the first year of college there was variation in ethnic identity exploration and commitment. Thus, this phase of development is often a time when young people find themselves in shifting contexts that promote further examination and negotiation of their ethnic/racial identities (Phinney, 2008).

Family is the most important proximal social context that guides ethnic and racial identity formation (Umaña-Taylor et al., 2013), and plays an important overall part in the identity development process of Latinx emerging adults. Researchers reported that parental involvement, parent–adolescent communication, and family support were positively related to identity coherence and negatively related to identity confusion among Latinx emerging adults (Schwartz, Pantin, et al., 2005). As Latinx emerging adults engage in these patterns of family responsibilities and support, they also experience changes in their ethnic identities while integrating these cultural values into their schema of sense of self (Syed et al., 2007; Syed & Azmitia, 2009). As they navigate patterns of family engagement alongside their own individual career and college decisions, family values can further shape these identity domains (Fuligni & Pedersen, 2002). Perceived discrepancy and conflict between emerging adults' own desires alongside those of their family members often require the renegotiation of personal and cultural identities (Azmitia et al., 2008). Taken together, family engagement and identity development are intertwined especially for Latinx immigrant-origin emerging adults coming of age.

The Present Study

To address these critical gaps in the literature, the present study examines the identity development and family engagement of Latinx immigrant-origin emerging adults through mixed methods. Specifically, this study explores the following questions: (1) what are broad the patterns of family engagement (i.e., family obligation and family support) in Latinx immigrant-origin emerging adults' lives? and (2) how do Latinx immigrant-origin emerging adults make sense of their hyphenated identities?

Method

Participants and Procedures

Data were collected regarding the family engagement among Latinx immigrant-origin emerging adults through an online survey ($N = 57$) and qualitatively explored the role of family in their identity development through in-depth idiographic case studies ($n = 2$). Participants' ages ranged from 19 to 28 years with a mean age of

20.4 years (*SD* = 2.2 years). The majority of our sample identified as female (*n* = 42). While our sample was multi-ethnic, we chose to limit our analyses to participants who self-identified as Latinx, which was the largest sub-group in our sample. Focusing our analyses on Latinx participants allowed us to understand within-group differences, rather than attempt to make claims regarding between-group differences with small sample sizes. While immigrant-origin youth can be either first- or second-generation, the majority of our Latinx sample were second-generation (86%) born in the U.S. to immigrant parents with 14% being first-generation who were born outside the U.S. and migrated earlier in their lives. The majority of participants' families in this group came from Mexico and a smaller group from Peru, Colombia, and Ecuador. The majority of participants reported that they were living with at least one family member (89.5%) and 10.5% reported living by themselves or with roommate(s) (See Table 5.1).

Methodological Approach

We employed a Participatory Action Research (PAR) framework to guide all aspects of this study (Stoudt et al., 2012; Torre et al., 2012). This framework positions research as an inclusive process among individuals that have a diversity of viewpoints centered in socio-political historical cultural context. Moreover, PAR fosters an interactive link between researchers and participants where expertise is shared; requires community involvement in methodological decisions to create meaningful goals to address community needs (Mertens, 2003). Creating a PAR Student Advisory Board at the inception of the project in order to inform all methodological and analytic decisions has been an effective way of utilizing PAR in developmental research (Katsiaficas et al., 2016). The student advisory board for this study was

Table 5.1 Descriptive statistics on demographic characteristics (*N* = 57)

	Frequency/Mean	SD/%
Characteristics		
Age (years)	20.4	2.2
Gender		
Male	42	73.7
Female	14	24.6
Other	1	1.8
Country of birth		
Born in U.S.	49	86.0
Born outside U.S.	8	14.0
Resident with		
By myself or with roommate	6	10.5
With family member	51	89.5

comprised of seven undergraduate students, all of whom with one exception, identified as immigrant-origin. With the guidance from the principal investigator and three graduate students, the PAR student advisory board members shaped the conversations and research questions that they felt were necessary to explore regarding the experiences of immigrant-origin emerging adults on campus. The goals of this overall project included the following: collect and analyze data regarding supporting the needs of immigrant-origin students to foster civic engagement, provide creative outlets for the dissemination of research findings by the PAR Student Advisory Board members, and showcase recommendations to better support the needs of immigrant-origin students on campus.

This study utilizes a mixed-methods design where quantitative and qualitative data were collected separately. A larger number of participants took part in the quantitative portion of the study, with a subset participating in the qualitative component of the study. Given the sensitivity of collecting data with immigrant-origin participants (many of whom were undocumented) under the current political administration, no identifying markers were collected, which resulted in data across methods not being able to be linked. Through these multiple methods, we aim to show a more complete picture of identity development in the context of family for immigrant-origin Latinx emerging adults.

Patterns of Family Engagement

To gain greater insights into the broad patterns of family engagement (i.e., family obligation and family support) in Latinx immigrant-origin emerging adults' lives, we conducted an online survey anonymously through the Qualtrics platform from April to August of 2018 at a large public university that serves a variety of immigrant communities in the Midwest of the U.S. Survey participants were recruited through a variety of outreach efforts such as posting flyers with the project website link on several message boards on campus and through campus list serves.

We utilized a number of measures to understand the patterns of family engagement. A version of the Family Responsibilities Index was adapted for this study (see Katsiaficas, 2017; Katsiaficas et al., 2018 for details). The six-item measure asked participants whether they engaged in the following activities in the past month for their family members: paying bills or expenses, errands or practical tasks, tutoring or homework help, translating, advice and solving problems, as well as a space of additional domains for which they wrote in their responses. The Family Assistance Checklist tapped into the types of assistance emerging adults received from their family members in the past month (see Katsiaficas et al., 2018 for details). On this six-item scale, participants indicated whether they had received help in the following domains: paying expenses, paying tuition, errands or practical tasks, tutoring or homework help, advice and solving problems as well as a space of additional domains for which they wrote in their responses.

Descriptive quantitative analyses were conducted to assess the activities participants engaged in and received from families, including frequencies to create a descriptive portrait of the ways in which Latinx emerging adults were engaged with their families. The results were then integrated with the qualitative verbal and visual narratives to add texture to these in-depth case studies.

Perceptions of Hyphenated Identities

To understand the ways in which Latinx immigrant-origin emerging adults make sense of their hyphenated identities, we drew upon individual interviews. Graduate research assistants, who were trained in these qualitative methodologies and had strong personal and professional ties to Latinx immigrant-origin communities, conducted the individual interviews for this study. Participants were recruited via email from different student and university organizations as well as through personal contacts. Undergraduate students who were enrolled at the university site, between the ages of 18 and 34, and had at least one foreign-born parent were recruited. All participants reported taking part in the quantitative portion of the study, however as noted above, these data were not linked to ensure participant safety.

Pluralistic narratives (Verbal and visual narratives; Katsiaficas et al., 2011, 2016) were collected through semi-structured interviews with participants. As part of the interview protocol (verbal narrative), participants were asked to complete an identity map (visual narrative) in which they were asked to "draw their multiple selves" (see Katsiaficas et al., 2011, 2016, 2018 for further details). Mapping techniques allow participants to express a visual narrative of self and relationships that may be "preverbal, affect-laden, metaphoric and/or relational" (Katsiaficas et al., 2011, p.123). After drawing, interviewers asked participants for specific interpretations of the participant's drawing to understand deeply into the narratives regarding the ways in which their identities played a role in immigrant-origin emerging adults' lives.

For the qualitative analysis, we analyzed the interview transcript (verbal narrative), and the identity map (visual narrative), by discussing themes and drawing connections between the data points for each case to identify how immigrant-origin emerging adults understood their family engagement and their identities. By placing both the visual and verbal narratives in dialogue with one another as part of the shared pluralistic analytic conversations, the researchers were able to develop in-depth descriptions of the multidimensional nature of the ways in which family engagement shapes immigrant-origin emerging adults' identities within immigrant families (Frost, 2009; Katsiaficas et al., 2011).

Understanding Family Engagement

The descriptive quantitative results show broad patterns of family engagement among Latinx immigrant families by the ways Latinx emerging adults are giving and receiving supports from their family members to help frame the in-depth case studies. Table 5.2 provides descriptive statistics of the types of support participants *received* from their family members. The majority of participants received financial supports from their families (64.9% of respondents). In addition, nearly half of participants reported that they received practical help from family members (49.1%). Only about a third of participants (38.6%) reported that they received support in helping to solve problems or getting advice from family members. Under a quarter reported receiving assistance from family members in paying tuition and very few participants reported receiving help with homework from family members. These findings reflect how the family supports that Latinx emerging adults receive are confined to either economic assistance or home-based help. Emerging adults may find it difficult to obtain practical assistance related to navigating institutional systems such as applying to college or searching for financial aid as they transition from high school to college, especially if they are the first in the families to do so.

Our findings also revealed that the proportion of family obligations were overall higher than the proportion of family support (see Table 5.3). For example, participants reported much higher rates of providing support to family members regarding errands (84.2%) than they received support in this domain (49.1%). While we were not able to gauge the significance or effect size of these differences, future work should use continuous variables to determine the size of these differences. These findings, however, point to the ways in which Latinx emerging

Table 5.2 Percentages of types of family support ($N = 57$)

Family Engagement Family Support	*N*	%
Received		
Paying my expenses	37	64.9
Helping me errands or practical tasks	28	49.1
Helping me to solve the problems/ give advice	22	38.6
Paying for tuition	13	22.8
Tutoring or homework help	2	3.5
Other		
Provided/family obligation	2	3.5

Table 5.3 Percentages of types of family support participants provided ($N = 57$)

Family Engagement Family Obligation	*N*	%
Helping my family with errands, household chores	48	84.2
Translating	45	78.9
Giving advice	32	56.1
Tutoring or helping my family members with homework/classes	31	54.4
Helping pay family's bills or expenses	26	45.6
Other	11	19.3

Table 5.4 Descriptive statistics on source to cover tuition and fees ($N = 57$)

	Family (Parent, relatives, etc.)		My Own (Savings from work, work study, other income)		Aid (Loan, scholarships, etc.)	
	N	%	*N*	%	*N*	%
None	22	38.6	12	21.1	8	14
1–25%	21	36.8	18	31.6	4	7
26–50%	7	12.3	10	17.5	8	14
51–75%	2	3.5	3	5.3	7	12.3
76–100%	5	8.8	14	24.6	30	52.6
Total	57	100	57	100	57	100

adults may often be more involved in supporting their families rather than receiving support from their families. Also, participants reported that they assisted their family by translating and interpreting as the second most frequent obligation among participants (78.9%), followed by giving advice to family members (56.1%) and tutoring family members (54.4%). Financial obligations such as "helping pay family's bills or expenses" were also undertaken by 45.6% of participants. In addition, there is also the financial obligation that they are responsible for their tuition and fees as emerging adults. Thus, the majority of students stated that they relied on themselves by working or savings from work (29.9%) or Aid such as loan or scholarships (64.9%) for over 50% of their tuition rather than family source (12.3%) (see Table 5.4).

Taken together, these findings highlight how Latinx emerging adults and their families are supporting one another, often in mutually constitutive ways that make use of the resources most available to parents and their emerging adult children. These broad patterns will be further illuminated through the case studies presented below.

Two Portraits of Latinx Immigrant-Origin Emerging Adults

Our qualitative data provided a more comprehensive exploration into the broad quantitative patterns reported above. We present two case studies that bring to life these patterns of family engagement and how family engagement relates to identities. While there were no identifiable group differences in gender in our quantitative data, we selected a case from a young woman and a young man, as the extant literature suggests their experiences may significantly differ. These cases were selected if they matched these demographic criteria and their narratives pertained to the intersection of family and identity development. This idiographic approach to understanding identity development helps to shed light on the unique experiences of participants through in-depth person-centered analysis rather than to provide generalizable patterns across aggregated data of participants (Hammack et al., 2009).

Carmen, Parallel Identities at the Hyphen

Carmen is a 22-year-old woman who was born in U.S. and living in a large urban center it the Midwestern U.S. She attended a local public university as she lived with her father and mother who migrated from Colombia as well as her younger sister who was also born in the U.S. Her family obligations and family support played an important role in how she selected her current college. When asked about how she made her decision to attend her current college, she responded,

> Cost is number one. I didn't want to leave my house. I think it's mostly because of cost. Also, because in, where my mom is from, I say where I'm from, but in reality, I wasn't born over there, but that's how I feel. Like it's normal when I go back to Colombia where students to live with their parents. So, I feel like I'm more to that culture than rather American like leaving home when they're eighteen, and you know, that was another reason. I wanted to really, um, enjoy my parents cause with all these things happening, I want to like be with my parents as much as possible as I can cause I know right after college, work, maybe move out of my house with someone, so I wanted to really enjoy as much of my life with my parents.

For Carmen, we observe in her narrative the ways in which her family values come into contrast with the "American" values of leaving home. She describes that she wanted to live with her family, even though she is an adult, since she wanted to save money and receive more emotional support, a feature common in her heritage culture, and one she notes as being distinct from American culture. At the same time, it is apparent that she has financial responsibilities for her family. She described her sense of collective contribution, reciprocating the supports she receives in the ways that she can also provide to her family. The reciprocal nature of this contribution underscores the importance of her family as it pertains to her experiences and identity as a college student. When asked about her family responsibilities she notes,

> Well my parents do give me like food and pay my phone bill, but there's little things that they really can't give me … But I do have like the responsibility of like giving them back what they give me in my house, and that has tied me a lot from like not being able to join some clubs because of the timing that they're meeting. Because I have to go back at home … literally five o'clock my dad comes home. And, I would really think he deserves coming home to a hot meal because he works like eight hours standing up, washing dishes at a hotel. So, I really think he deserves like going home and finding food. [Note. She pauses to hold back tears]

Carmen's narrative highlights her understanding her parents' hard work and sacrifices to support the children in the family, and illuminates the ways in which this understanding shapes how she engages in college extracurricular activities and ultimately orients herself to her role as a college student, who puts her family first.

Further in her narrative, we hear a longing for Colombia and a sense of complicated belonging in the U.S. As Carmen emerges into adulthood, she poignantly describes a shift within her identities that coincides with the changing socio-political landscape of the U.S. during the Trump administration and perhaps shows the ways in which she is reforming her identity in the current climate of the U.S.

…Although I was not born there [Colombia], I identify more with my roots, and I really care about that. Because America... they don't want, they don't want us here. I don't feel like they want me here. So, why, why-you now *agradecida* with this country cause like you know an education, a bachelor's, but ideally if I had the opportunity to be economically well in Colombia, I would be there.

Through Carmen's description, we observe the difficulties of living here but longing to be there, while still knowing she would not have the same opportunities. She showed her strong orientation toward Colombia where she felt a deeper sense of belonging. She explained that she went to Colombia every two years to see her grandparents, and her parents purchased a house in Columbia to go back to in the future. These activities gave way to shared family values. She notes,

I think my parents did a good job and we decided to follow, because that's deciding because we decided to follow, the Colombian culture, cause I know some other Latinos here, their parents listen to music, but they decide to hear American rap. Like I decided to follow my parent's path in their culture and in their food and how they talk.

In these ways, Carmen is describing a deep desire to protect the connections she has with Colombia in ways that may significantly differ from her peers who adopt a more "American" orientation to music. This desire influences her identity as a Latina in the U.S. who "follows" closely in Colombian traditions and cultural practices.

Her strong family ties also influenced her future plans about where she wants to live. These strong family engagement values and her aspirations are represented in her Identity Map (see Fig. 5.1). When asked to draw her identity during the mapping exercise, she drew herself, her boyfriend, and her family members as the main focus of her identity map. This visual narrative helps to open a conversation about her identities, and reveals her sense of belonging across these two nations.

My boyfriend is there [Colombia], actually, I'm thinking of like maybe like moving back… [sighs], my mom came here when she was really, I mean not young, she got married. The point is, my mom was here by herself with her nine brothers and sisters in Colombia with my grandparents, and I know how that feels. Not seeing, I mean, *gracias a Dios*, I went there every two years. I think that's better than other cases of other immigrant families. I mean I feel really blessed I did see them, my grandparents every two years, but I still feel like I missed a lot like of parties, events, and ironically, I went there every two years. I don't want my kids to, cause my parents, they already purchased their house over there, an apartment. I know they're going to come over there. I don't know if they're going to come back. The point is I don't want my kids to go through that. One of the reasons why I would strongly think of moving back. I think it's more important than having like money, and like I talk a lot about it with my boyfriend, and he's like, I study here in public schools, good education, and he's got so many things to come at me. He has free education. But, you know, what if our kids will come out dead and like *drugaditos*. Cause reality here I saw so many drugs, so many, I got offered, but I didn't take that path, and it's more than I see it over there. More bullying…

Carmen explains that these biennial visits to Colombia helped to keep her connected to her family while also revealing some of the drawbacks of raising a family in the U.S. Her narrative illuminates the ways her identities intersect with her fears and desires for her future and that of her potential children.

Fig. 5.1 Carmen's identity map

Furthermore, she put her passport in her identity map to explain her career aspiration to work in the embassy to connect Colombia and the U.S. She explains,

> I wrote passport 'cause I have the U.S. passport. I hope to one day work in an Embassy, or like passport agent or something like that 'cause I feel like traveling has been, has connected like two of my worlds. Like the U.S. and Colombia, which I would really like to do something like that, like representing either like America. I can't do a Colombian one cause, you know, I was born here. I tried that, but I'm really open to also living in other countries. I don't know what's, but that's I want something related…

This passport in many ways might represent the hyphen at which her identities come together; a marker of her "Americaness" that will also afford her the opportunity to travel outside of a place she has complicated feelings of belonging.

The way she narrates her identities at the hyphen becomes further complicated in the verbal narrative when she was asked about potential conflicts with the American culture. She stated,

> A lot [laughs]. I mean it's difficult because when I'm here I identify myself as Colombian because I feel like I don't fit in with American people. And then when I'm in Colombia, they're like *Gringa* [A female native speaker of English]. So I'm like what the hell? Like what am I? I mean I love both countries, like they're both give me something that the other one can't.

Through her verbal narrative, there is a sense of a gap in-between being Colombian and an American, a space she depicts in her map as floating above her head with the Colombian flag in a heart, a picture of the globe and her passport between the two. She goes on to discuss the idea of belonging in greater depth, when the interviewer asks her about where she sees herself living in the future. She responds,

> Miami. I mean I went there not long, but the little time I was there, I mean it's not just because there's, I mean there are more Colombians, but there's like Cubans and Argentineans, and I thought I fit in more with them than I do with the Mexican people here in Chicago. So that's what I mean by my university community cause the majority of the Latinos are Mexican. I feel like a minority in that minority in my classrooms.

This sense of navigating the hyphen presents two parallel identities that while not in conflict, do not seem to provide much overlap. Given that one's identity formation is developed with the socialization process of the family, the ethnic community, and larger settings (Phinney et al., 2001), her parallel identity development might be related to the segmentation of these contextual messages. This is closely related to the multiple messages she receives about her sense of belonging within her environment. She reported she felt excluded on campus because of the majority of Mexican and Puerto Rican students there. The feeling of being a minority among her Latinx peers may also influence Carmen's identity development.

Eric, Giving Back to his Community

Eric was born in the U.S. and lives in a well-known Latinx enclave in a large urban center in the Midwest U.S. He commutes to the university which is close to his house. As a sophomore, he aims to be a teacher and received a prestigious state-level scholarship to further this goal. His aspirations for the future are well represented in his visual narrative (see Fig. 5.2). He drew himself wearing a shirt, black pants, and brown shoes, with an active stance and a smile on his face. He is depicted standing between a math book and backpack for studying math class as a college student with his "Prestigious Teacher Training Scholarship" [pseudonym to protect participant's identity]. On his other side, stands the flags of Mexico and the U.S. city where he currently resides. We could see his identity is tied not only to his current status as a college student but also to his future career aspirations as a math teacher.

Given that Carmen's identity was more situated within her close relationships, Eric emphasized his identity through his career. These differences highlight the traditional gender roles and expectations among Latinx emerging adults that may shape the ways in which they depict and understand their identities. Even though they both drew their future on their identity maps, Carmen anticipated her future as being related with her romantic and family relationships passively, whereas Eric connected his current status with his future career in greater detail. These differences also appeared in their family obligations. Eric described,

Fig. 5.2 Eric's identity map

> I have two brothers. One's older. He's still in school as well. He's only a year older. Um, and then one younger, and he's a freshman. But my older brother does not go to school here. He actually moved away to go to school. My mom and dad also live in the household as well as one of my cousins. I would say my role in the household mostly would be like since my older brother's not there, I'm like the older brother to my little brother. I'm also the ones who's probably the responsible one in the aspect of paperwork, like for like university stuff specifically, too. When it comes to me and my little brother, I am the one who makes sure we both fill out, like FAFSA, and the financial aid. And I go file taxes. We gotta do taxes because then we got fill out FAFSA for next year. Where are the W2s? And all that stuff. I am trying to take the role of like, trying to take some of the things away from my parents in the aspect of letting them not to worry so much on like, like let's say taxes. Just because I am on top of what I'm doing on taxes. And like by me being on top of it, it helps them because then they don't have to be nagging me about where are your W2s forms? I already have them collected. I keep them. I'd say, I also work. And then I'm doing observations also because I'm in my classes, we have to observe. Um, so I'm just those things combined, in addition with homework, like those things combined just make me very busy. I don't have a lot of free time all the time. When I spend it doing things I like to do. So, I would say it stops me in the aspect of sometimes by the end of the day, I'm just kind of tired, and I just want to go home and chill.

Eric's obligations for his family are tied to navigating the financial realities of being a student for both himself and his brother, including completing the taxes for his family. His identity as an older brother and financial steward in his family impacts his identity as a student, as the two are closely related.

Furthermore, Eric's responsibility toward his family expanded to his community. During the interview, Eric mentioned receiving clear messages about the value of collective contribution that is promoted in the community.

> I am a firm believer in helping those around you as well, trying to elevate not just yourself but the people around you should try to elevate you as well and you to them. And the type of people who don't reciprocate that type of mentality are the people not in my community.

> I'm not trying to say they're not welcome, I want to help everyone eventually. But as far as like trying to get to where I need to be to put myself in the position to help people, I try to surround myself around people who have goals, as well, and try to push each other.

He further mentions that he would like to teach in a neighborhood school that serves a majority of Latinx students as an extension of his work with his family. In this way, his ethnic identity based on his family engagement impacts his career identity.

Eric is focused on giving back to his family and community by being a teacher. His community membership and his desire to give back to this community are important to his identity. This is depicted in his picture of the city's flag which is related to feelings of belonging and attitudes toward the larger society. Moreover, he described how the depiction of the Mexican flag has a large impact on his identity as well,

> I put the Mexican flag just because it's big on my day to day life. I would say in the fact that from the food I eat at home, to Spanish, to my parents, to my grandparents, to like our name in our household like, they're all Mexican based names. I would say this is probably the biggest thing.

His ethnic identity was developed within his family context and by broader extension his community. His ethnic and local community identities were entertwined and in turn informed his career identity. For Eric, the hyphen acts almost as a loop highlighting the ways in which he has received and continues to give back to his community.

Discussion and Conclusions

Across these mixed-methods data, we come to understand how immigrant-origin Latinx emerging adults have developed their multiple identities within their families, and by extension their community and cultural contexts. While the majority of extant research on emerging adults has focused on identity development in the domains of love, work, religion, and politics (see Schwartz, Pantin, et al., 2005; Schwartz et al., 2013 for review), few studies have examined the ways in which families play a role in identity development during this developmental period. Yet, our results indicate that there are not only high levels of engagement with families for immigrant-origin emerging adults in our study, but family played a significant role in shaping their multiple identities.

Specifically, results revealed that participants were highly engaged in supporting their families through a variety of activities and also reported receiving a significant amount of support from their families. These findings extend the work of family obligations (Fuligni et al., 1999) by examining the bi-directional contributions between family and students, referred to as collective contribution (Katsiaficas et al., 2018). From the quantitative data, we found that the reciprocal relationship of family engagement among Latinx immigrant-origin emerging adults by engaging in family obligations and receiving family supports. Specifically, the results showed that participants relied on financial support from their family, and that concurrently they provided daily assistance such as household chores and translating in return. While the quantitative data did not highlight the financial contributions of these emerging adults, we saw through their narratives how financial responsibilities played an important role in their lives. Carmen and Eric reported that they choose their current university to save the tuition and dormitory costs. Specifically, Eric showed a concrete future to give back to his community as being a teacher. Considering their current status of college students, we could see their heavy financial responsibilities to repay their families in the future.

This interdependence between emerging adults and their families is often overlooked by current theories of this developmental period. The theory of emerging adulthood suggests that as young people emerge into adulthood, they become increasingly independent and self-focused (Arnett, 2000). Yet, these results highlight the ways in which emerging adulthood can also be a period of increased social responsibilities, where young people are able to contribute in new and profound ways to their families as they may not have the opportunity to do so prior. These social responsibilities were powerful contributors to these emerging adults' identity formation processes.

We also found that the development of ethnic identities were intertwined with these social responsibilities through both their verbal and visual narratives. This work extends the work of Syed and Azmitia (2009) who found that ethnic identity exploration tends to increase during the college-going years particularly for ethnic minority emerging adults. The narratives presented in the results help to further illuminate how families play an integral role in this process. Carmen demonstrated a strong inclination toward her heritage culture as a Colombian. She followed Colombian culture since her parents adhered to their heritage culture in the U.S. and experienced it by visiting Colombia every 2 years. Eric also described the ways that his Mexican heritage was rooted in his identity by way of his community, including the food, language, and experiences he shared with his family. The daily experiences of their heritage culture from their immigrant family context supported and shaped their developing ethnic identities. For Carmen, we saw the ways the hyphen became a place she occupied between her multiple selves, as she depicted the complications of the hyphen through her future aspirations, her passport, and her future family goals. Particularly for Carmen, we saw in her narratives how the current socio-political environment sparked a reformation of her identities in response to a shifting sense of belonging in the U.S. For Eric, on the other hand, there seemed to be a more seamless transition between these multiple identities. Eric narrated how his family identity shaped his future career goals to give back to his community. The hyphens for Eric seemed to be an integration of his family, career and ethnic identities as he found ways to expand his family cultural values into his neighborhood and community. Taken together, the Hyphenated Selves framework (Fine & Sirin, 2007) allowed for rich narratives of identity, family, and socio-political context to weave together to present a more holistic understanding of identity development during this particular historical moment.

The case studies demonstrated how these reciprocal forms of family support played a role in students' identities. Taken together, these findings point to ways that family are integral members in shaping identities and college experiences for immigrant-origin emerging adults and have important implications for practice and policy. In particular, it is critical for higher education institutions to recognize the important role that families play in shaping identities of emerging adult students but also influence their future career aspirations. Further, recognizing the ways broad messages about immigrant families from the socio-political context can impact identity development and the engagement of emerging adults in society is a critical point of intervention. While many immigrant-origin Latinx emerging adults are encountering messages that they do not belong in U.S. society, they simultaneously represent key bridging agents between their families, communities, and American social institutions. Recognizing the potential of immigrant-origin Latinx emerging adults can help policy makers to recognize and harness the power these young people possess.

It is important to note key limitations of these data and findings. Securing participants' identifiable data were paramount in this study and therefore anonymous data collection was a key element. As such, this did not allow for our quantitative and qualitative data to be linked, which would have allowed us to make inferences not possible in the current analysis. Furthermore, while beyond the scope of this study, data collection that included a larger sample might also help to capture the variation in experiences during this particular socio-political moment.

The Hyphenated Selves framework allowed for us to capture thick descriptions of the multiple ways in which these dynamic contexts shaped the identities of these emerging adults. The pluralistic narratives alongside the idiographic analytic approach allow for theorizing about the role of family, community, and socio-political contexts as they dynamically relate to identity formation and reformation. In both of these cases, these frameworks and methodologies helped to start conversations about these processes, however, further research is necessary to more deeply understand these processes for Latinx emerging adults. Specifically, examining how these identities are shaped across time and across contexts would further illuminate the role of family in shaping identity during this developmental period. Future research should examine further the dynamic ways in which families shape the identities of emerging adults. While this study is an important step in that direction, larger samples and longitudinal methods would help to examine these phenomena in greater depth and over time.

References

Arnett, J. J. (2000). Emerging adulthood: A theory of development from the late teens through the twenties. *American Psychologist, 55*(5), 469–480. https://doi.org/10.1037/0003-066X.55.5.469

Arnett, J. J. (2004). *Emerging adulthood: The winding road from the late teens through the twenties*. Oxford University Press.

Arnett, J. J. (2008). The neglected 95%: Why American psychology needs to become less American. *American Psychologist, 63*(7), 602–614. https://doi.org/10.1037/0003-066X.63.7.602
Azmitia, M., Syed, M., & Radmacher, K. (2008). On the intersection of personal and social identities: Introduction and evidence from a longitudinal study of emerging adults. In M. Azmitia, M. Syed, & K. Radmacher (Eds.), *New directions for child and adolescent development: Intersections between personal and social identity* (pp. 1–16). Jossey-Bass.
Baer, J. C., & Schmitz, M. F. (2007). Ethnic differences in trajectories of family cohesion for Mexican American and non-Hispanic white adolescents. *Journal of Youth and Adolescence, 36*(4), 583–592. https://doi.org/10.1007/s10964-007-9177-3
Batalova, J., & Fix, M. (2011). *Up for grabs: The gains and prospects of first- and second generation youth adults*. Migration Policy Institute.
Cervantes, J. M., Minero, L. P., & Brito, E. (2015). Tales of survival 101 for undocumented Latina/o immigrant university students: Commentary and recommendations from qualitative interviews. *Journal of Latina/o Psychology, 3*(4), 224–238. https://doi.org/10.1037/lat0000032
Cooper, C. R., Jackson, J. F., Azmitia, M., & Lopez, E. M. (1998). Multiple selves, multiple worlds: Three useful strategies for research with ethnic minority youth on identity, relationships, and opportunity structures. In V. C. McLoyd & L. Steinberg (Eds.), *Studying minority adolescents: Conceptual, methodological, and theoretical issues* (pp. 111–125). Erlbaum.
Côté, J. E. (2006). Emerging adulthood as an institutionalized moratorium: Risks and benefits to identity formation. In J. J. Arnett & J. L. Tanner (Eds.), *Emerging adults in America: Coming of age in the 21st century* (pp. 85–116). American Psychological Association. https://doi.org/10.1037/11381-004
Deaux, K., & Perkins, T. S. (2001). The kaleidoscopic self. In C. Sedikides & M. B. Brewer (Eds.), *Individual self, relational self, collective self; individual self, relational self, collective self* (pp. 299–313). Psychology Press.
Fine, M. (1994). Working the hyphens: Reinventing self and other in qualitative research. In N. K. Denzin & Y. S. Lincoln (Eds.), *Handbook of qualitative research: Handbook of qualitative research* (pp. 70–82). Sage.
Fine, M., & Sirin, S. R. (2007). Theorizing hyphenated selves: Researching youth development in and across contentious political contexts. *Social and Personality Psychology Compass, 1*, 6–38. https://doi.org/10.1111/j.1751-9004.2007.00032.x
Frost, N. (2009). "Do you know what I mean?": The use of a pluralistic narrative analysis approach in the interpretation of an interview. *Qualitative Research, 9*(1), 9–29. https://doi.org/10.1177/1468794108094867
Fuligni, A. J., Tseng, V., & Lam, M. (1999). Attitudes toward family obligations among American adolescents with Asian, Latin American, and European backgrounds. *Child Development, 70*(4), 1030–1044. https://doi.org/10.1111/1467-8624.00075
Fuligni, A. J., & Pedersen, S. (2002). Family obligation and the transition to young adulthood. *Developmental Psychology, 38*(5), 856. https://doi.org/10.1037/0012-1649.38.5.856
Fuligni, A., Witkow, M., & Garcia, C. (2005). Ethnic identity and the academic adjustment of adolescents from Mexican, Chinese, and European backgrounds. *Developmental Psychology, 41*, 799–811. https://doi.org/10.1037/0012-1649.41.5.799
Galliher, R. V., McLean, K. C., & Syed, M. (2017). An integrated developmental model for studying identity content in context. *Developmental Psychology, 53*(11), 2011–2022. https://doi.org/10.1037/dev0000299
Gandara, P. (1995). *Over the ivy walls: The educational mobility of low-income Chicanos*. State University of New York Press.
Goldscheider, F., & Goldscheider, C. (1999). *The changing transition to adulthood: Leaving and returning home*. Sage.
Guan, S. A., & Fuligni, A. J. (2016). Changes in parent, sibling, and peer support during the transition to young adulthood. *Journal of Research on Adolescence, 26*(2), 286–299. https://doi.org/10.1111/jora.12191

Hammack, P. L., Thompson, E. M., & Pilecki, A. (2009). Configurations of identity among sexual minority youth: Context, desire, and narrative. *Journal of Youth and Adolescence, 38*(7), 867–883. https://doi.org/10.1007/s10964-008-9342-3

Hooghe, M., & Dassonneville, R. (2018). Explaining the trump vote: The effect of racist resentment and anti-immigrant sentiments. *PS, Political Science & Politics, 51*(3), 528–534. https://doi.org/10.1017/S1049096518000367

Hurtado, S., Carter, D., & Spuler, A. (1996). Latino student transition to college: Assessing difficulties and factors in successful college adjustment. *Research in Higher Education, 37*(2), 135–157. http://www.jstor.org/stable/40196170

Katsiaficas, D., Alcantar, C. M., Hernandez, E., Samayoa, E., Gutierrez, M. N., Texis, O. R., & Williams, Z. (2016). Important theoretical and methodological turning points for understanding contribution with undocumented undergraduates. *Qualitative Psychology, 3*(1), 7–25. https://doi.org/10.1037/qup0000043

Katsiaficas, D., Futch, V. A., Fine, M., & Sirin, S. R. (2011). Everyday hyphens: Exploring youth identities with methodological and analytic pluralism. *Qualitative Research in Psychology, 8*(2), 120–139. https://doi.org/10.1080/14780887.2011.572743

Katsiaficas, D., Hernandez, E., Alcantar, C. M., Samayoa, E., Gutierrez, M. N., & Williams, Z. (2018). "We'll get through this together": Collective contribution in the lives of Latino undocumented undergraduates. *Teachers College Record, 120*, 120306.

Katsiaficas, D. (2017). "I know I'm an adult when…I can care for myself and others": The role of social responsibilities in emerging adulthood for community college students. Emerging Adulthood, 5(6), 392–405. https://doi-org.proxy.cc.uic.edu/10.1177/2167696817698301

Lanuza, Y. R. (2017). Who helps? Immigrant-native differences in patterns of homework assistance. *Sociological Perspectives, 60*(2), 293–314. https://doi.org/10.1177/0731121416636086

López, M. H., Morin, R., & Taylor, P. (2010). Illegal immigration backlash worries, divides Latinos. *Pew Research Center.* http://pewhispanic.org/files/reports/128.pdf

Louie, V. (2012). *Keeping the immigrant bargain: The costs and rewards of success in America.* Sage. http://www.jstor.org/stable/10.7758/9781610447799

Mertens, D. M. (2003). Mixed methods and the politics of human research: The transformative-emancipatory perspective. In A. Tashakkori & C. Teddlie (Eds.), *Handbook of mixed methods in social and behavioral research* (pp. 135–164). Sage.

Morales, A., & Hanson, W. E. (2005). Language brokering: An integrative review of the literature. *Hispanic Journal of Behavioral Sciences, 27*(4), 471–503. https://doi.org/10.1177/0739986305281333

Muñoz, S. M., & Maldonado, M. M. (2012). Counter stories of college persistence by undocumented Mexicana students: Navigating race, class, gender, and legal status. *International Journal of Qualitative Studies in Education, 25*(3), 293–315. https://doi.org/10.1080/09518398.2010.529850

Opotow, S. (1990). Moral exclusion and injustice: An introduction. *Journal of Social Issues, 46*(1), 1–20. https://doi.org/10.1111/j.1540-4560.1990.tb00268.x

Parrott, T. M., & Bengtson, V. L. (1999). The effects of earlier intergenerational affection, normative expectations and family conflict on contemporary exchanges of help and support. *Research on Aging, 21*(1), 73–105. https://doi.org/10.1177/0164027599211004

Phinney, J. S. (2008). Ethnic identity exploration in emerging adulthood. In D. L. Browning (Ed.), *Adolescent identities: A collection of readings* (pp. 47–66). The Analytic Press/Taylor & Francis Group.

Phinney, J. S., Horenczyk, G., Liebkind, K., & Vedder, P. (2001). Ethnic identity, immigration, and well-being: An interactional perspective. *Journal of Social Issues, 57*(3), 493–510. https://doi.org/10.1111/0022-4537.00225

Phinney, J. S., Ong, A., & Madden, T. (2000). Cultural values and intergenerational value discrepancies in immigrant and non-immigrant families. *Child Development, 71*(2), 528–539. https://doi.org/10.1111/1467-8624.00162

Rumbaut, R. G., & Komaie, G. (2010). Immigration and adult transitions. *The Future of Children, 20*(1), 43–66. https://doi.org/10.1353/foc.0.0046
Saenz, V. B., & Ponjuan, L. (2011). *Men of color: Ensuring the academic success of Latino males in higher education*. Institute for Higher Education Policy. http://www.ihep.org/sites/default/files/uploads/docs/pubs/brief_men_of_color_latinos.pdf
Schwartz, S. J., Côté, J. E., & Arnett, J. J. (2005). Identity and agency in emerging adulthood: Two developmental routes in the individualization process. *Youth & Society, 37*(2), 201–229. https://doi.org/10.1177/0044118X05275965
Schwartz, S. J., Pantin, H., Prado, G., Sullivan, S., & Szapocznik, J. (2005). Family functioning, identity, and problem behavior in Hispanic immigrant early adolescents. *The Journal of Early Adolescence, 25*(4), 392–420. https://doi.org/10.1177/0272431605279843
Schwartz, S. J., Rodriguez, L., Weisskirch, R. S., Zamboanga, L. B., & Pantin, H. M. (2012). Personal, ethnic, and cultural identity in urban youth: Links with risk and resilience. In G. Creasey & P. A. Jarvis (Eds.), *Adolescent development and school achievement in urban communities: Resilience in the neighborhood* (pp. 228–238). Routledge.
Schwartz, S. J., Zamboanga, B. L., Luyckx, K., Meca, A., & Ritchie, R. A. (2013). Identity in emerging adulthood: Reviewing the field and looking forward. *Emerging Adulthood, 1*(2), 96–113. https://doi.org/10.1177/2167696813479781
Sirin, S. R., & Fine, M. (2008). *Muslim American youth: Understanding hyphenated identities through multiple methods*. New York University Press.
Sirina, S. R., Bikmen, N., Mir, M., Fine, M., Zaal, M., & Katsiaficas, D. (2008). Exploring dual identification among Muslim-American emerging adults: A mixed methods study. *Journal of Adolescence, 31*(2), 259–279. https://doi.org/10.1016/j.adolescence.2007.10.009
Smith, R. (2006). *Mexican New York: Transnational lives of new immigrants*. University of California Press.
Spencer, M. B. (1999). Social and cultural influences on school adjustment: The application of an identity-focused cultural ecological perspective. *Educational Psychologist, 34*(1), 43–57. https://doi.org/10.1207/s15326985ep3401_4
Stoudt, B. G., Fox, M., & Fine, M. (2012). Contesting privilege with critical participatory action research. *Journal of Social Issues, 68*(1), 178–193. https://doi.org/10.1111/j.1540-4560.2011.01743.x
Suárez-Orozco, C. (2004). Formulating identity in a globalized world. In M. Suárez-Orozco & D. B. Qin-Hilliard (Eds.), *Globalization: Culture and education in the new millennium* (pp. 173–202). University of California Press.
Suárez-Orozco, C., Motti-Stefanidi, F., Marks, A., & Katsiaficas, D. (2018). An integrative risk and resilience model for understanding the adaptation of immigrant-origin children and youth. *American Psychologist, 73*(6), 781–796. https://doi.org/10.1037/amp0000265
Suárez-Orozco, C., Suárez-Orozco, M., & Todorova, I. (2008). *Learning a new land: Immigrant students in American society*. Harvard University Press.
Sue, D. W., Capodilupo, C. M., Torino, G. C., Bucceri, J. M., Holder, A. M. B., Nadal, K. L., & Esquilin, M. (2007). Racial microaggressions in everyday life. *American Psychologist, 62*, 271–286. https://doi.org/10.1037/0003-066X.62.4.271
Syed, M., & Azmitia, M. (2009). Longitudinal trajectories of ethnic identity during the college years. *Journal of Research on Adolescence, 19*(4), 601–624. https://doi.org/10.1111/j.1532-7795.2009.00609.x
Syed, M., Azmitia, M., & Phinney, J. S. (2007). Stability and change in ethnic identity among Latino emerging adults in two contexts. *Identity: An International Journal of Theory and Research, 7*(2), 155–178. https://doi.org/10.1080/15283480701326117
Torre, M. E., Fine, M., Stoudt, B. G., & Fox, M. (2012). Critical participatory action research as public science. In H. Cooper, P. M. Camic, D. L. Long, A. T. Panter, D. Rindskopf, & K. J. Sher (Eds.), *APA handbook of research methods in psychology* (*Research designs: Quantitative, qualitative, neuropsychological, and biological*) (Vol. 2, pp. 171–184). American Psychological Association. https://doi.org/10.1037/13620-011

Tseng, V. (2004). Family interdependence and academic adjustment in college: Youth from immigrant and U.S.-born families. *Child Development, 75*(3), 966–983. https://doi.org/10.1111/j.1467-8624.2004.00717.x

U. S. Census Bureau. (2013). *Current population survey, annual social and economic supplement.* https://www-census-gov.proxy.cc.uic.edu/programs-surveys/acs/about.html

Umaña-Taylor, A. J., Zeiders, K. H., & Updegraff, K. A. (2013). Family ethnic socialization and ethnic identity: A family-driven, youth-driven, or reciprocal process? *Journal of Family Psychology, 27*(1), 137–146. https://doi.org/10.1037/a0031105

Vega, W. (1990). Hispanic families in the 1980s: A decade of research. *Journal of Marriage and Family, 52*(4), 1015–1024. https://doi.org/10.2307/353316

Chapter 6
Buffering Against Parent-Child Conflict: Exploring the Role of Biculturalism in the Relationship Between South Asian College Students and Their Parents

Vrinda Kalia, Pankhuri Aggarwal, and Vaishali V. Raval

More than 20 million Asian immigrants live in the U.S. (Lopez et al., 2017). According to the Center for American Progress, Asians are the fastest-growing immigrant population in the U.S. (Kieu, 2013), with 66% of foreign-born adult residents in the country identifying as Asian (Malik, 2015). Asian immigrants are also a diverse group. The largest proportion of Asian immigrants, approximately 22%, identify as Chinese, whereas the smallest proportion come from Japan, approximately 7% (Kieu, 2013). Asian Indians are the second largest group of Asian immigrants and constitute about 20% of the total Asian population (Lopez et al., 2017). Although developmental theories have long acknowledged that contextual forces shape individual development (Bronfenbrenner, 1979; Vygotsky, 1978), it is only recently that developmental researchers have begun to examine the role of immigration and emigration as developmental contexts that contribute to individual differences in developmental outcomes (Mistry et al., 2016). In view of the fact that Asian immigrants account for a significant portion of immigrants in the U.S. it is imperative that we study developmental outcomes for this population within the context of immigration.

According to Erikson (1968), identity development results from the interaction between individual characteristics and the socio-cultural context within which the individual is trying to establish their role as a member of the community. Thus, some have claimed that "the person and the person's social world coauthor identity" (McAdams, 2001, p. 116). As part of the developmental process of acquiring an identity, the individual must achieve some separation from their parents and establish their own individual characteristics. Since parents serve as a bridge between the macro and micro contexts in a child's search for identity, they can either support

V. Kalia (✉) · P. Aggarwal · V. V. Raval
Miami University, Oxford, OH, USA
e-mail: kaliav@miamioh.edu; aggarwp@miamioh.edu; ravalvv@miamioh.edu

D. J. Johnson et al. (eds.), *Re/Formation and Identity*, Advances in Immigrant Family Research, https://doi.org/10.1007/978-3-030-86426-2_6

their child's need for autonomy and individuation or stifle it by curtailing exploration (Koepke & Denissen, 2012). The challenges associated with acculturation can influence the role that parents play in identity development and increase intergenerational conflict in immigrant families (Tardiff-Williams & Fisher, 2009).

Building on the interactional framework of identity development proposed by Erikson (1968), in the present chapter we provide a brief overview of the racialized immigration history of the U.S., as race has always played a predominant role in immigration in the country (Bhatia & Ram, 2001). We use the racialized history of Asian immigration to justify the examination of the parent-child relationship in identity reformation of South Asians using a postcolonial feminist lens. In view of contemporary research showing that family plays a crucial role in the development of ethnic identity (Syed et al., 2007) and family conflict in immigrant families is negatively associated with psychological adjustment (Juang et al., 2012), our primary focus will be on the relation between family conflict and the reconstruction of immigrant identity.

Racialized Immigration History of the United States

In an attempt to address labor shortages resulting from the Civil War, the United States Congress enacted the immigration act of 1864. Consequently, between the early nineteenth and twentieth centuries, the U.S. added approximately 35 million immigrants to its citizenry (Calavita, 2020). Yet, by the late nineteenth century, as the number of non-European immigrants began to rise, faith in the immigrant's ability to assimilate with American culture began to fade (Lehtinen, 2002). Within this environment of skepticism and hostility toward immigrants, the immigration station was opened at Ellis Island. All individuals intending to immigrate had to be approved for entry by government officials from the Department of Commerce and Labor and medical physicians. A key aspect of the approval process was determining whether the intended immigrants were "feeble-minded" or had mental deficiencies so as to prevent those who were ineducable from entering the U.S. (Richardson, 2003). Eventually, Asians, except those from Japan, were barred from immigrating to the U.S. (Lehtinen, 2002). The culmination of this race-based nativism, favoring Northern and Western Europeans over others, was a series of legislations that were promoted and supported by eugenists. These laws irrevocably institutionalized racial quotas for immigrants based on country of origin (Ngai, 1999).

When considered through the lens of the history of immigration in the U.S., it is perhaps unsurprising that racial and ethnic discrimination and prejudice are normative experiences for Asian immigrants (Bhatia & Ram, 2001). The experience of prejudice and discrimination can take many forms, including that of systemic barriers that homogenize the Asian immigrant experience by silencing differences among Asians (King, 2000). The myth of the 'model minority' is an example of the imposed institutional homogenizing that has obscured both the financial (Nam et al., 2015) and mental health issues that Asian immigrants struggle with, which are by-products of the stress associated with immigration (Miller et al., 2011).

However, it is not just institutionalized and overt racism that can cause damage (Sue & Sue, 2003). Implicit stereotyping, in the form of microaggressions, has the potential to be harmful as well (Sue et al., 2009). For instance, Benner and Kim (2009) report on a longitudinal study with Chinese adolescents showing that experiences with daily discrimination predicted symptoms of depression, reduced engagement in school, and lower grades. Interestingly, they also found that individual differences in cultural orientation toward American culture moderated the relationship between experiences with discrimination and developmental outcomes. Specifically, Chinese adolescents who exhibited increased orientation toward American culture (e.g., I believe in mainstream American values) were also more likely to experience negative effects associated with discrimination. Benner and Kim (2009) explained this finding in terms of cognitive dissonance. Despite the belief that they were American, Chinese adolescents who expressed high levels of American orientation were reminded through daily acts of discrimination of their 'foreigner' status. These reminders through acts of microaggression and discrimination contradicted their view of themselves and caused them emotional distress (Benner & Kim, 2009). In effect, Benner and Kim (2009) demonstrated that Chinese adolescents' integration into the host culture, as traditionally defined by Berry's (1997) model of acculturation, was having a negative impact on developmental outcomes. As other researchers have highlighted previously (Boski, 2008) and we argue below, Berry's model provides an appealing but overly simplistic account of the complex process of the immigrant's engagement with the host culture.

Acculturation Approach to Identity

In his seminal work, Berry (1997) characterized acculturation as a process through which two cultures come in contact and adapt around one another. At the individual level, Berry (1997; Berry & Sam, 1996; Sam & Berry, 2010) identified four acculturative strategies that enable this process of adaptation to the host culture. The most successful strategy is integration, which is characterized by a willingness exhibited by immigrants to retain the values of their culture of origin and integrate them with the values of the host culture that they adopt. The least successful strategy is marginalization. Marginalization is defined by a lack of interest in maintaining the values of the culture of origin and lack of engaging with the host culture, which can result in rejection and isolation from both cultures. The two other acculturation strategies, which are less successful than integration but not as damaging as marginalization, are assimilation (immigrant adopts, not always voluntarily, their host culture's values in order to fit in) and separation (individual values their culture of origin and avoids the dominant culture) (Sam & Berry, 2010).

The underlying premise of Berry's work is that the individual's acculturative processes in the host culture are motivated by two factors: (1) the extent to which the immigrant is allowed to and wants to retain their culture of origin; and (2) the extent to which the immigrant is allowed to and wants to identify with and engage with their host culture (Benet-Martínez et al., 2006). The model of acculturation proposed by Berry has had a profound impact on research conducted in cross-cultural psychology and our understanding of immigration through a psychological lens (Bhatia & Ram, 2001).

Although Berry mentions the possibility of biculturalism as an outcome of the acculturative process (Berry, 1997; Berry & Sam, 1996), his theorizing about acculturation processes has two major problems. First, Berry insists on a universalist stance on the processes underlying acculturation (Bhatia & Ram, 2001). While acknowledging that social contexts within the host culture could influence the type of acculturative strategies that an individual employed (Berry, 1997), the acculturation model also proposes that processes that operate to enable acculturation are the same regardless of the immigrant group (Bhatia & Ram, 2001). By universalizing the psychological processes that support acculturation, Berry's model silences the historical and political realities that determine individual differences in the immigrant's experience within the host culture (Bhatia & Ram, 2001).

Second, Berry (1997) proposes a linear model of culture (Bhatia, 2002). In Berry's vision of immigration, an individual lives in culture A (which is their culture of origin) and then moves to culture B (which is their host culture) in a linear fashion. However, this model ignores the reality of intercultural contact and immigration in a globalized world (Nguyen & Benet-Martínez, 2013), as well as the experiences of immigrants who may be multinational or have hyphenated identities (Bhatia & Ram, 2001). For these individuals, acculturation is better characterized as 'mixing' of cultures (Bhatia, 2002). And it is a dynamic experience that requires continuous engagement with the process of reforming and negotiating their identity in changing cultural landscapes (Bhatia, 2002).

Postcolonial Feminist Views on Immigrant Identity and Biculturalism

Postcolonial scholars were the first to identify the ways in which historical, political, and social institutions create novel experiences of the 'Third-World' immigrant in 'First World' contexts (Bhatia, 2002). For instance, Mohanty (1991) has argued that the racialized immigration policies of the U.S. have created a class of individuals known as immigrant aliens, who, regardless of the time lived in the country, will always be foreign or 'other.' By highlighting the perspective of the minority immigrant community on immigration policies created and enforced by the majority, this line of work questions one of the assumptions implicit in Berry's model of acculturation: that both the majority and the minority community have equal power status (Bhatia, 2002).

More importantly, the postcolonial perspective shines much-needed light on the conflict and dynamism that underlie the process of identity negotiation in immigrant communities (Hegde, 1998). In particular, the work of feminist writer Gloria Anzaldúa provides an account of identity that is centered on people's social locations with an appreciation for historical, societal, and material processes that permeate everyday life experiences to give shape to individual identities (Barcinski & Kalia, 2005). Anzaldúa (1997) turns away from the modern transcendental and unitary self, which is assumed to remain fixed and relatively unchanging once it has developed. In its place she proposes the *mestiza* consciousness, which describes the sense of self that emerges from experience of living in more than one culture. Because individuals who embody more than one culture cross cultural boundaries with regularity in their daily life (e.g., eating Indian food at home and hot dogs in school), Anzaldúa (1997) believes they have a sense of self that is mixed, not fully belonging to either culture, and fluid. This mestiza consciousness can help these individuals make sense of the chaotic and multiple experiences of hyphenated identities (Barcinski & Kalia, 2005).

Anzaldúa (1997) describes a developed tolerance for ambiguity and contradictions as a key feature of the hyphenated identity. The mestiza, Anzaldúa proposes, is open to the idea that the experiences and values of the two identities that constitute the mestiza consciousness will sometimes clash and at other times complement each other. For instance, Asian Indian adolescents may feel torn between obligations to their family and the desire to spend time with their friends. The mestiza consciousness shows its capacity for tolerance and fluidity by successfully inhabiting different realities (e.g., fulfilling family obligations and finding ways to connect with friends) and absorbing contradictory or opposing messages (e.g., family is more important than friends vs. friends are more important than family). In order to successfully engage in constant negotiations between the broader societal and historical frameworks that determine the complexities of this daily reality, the hyphenated identity must remain fluid and non-fixed (i.e., able to change based on contextual demands).

Another aspect of the hyphenated identity highlighted by Anzaldúa (1997) is the notion of living in the borderlands. Even though the term borderland was initially used to refer to the experience of Chicana women who lived in the border regions between the U.S. and Mexico, the borderlands do not refer to any specific place in Anzaldúa's theorizing. Instead, borderlands are spaces where the hyphenated identity may experience a confluence of ideas and values from both cultures. This is a space of contradiction and struggle that is available to the mestiza consciousness. For instance, an Asian Indian adolescent may feel that their culture of origin is important to them but also keeps them from fully assimilating into their host culture. Celebratory rituals that are valued by either culture and language can become the borderlands. Since the conflicts between two cultures are commonplace the only way for a hyphenated identity to survive the borderlands is to be flexible (Barcinski & Kalia, 2005).

Resonating with the ideas proposed by postcolonial feminist writings, researchers in cross-cultural psychology advanced the theory of biculturalism in order to characterize psychological dimensions of hyphenated identities. Within psychology, individuals with hyphenated identities are known as 'bicultural' and are viewed as having two distinct or even conflicting 'minds.' Consistent with the notion of the mestiza, bicultural individuals have two frames of reference or two ways of thinking about and viewing their world. Substantial empirical evidence has been provided to show that bicultural individuals have access to dual cognitive frameworks (Cheng et al., 2014). For instance, Ramírez-Esparza et al. (2006) found that when given a personality test, Spanish-English bilinguals in the U.S. and Mexico exhibited more extraverted, agreeable, and conscientious personality traits when tested in English than in Spanish. Based on these findings, the authors concluded that language activated a shift in the biculturals' personality by switching their frame of reference.

Benet-Martínez et al. (2002) have proposed a method for separating bicultural individuals into two groups based on subjective perceptions of the ways in which their dual identities interact with one another. Those individuals who identify as blended biculturals tended to view their two cultures as being concordant or in harmony with one another. Consequently, blended biculturals found it easy to switch fluidly between their two cultures in the presence of relevant cultural cues. This ability to switch between cultural meaning systems, when primed, is known as cultural frame switching (Hong et al., 2000). As a result, the blended bicultural is able to meet the demands of complex social contexts without compromising their identity (Sirikantraporn, 2013). The alternate bicultural, on the other hand, believed that their two cultures are in opposition with one another. Because the two cultures are seen as conflicting, with disparate ideas and beliefs, the experience of having two cultures is seen as a burden to the alternate bicultural (Cheng et al., 2014). Consequently, they can only activate schemas pertinent to one culture or another, in the presence of culturally relevant cues, but cannot engage with both at the same time (Benet-Martínez & Haritatos, 2005).

There is some empirical evidence to indicate that there are psychological benefits associated with being a blended bicultural (Cheng et al., 2008). For instance, viewing one's dual cultures as being harmonious or blended may enhance an individual's cognitive abilities. Specifically, Saad et al. (2013) have shown that Chinese American biculturals who perceived their two cultures as being compatible and blended exhibited more originality and creativity in cultural contexts where both their cultures are salient in comparison to monocultural contexts. The benefits associated with being bicultural extend beyond cognitive abilities. For example, Sirikantraporn (2013) has demonstrated that Southeast Asian adolescents who identified as bicultural were more resilient, despite exposure to domestic violence at home, than those who identified as mostly Asian or mostly American.

Development of Bicultural Identity and Family Conflict in South Asian Immigrants

Erikson (1968) suggested that the primary goal of transitioning from an adolescent to adult was to establish a stable identity. As part of a stable identity an individual must achieve some detachment from their parents and also acquire their own value system (Erikson, 1968). Since parents bridge the divide between micro and macro contexts in a child's life, their individual value system often develops in juxtaposition to parental ideas and values (Marcia, 1966). Thus, before individuals make a commitment to any value system, they spend some time exploring and evaluating the values of their parents and comparing them with other available ideologies. The process of acculturation can exacerbate the divergence of values and perspectives in immigrant families (Tardiff-Williams & Fisher, 2009).

Past research has shown that acculturation can influence family relationships in Asian immigrant communities (Ho & Birman, 2010). In this chapter, we focus on the role of acculturation practices in family conflict among South Asians in the U.S. South Asia consists of a cluster of countries including Afghanistan, Bangladesh, Bhutan, India, Pakistan, Myanmar, Nepal, and Sri Lanka. Each South Asian country has its own rich linguistic, religious, and cultural heritage. South Asians have been migrating to the U.S. since the late nineteenth century (Deepak, 2005) and are part of a rapidly expanding immigrant group in the country (Kieu, 2013), but are relatively understudied (Shariff, 2009).

Child-rearing practices among South Asians emphasize close familial bonds and the expectation that each member contributes toward the honor of the whole family (Jambunathan & Counselman, 2002). However, physical and psychological stress associated with immigration can put a strain on family bonds. The chances of conflict are further enhanced when the immigrant comes from a culture where children are expected to remain at home until they marry and to follow parental advice when dating, considering marriage, or making career choices (Dasgupta, 1998). As a result, parent—offspring conflicts in South Asian immigrant families are more likely to center on differing cultural values regarding family roles and independence in important life decisions.

Empirical evidence indicates that Asian Indian immigrant parents who either reported using separation (i.e., individual rejects the dominant culture) or marginalization (i.e., disconnection from both host culture and heritage culture) as dominant strategies for acculturation were more likely to have conflicts with their children (Farver et al., 2002; Farver et al., 2007). Additionally, parent-child conflict was more likely in families that used separation and marginalization acculturation strategies predominantly (Farver et al., 2002). Conversely, Asian Indian parents and children reporting integrated or assimilated styles of acculturation were less likely to report conflict, suggesting a protective role of bicultural identity.

Some of the literature suggests that for South Asian immigrants, the social, cultural, and historical contexts prior to migration (e.g., familiarity with English language and exposure to different cultural worldviews through years of colonial rule) may facilitate the development of bicultural identity (Ibrahim et al., 1997). For example, Indian immigrant parents in the U.S. reported being mindful of the importance of bicultural competence for their children and adapted their parenting behaviors to help their children navigate differing cultural spheres (Inman et al., 2007). There is some evidence that South Asian immigrant parents thoughtfully retain values and beliefs reflecting their cultural heritage while being open to incorporating characteristics of mainstream parenting from North American cultures (Farver et al., 2007; Inman et al., 2007).

On the other hand, there is also evidence to indicate that South Asian parents emphasize the maintenance of their culture (Farver et al., 2002). For example, Indian immigrant mothers described their efforts to maintain 'an Indian home' where Indian languages are spoken, Indian food is served, and children are exposed to books, music, and films representing their cultural heritage. Interestingly, while describing efforts to maintain their culture, these mothers also characterized their parenting as flexible and autonomy-promoting to help their children adapt successfully in the broader American society. Nevertheless, some children perceived their mothers as less flexible and hoped that they would develop increased familiarity with American culture (Raj et al., 2021). Therefore, it is perhaps unsurprising that numerous studies have observed that South Asian immigrants struggle with identity development due to their parents' enforcement of traditional values (Farver et al., 2002; Shariff, 2009).

The Present Study

In view of the fact that an individual's identity develops in interaction with immediate others and their macro environment (Koepke & Denissen, 2012), it is relevant to consider a transactional model for the development of a bicultural identity. A transactional model would incorporate reciprocal interactions between parenting and children's bicultural identity such that parenting goals and practices contribute to the development of bicultural identity, and bicultural identity contributes to ongoing interactions between parents and children including parent-child conflict. If we accept that there is variability in acculturative phenomena across groups then we must also acknowledge that there are likely differences in the ways in which acculturative processes influence individuals at the interpersonal level (Tardiff-Williams & Fisher, 2009). Thus, an examination of the relationship between bicultural identity identification and intergenerational conflict could provide a more nuanced understanding of the ways in which acculturative processes influence interpersonal relations in families.

In addition to providing insight about the role of family interaction in bicultural identity development, it is relevant to examine family conflict in South Asian American families, as it has been identified as a risk factor for depression and psychological distress among Asian Americans (Bahrassa et al., 2011). Since there is some empirical evidence to suggest that South Asians are socialized to view the needs of their family above their individual needs and goals (Masood et al., 2009), this line of inquiry may be particularly relevant for South Asians immigrants in the U.S. In contrast to the heritage culture of South Asians, the U.S. is an individualistic culture that values the development of an individual's independence and autonomy (Markus & Kitayama, 1991). Thus, it is likely that the distance between host and heritage culture could be fertile ground for family conflict.

The primary objective of this study was to examine the association between biculturalism and parent-child conflict in South Asian American emerging adults. Arnett (2007) defined emerging adulthood as the age period between the late teens and early twenties. Emerging adulthood serves as a transition between adolescence and adulthood and is characterized by explorations associated with identity development. As such, emerging adulthood is likely to be the period when children of South Asian immigrants would engage in activities that would help them explore their dual identities.

Based on prior research (Huynh et al., 2011), our primary expectation was that individuals who reported higher ratings on blended biculturalism would report lower levels of conflict with the family. Additionally, as previous research has shown that gender differences exist in ethnic identity development among South Asians (Dion & Dion, 2001), we also sought to examine whether the reported conflict between emerging adult South Asian Americans differed as a function of the emerging adult's gender. Finally, since past research has demonstrated that biculturalism is associated with resilience (Huynh et al., 2011), we speculated that individuals identifying as biculturals (harmony or blended) would likely also report using adaptive emotion regulation strategies and have fewer adjustment problems in college.

Method

Participants

We recruited a sample of 52 (female = 53.8%; M_{age} = 23.08 years) South Asian/South Asian American emerging adults in the U.S. Table 6.1 summarizes the sociodemographic characteristics of the sample. On average, participants had spent 10 years (range = 1–23 years) in the U.S. and their parents had spent 12 years (range = 0–37 years). Around half of the participants (45.1%) were graduate students, with the remaining being undergraduate. More than half (60.9%) of the graduate students were in their first year of schooling and 35.7% of the undergraduate students were in their senior year.

Table 6.1 Socio-demographic details of the sample

Variable	Categories (percentage of participants)
Participants' ethnic identity	Indian/Indian American (84.6%); Bangladeshi/Bangladeshi American (9.6%); Nepalese/Nepalese American (3.8%); Pakistani/Pakistani American (1.9%)
Participants' religion	Hinduism (60.8%); Islam (15.7%); Christianity (7.8%); other (15.7%)
Participants' relationship status	Single (67.3%); in a relationship or engaged (25%); married/ common-law (7.7%)
Participants' living arrangement	Live on their own (55.8%); in a resident hall (25%); at home with parents (19.2%)
Participants' country of origin	India (51.9%); U.S. (28.8%); Bangladesh (7.7%); Nepal (3.8%); Pakistan (1.9%); Singapore (1.9%); Middle East (1.9%)
Country where participants spent a majority of their childhood	India (49%); U.S. (35.3%); Bangladesh (5.9%); Nepal (2%); Pakistan (2%); China (2%); Singapore (2%); Middle East (2%).
Participants' parents' country of origin	India (84.3% mothers, 82.4% fathers); Bangladesh (7.8% mothers, 9.8% fathers); Nepal (3.9% mothers, 3.9% fathers); Pakistan (3.9% mothers, 2% fathers); U.S. (2% fathers)
Participants' parents' marital status	Married/common-law (94.2%); divorced (1.9%); widowed (1.9%); never married (1.9%)
Participants' parents' education	<Grade 12 (14% mothers, 10% fathers); high school completed (18% mothers, 6% fathers); Bachelor's degree (48% mothers, 30% fathers); graduate/professional degree (20% mothers, 54% fathers)
Participants' parents' occupation	Homemaker/unemployed (40% mothers, 7.9% fathers); labor work/ skilled trade (12.5% mothers, 13.2% fathers); secretarial/clerical (12.5% mothers, 2.6% fathers); managerial/professional (30% mothers, 63.2% fathers); retired (5% mothers, 13.2% fathers);
Participants' parents' annual income	$50,000 (38%); $51,000–$100,000 (26%); $101,000–$150,000 (24%); more than $200,000 (12%)

Procedure and Measures

South Asian student organizations at universities and colleges in the U.S. were contacted and asked to circulate the Qualtrics link to the survey among their members. Flyers containing information about research participation were also posted on the Facebook page of the Culture, Affect & RElationships lab (CARE), Miami University. Inclusion criteria included (1) 18 years or older; (2) identify as South Asians or South Asian Americans (whose families are from India, Pakistan, Nepal, Sri Lanka, and Bangladesh); and (3) currently attending college (i.e., for the pursuit of an undergraduate degree, graduate degree, professional degree). Participants completed the survey online (duration: 30-35 min) at a location that was convenient to them. After the completion of the survey, participants were asked to provide their email address using a separate Qualtrics survey link, in case they wanted to enter their names into a raffle to win $100.

Demographics Questionnaire. Participants were asked questions about their date of birth, gender, ethnic identity, relationship status, religion, living situation, college/university status, year in the program, country of origin, and country in which a majority of their childhood years were spent. Participants were also asked to report their parents' marital status, education, occupation, and annual income.

Bicultural Identity Integration Scale. Bicultural identity was assessed using the Bicultural Identity Integration Scale (BII, Benet-Martínez et al., 2002). The BII consists of 19 questions on a 10-point Likert scale (1 = strongly disagree; 10 = strongly agree) adapted for the South Asian context (e.g., "I feel torn between the American and South Asian cultures"; "I relate better to a combined culture than to American culture or South Asian culture alone"). The BII provides scores on two dimensions of bicultural identity: cultural harmony (vs. conflict) and cultural blendedness (vs. compartmentalization). Cultural harmony (10 items; $\alpha = 0.86$) is a measure of the degree of clash or tension versus compatibility perceived between the two cultures. Cultural blendedness (9 items; $\alpha = 0.88$) is a measure of the degree of dissociation versus overlap perceived between the two cultures (Huynh et al., 2011).

Intergenerational Conflict Inventory. Parent-child conflict was assessed using the Intergenerational Conflict Inventory (ICI, Chung, 2001). The ICI consists of 24 questions on a 6-point Likert scale (1 = no conflict over this issue, 6 = a lot of conflict over this issue) that provide information about 3 domains of conflict: family expectations (11 items; $\alpha = 0.90$), education and career (10 items; $\alpha = 0.91$), and dating and marriage (3 items; $\alpha = 0.82$). Examples of statements include "your desire for greater independence and autonomy," "whom to marry," and "how much time to spend on studying." Two statements ($\alpha = 0.87$) on financial conflict (i.e., opening your own bank account and seeking financial independence) were added to the inventory to adapt the ICI to a South Asian sample.

Emotion Regulation Questionnaire. Habitual use of emotion regulation strategies was assessed using the Emotion Regulation Questionnaire (ERQ, Gross & John, 2003). The ERQ is a measure of individual differences in the habitual use of two emotion regulation strategies, namely cognitive reappraisal and expressive suppression (Gross & John, 2003). The cognitive appraisal subscale comprises 6 items ($\alpha = 0.86$; e.g., "When I'm faced with a stressful situation, I make myself think about it in a way that helps me stay calm") and the expressive suppression subscale comprises 4 items ($\alpha = 0.78$; e.g., "I control my emotions by not expressing them"). Participants rate each of the 10 statements on a 7-point scale (1 = strongly disagree, 7 = strongly agree).

College Adjustment Scale. College adjustment was assessed using the College Adjustment Scale (CAS, Anton & Reed, 1991). The CAS is a 36-item measure of psychological and adjustment problems among college students. Each participant completed three of the nine subscales of the CAS: Anxiety ($\alpha = 0.94$), Family Problems ($\alpha = 0.89$), and Academic Problems ($\alpha = 0.86$) on a 4-point rating scale (1 = statement is false or not at all true, 4 = the statement is very true). Lower scores on the CAS indicate better adjustment. Examples of statements include "I often feel afraid but don't know why," "My family doesn't understand me," and "I have poor study skills."

Results

Preliminary Analyses

All variables were normally distributed. Overall, 27.97% of the data were missing (range: 0–46.1%). Variables with most missingness included the emotion regulation and college adjustment scales (44.2–46.1%). Missing data were handled with multiple imputation (n = 40), and a single dataset was created by averaging across imputed datasets with no missing data. The mean scores obtained by participants on each of the measures are summarized in Table 6.2. One-way ANOVA demonstrated no significant difference in ratings on biculturalism, reported conflict, emotion regulation, and college adjustment as a function of participants' gender (all $ps > 0.10$). Item-level analysis of the ICI indicated that participants reported relatively fewer conflicts with their parents on how much time they should spend in recreation (M = 2.13, SD = 1.30) and most conflicts regarding whom to date (M = 4.05, SD = 2.03).

Table 6.2 Gender-wise distribution of mean scores obtained by participants

Variable	Males (N = 24)	Females (N = 28)
BII Subscales (Benet-Martínez et al., 2002)	Mean	Mean
Bicultural harmony	6.18	5.91
Bicultural Blendedness	6.82	6.58
ICI Subscales (Chung, 2001)		
Family expectations	31.70	32.18
Education and career	33.64	30.86
Dating and marriage	11.39	10.53
Finance	7.58	7.87
ERQ Subscales (Gross & John, 2003)		
Cognitive reappraisal	28.97	28.91
Emotion suppression	19.13	15.79
CAS Subscales (Anton & Reed, 1991)		
Anxiety	25.11	26.68
Family problems	27.26	24.33
Academic problems	25.41	24.54

Bivariate Associations Among Biculturalism, Intergenerational Conflict, and Well-being

Pearson correlations were conducted to examine the relations between biculturalism, reported parent-child conflict (both domain and item wise), emotion regulation, college adjustment, and years of living in the U.S. (both self and parents). The correlation coefficients are summarized in Tables 6.3 and 6.4.

When participants reported increased compatibility between the two cultures, they were also more likely to report perceived overlap between cultures ($p < 0.01$; see Table 6.3 for coefficients). The more compatible the participants perceived their cultures to be, the less likely they were to report feelings of anxiety ($p < 0.05$). When participants reported higher levels of overlap between their cultures, they were more likely to report cognitive appraisals, an adaptive emotion regulation strategy ($p < 0.05$), and less likely to report academic problems ($p < 0.05$). The more conflicts the participants reported related the domains of family expectations and education and career, the more likely they were to report anxiety ($p < 0.05$). When participants reported more conflicts based on family expectations, education and career, and dating and marriage, they were also more likely to report family problems. The two added questions regarding financial conflict were not significantly related to conflicts based on family expectations, education and career, and dating and marriage ($p > 0.05$).

Our item-level analyses indicated that when participants reported higher bicultural blendedness, they were less likely to report conflicts with their parents about learning their own South Asian language ($p < 0.01$; see Table 6.4 for coefficients). Contrary to what one might expect, when participants reported higher bicultural

Table 6.3 Correlation coefficients between biculturalism, reported intergenerational conflict (domain-wise), emotion regulation, and college adjustment

	1	2	3	4	5	6	7	8	9	10	11
1	–	0.50**	−0.19	−0.09	−0.18	−0.00	0.23	0.09	−0.33*	−0.20	−0.27
2		–	−0.09	0.05	0.00	0.08	0.31*	0.09	−0.25	−0.11	−0.36*
3			–	0.72**	0.65**	0.27	−0.00	0.11	0.42**	0.51*	0.35*
4				–	0.71**	0.38**	0.01	0.16	0.33*	0.48*	0.30*
5					–	0.59**	0.05	0.14	0.23	0.38*	0.21
6						–	0.05	−0.12	−0.07	0.10	−0.04
7							–	0.32*	0.08	−0.09	−0.05
8								–	0.41**	0.24	0.23
9									–	0.57**	0.52**
10										–	0.45**
11											–

Note. *$p < 0.05$; **$p < 0.01$; 1 = BII Harmony; 2 = BII Blendedness; 3 = ICI Family Expectations; 4 = ICI Education and Career; 5 = ICI Dating and Marriage; 6 = ICI Finance; 7 = ERQ Cognitive Reappraisal; 8 = ERQ Emotion Suppression; 9 = CAS Anxiety; 10 = CAS Family Problems; 11 = CAS Academic Problems

Table 6.4 Correlation coefficients between biculturalism, years of immigration, and reported intergenerational conflict (item-wise)

Variable	BII - Harmony	BII - Blendedness	Years in US (Self)	Years in US (Parent)
Lack of communication with your parent	−0.13	−0.05	0.23	0.22
Your desire for greater independence and	−0.16	0.12	0.29	0.33*
autonomy	−0.02	−0.10	0.10	0.24
Following cultural traditions	−0.21	−0.37**	0.16	0.27
Pressure to learn one's own South Asian	−0.06	−0.10	0.29	0.40**
language	0.10	−0.03	0.14	0.25
Expectations based on being male or	−0.23	−0.17	0.28	0.31*
female	−0.17	−0.02	0.17	0.22
Expectations based on birth order	−0.23	−0.08	−0.06	−0.07
Family relationships being too close	−0.18	0.04	0.22	0.23
Family relationships being too distant	−0.05	0.14	0.22	0.27
How much time to spend with family	−0.11	−0.03	0.36*	0.45**
How much to help around the house	−0.02	0.06	−0.04	−0.01
How much time to help out in the family	−0.16	0.03	0.11	0.21
business	−0.11	0.07	0.22	0.30
How much time to spend on studying	−0.27	−0.17	0.07	0.10
How much time to spend on recreation	0.23	0.28*	0.08	0.18
How much time to spend on sports	−0.04	0.10	0.16	0.27
How much time to spend on practicing	−0.07	−0.03	0.09	0.18
music	−0.02	0.04	0.31*	0.43**
Importance of academic achievement	−0.04	−0.04	0.19	0.27
Emphasis on materialism and success	−0.20	−0.09	0.15	0.20
Which school to attend	−0.17	−0.02	0.25	0.29*
What to major in at college	−0.08	0.12	0.02	0.01
Which career to pursue	0.04	0.09	0.09	0.15
Being compared to others	−0.06	0.04	0.08	0.11
When to marry				
Whom to date				
Whom to marry				
Opening your own bank account				
Seeking financial independence				

Note. $^{*}p < 0.05$. $^{**}p < 0.01$

blendedness, they reported more conflicts emphasizing materialism and success ($p < 0.05$). Although not statistically significant, the more the participants reported harmony and compatibility between cultures, the less likely they were to report intergenerational conflicts across different topics.

Our item-level analyses indicated that the more years the participants reported spending in the US, the more likely they were to report conflicts related to the time they spent studying and which career to pursue ($p < 0.05$; see Table 6.4). The more years the participants reported their parents spent in the US, the more likely they were to report conflicts based on gender-based expectations, time to be spent in studying, which career to pursue, desire for greater independence and autonomy, family relationships being too close, and whom to date (all p's < 0.05).

Correlation coefficients between the variables of interest were also calculated on the original dataset (without imputations). Each coefficient pair was converted into a z-score using Fisher's r-to-z transformation scale and p values were computed to test significance (Steiger, 1980). All p values were greater than 0.2 indicating no statistically significant differences between the two sets of correlation coefficients (with and without multiple imputations).

Discussion

Erikson (1968) believed that a stable identity, which defines an individual's value system and their role in society, was a singular achievement of the adolescent phase in an individual's lifespan. However, as Arnett (2007) points out the transition between adolescence to adulthood post industrialization has extended past adolescence and into an individual's mid-twenties. Arnett called this period of transition, between adolescence and adulthood, emerging adulthood and highlighted the importance of identity exploration as a key process of developmental change.

The role of parents in identity development, in emerging adulthood, is that of mediators or socializing agents who help transfer social norms and expectations from the macro environment to their children. When viewed through this lens, parents and children can be seen as two identity systems in relation with each other and conflicts that arise between them as indicators of the tension that exists between two identity systems (Koepke & Denissen, 2012). Within the context of acculturation, both of these identity systems are independently, and with each other, engaged in the interactions that construct and transform their culture. As such, everyday interactions between family members, including conflicts, can become borderlands where cultural differences are given voice and identities are reconstructed (Tardiff-Williams & Fisher, 2009). The purpose of our study was to examine the relation between bicultural identity identification and family conflict, as well as well-being, in South Asian emerging adults in the US.

Biculturalism and Family Conflict

Consistent with our primary objective, we found some evidence that individuals who reported higher bicultural blendedness were also less likely to report conflicts with their parents about learning their own South Asian language. In effect, individuals who reported that they 'could not ignore' either their Asian or American side or felt connected to both their Asian and American culture at the same time were less likely to engage in conflict about learning their heritage language. Our data show that emerging adults who identify as blended biculturals have come to an understanding, with their parents, about heritage language maintenance. It could either be that the emerging adult has accepted that maintaining their heritage language is a part of their value system or that the parents have accepted that sustaining their heritage language within the socio-cultural context of the US is difficult. Since past research has shown that most Asian parents want their children to learn their heritage language (Inman et al., 2007; Li & Wen, 2015), this finding implies that blended biculturals would likely be open to maintaining their heritage language. Considering that heritage language loss is at the highest levels among Asians in the U.S. (Li & Wen, 2015), future research should examine whether biculturalism could be associated with bilingualism and heritage language learning in South Asians in the U.S.

Unexpectedly, we also found that when participants reported higher scores on bicultural blendedness, they also reported having more conflicts about materialism and success. We are not the first to report that South Asian parents and emerging adults have conflicts about success and materialism (Dugsin, 2001; Farver et al., 2007; Shariff, 2009). For instance, Segal (1991) has observed that Indian immigrants expect their children to be achievement driven and successful so they can uphold the standards of a 'model minority.' Segal believes that the answer to why South Asian parents push their children to achieve material success lies in their own experience of immigration. Due to structural barriers that make immigration to the U.S. difficult for South Asians, the ones who are successful in immigrating tend to score higher on intrinsic motivation and value success and achievement.

Individual autonomy is differentially defined across cultures; allocentric cultures expect individuals to perform their obligations to their group per culturally stipulated norms (Triandis, 1996). Because many South Asians are considered allocentric (Masood et al., 2009), children in South Asian families are socialized to develop a *we-self* (Roland, 1988). This sense of self prioritizes the needs of the family above their individual desires (Lou et al., 2012; Masood et al., 2009). Thus, it is possible that South Asian parents are driven by a desire to protect their own when they emphasize material success in their interaction with their children. Alternatively, it is also probable that South Asian parents prefer to maintain control over their children because they feel less agentic (Koepke & Denissen, 2012) in their host culture due to acculturative stress (Farver et al., 2007). Future research that conducts a systematic examination of the underlying motivations of family conflict about material success in South Asian immigrants would add to our understanding of family dynamics in this cultural group.

Our results indicated that reported family conflict increased with the number of years the participants reported spending in the U.S. In particular, participants reported more conflicts about time spent studying and career goals. This finding is consistent with prior evidence that South Asian immigrant families experience frequent family conflict within the domains of school and career choices (Giguère et al., 2010). The conflict may be due to the assimilation of our participants into American culture over time. The emerging adults may be increasingly endorsing norms more consistent with American culture than with South Asian culture, which their parents expect them to adhere to. Our work suggests that despite the stereotypically positive portrayal of South Asians as 'model minorities' who are motivated to excel in school, performance in academic and career domains may be accompanied by a psychological cost that is unseen and unacknowledged.

Previous research has also shown that gender norms, dating, and marriage are domains of significant conflict in South Asian families (Dasgupta, 1998; Kallivayalil, 2004; Kalia & Weatherall, 2009), which is concordant with our data. The longer the participants reported their parents spent in the U.S., the more likely they were to report conflicts regarding gender-based expectations and dating. This finding suggests that despite the number of years spent in the U.S. first-generation South Asian adults continue to endorse norms of their heritage culture. In contrast, South Asian emerging adults are more likely to endorse American norms about gender and dating, which are incompatible with South Asian norms (Dasgupta, 1998; Kalia & Weatherall, 2009). This incompatibility between the host and heritage cultures enhances the differences that exist between first- and second-generation immigrants, thereby increasing family conflict (Giguère et al., 2010).

Overall, on the question regarding the relationship between biculturalism and family conflict, our data indicate that biculturalism is not a panacea to reduce family conflict. This is particularly noteworthy for South Asian immigrants since past research has shown that quality of family life and interpersonal interactions with family members can influence their well-being (Dugsin, 2001; Farver et al., 2007; Masood et al., 2009). For instance, Dugsin has reported that intergenerational conflict regarding value systems resulted in self-harming behavior as well as use of drugs and alcohol in second-generation Indian immigrants. Concordant with previous reports we found that family conflict about family expectations as well as education and career was associated with increased anxiety and more academic problems. It is relevant to point out that although not statistically significant, the more our participants endorsed harmony and compatibility between their two cultures, the less likely they were to report intergenerational conflicts across different domains. However, our small sample size limits our ability to make any firm claims, and more research on the relationship between biculturalism and family conflict in South Asian immigrant communities is needed.

Biculturalism as a Protective Factor

We found that endorsement of harmony between cultures was associated with reports of reduced anxiety. Although our sample size is small, our findings are consistent with previous work showing that Asian immigrants who experience difficulty switching between their two cultures are also more likely to experience anxiety and depression (Hsu et al., 2012). There is some evidence to suggest that prior experience with prejudice, in particular, can enhance anxiety about social interactions that require intercultural contact (Mendoza-Denton et al., 2002). It remains unclear whether bicultural self-efficacy can serve as a buffer against anticipated anxiety about intergroup contact in immigrant communities. Certainly, our study bolsters the notion that blended biculturalism can aid South Asian immigrants in meeting the dynamic demands of their complex social contexts without compromising their well-being (Sirikantraporn, 2013).

Additionally, endorsement of blended biculturalism was associated with fewer academic problems and the increased use of cognitive reappraisal as a strategy to regulate emotions. Considering that effectively using cognitive reappraisal requires that the individual alter the way they think about an event in order to reduce its emotional impact (Gross, 1998), it is possible to speculate that the enhanced cognitive benefits associated with blended biculturalism (Cheng et al., 2008; Saad et al., 2013) could facilitate the use of cognitive reappraisal. Thus, it is perhaps unsurprising that we observed this positive association between cognitive reappraisal and blended biculturalism.

This latter finding is also interesting because South Asians come from a collectivist culture (Masood et al., 2009). Collectivist cultures tend to emphasize social cohesion (Markus & Kitayama, 1991), and expressive suppression (e.g., biting your tongue) as a strategy to regulate emotions is sometimes encouraged in order to avoid discord and achieve harmony (Arens et al., 2013). Consistent with this notion, there is some empirical evidence to suggest that expressive suppression is associated with fewer negative outcomes in collectivist cultures (Butler et al., 2007). For instance, Turkish immigrants in Germany who endorsed collectivist values experienced less emotional negativity if they employed expressive suppression as a strategy for regulating emotions (Arens et al., 2013). Our data, on the other hand, suggest that expressive suppression as a strategy for regulating emotions is less effective for South Asians since it is associated with increased levels of anxiety. Future research should examine whether habitual use of expressive suppression is associated with other mental health outcomes (e.g., depression).

Overall our study suggests that endorsement of blended and harmonious biculturalism is associated with resilience factors (i.e., reduced anxiety levels, adaptive emotion regulation strategies) in South Asian emerging adults. This provides further evidence that blended and harmonious biculturalism is implicated in positive outcomes for immigrants (Huynh et al., 2011).

Final Message

Consistent with the vast majority of the existing work on South Asian Americans (Masood et al., 2009), our sample was mostly Hindu and college educated. Despite this limitation, our data adds to the limited body of work that has studied this population. Our findings are limited by a small sample size, as well as data collected at a single point in time and from a single respondent. We recognize that shared method variance may have contributed to some of the significant associations we found, and that it is not possible to establish direction of effect with our data. Future research may build on our findings and examine blended biculturalism and its relation to family conflict, as well as well-being indices, in a larger sample, collecting data from emerging adults and parents across multiple time points.

By situating our exploration of identity reconstruction within the framework of biculturalism, we were able to incorporate some of the ways in which macro environmental forces can influence the daily, everyday interactions in immigrant families in the U.S. Although Erikson (1968) proposed that establishing a stable ego-identity that agentively synthesizes experiences is crucial for development, his theory does not account for the fact that an individual's sense of self is dynamic and rooted in the everyday interactions of their life (Koepke & Denissen, 2012). Immigrants who endorse blended and harmonious biculturalism show us that it is possible for an individual to develop more than one meaning system and process conflicting values at the same time (Saad et al., 2013). In fact, the ability of bicultural individuals to reform and reconstruct their ego-identity based on changing contextual demands may be associated with enhanced resilience.

References

Anton, W. D., & Reed, J. R. (1991). *CAS: College adjustment scales: Professional manual*. Psychological Assessment Resources.

Anzaldúa, G. (1997). La consciencia de la mestiza. Towards a new consciousness. In K. Conboy et al. (Eds.), *Writing the body: female embodiment and feminist theory* (pp. 233–247). Columbia University Press.

Arens, E. A., Balkir, N., & Barnow, S. (2013). Ethnic variation in emotion regulation: Do cultural differences end where psychopathology begins? *Journal of Cross-Cultural Psychology, 44*(3), 335–351. https://doi.org/10.1177/0022022112453314

Arnett, J. J. (2007). Emerging adulthood: What is it, and what is it good for? *Child Development Perspectives, 1*(2), 68–73. https://doi.org/10.1111/j.1750-8606.2007.00016.x

Bahrassa, N. F., Syed, M., Su, J., & Lee, R. M. (2011). Family conflict and academic performance of first-year Asian American undergraduates. *Cultural Diversity and Ethnic Minority Psychology, 17*(4), 415–426. https://doi.org/10.1037/a0024731

Barcinski, M., & Kalia, V. (2005). Extending the boundaries of the dialogical self: Speaking from within the feminist perspective. *Culture & Psychology, 11*(1), 101–109. https://doi.org/10.1177/1354067X05050752

Benet-Martínez, V., & Haritatos, J. (2005). Bicultural identity integration (BII): Components and psychosocial antecedents. *Journal of personality, 73*(4), 1015–1050. https://doi.org/10.1111/j.1467-6494.2005.00337.x

Benet-Martínez, V., Lee, F., & Leu, J. (2006). Biculturalism and cognitive complexity: Expertise in cultural representations. *Journal of Cross-Cultural Psychology, 37*(4), 386–407. https://doi.org/10.1177/0022022106288476

Benet-Martínez, V., Leu, J., Lee, F., & Morris, M. (2002). Negotiating biculturalism: Cultural frame switching in biculturals with oppositional versus compatible cultural identities. *Journal of Cross-Cultural Psychology, 33*, 492–516. https://doi.org/10.1177/0022022102033005005

Benner, A. D., & Kim, S. Y. (2009). Experiences of discrimination among Chinese American adolescents and the consequences for socioemotional and academic development. *Developmental Psychology, 45*(6), 1682–1694. https://doi.org/10.1037/a0016119

Berry, J. W. (1997). Immigration, acculturation, and adaptation. *Applied Psychology, 46*(1), 5–68. https://doi.org/10.1111/j.1464-0597.1997.tb01087.x

Berry, J., & Sam, D. (1996). Acculturation and adaptation. In J. Berry, M. Segall, & C. Kagitcibasi (Eds.), *Handbook of cross-cultural psychology: Social behavior and applications* (Vol. 3, pp. 291–325). Allyn & Bacon.

Bhatia, S. (2002). Acculturation, dialogical voices and the construction of the diasporic self. *Theory & Psychology, 12*(1), 55–77. https://doi.org/10.1177/0959354302121004

Bhatia, S., & Ram, A. (2001). Rethinking 'acculturation' in relation to diasporic cultures and postcolonial identities. *Human Development, 44*(1), 1–18. https://doi.org/10.1159/000057036

Boski, P. (2008). Five meanings of integration in acculturation research. *International Journal of Intercultural Relations, 32*(2), 142–153. https://doi.org/10.1016/j.ijintrel.2008.01.005

Bronfenbrenner, U. (1979). *The ecology of human development*. Harvard University Press.

Butler, E. A., Lee, T. L., & Gross, J. J. (2007). Emotion regulation and culture: Are the social consequences of emotion suppression culture-specific? *Emotion, 7*(1), 30–48. https://doi.org/10.1037/1528-3542.7.1.30

Calavita, K. (2020). *US immigration law and the control of labor: 1820–1924* (Vol. 31). Quid Pro Books.

Cheng, C. Y., Lee, F., Benet-Martínez, V., & Huynh, Q. L. (2014). Variations in multicultural experience: Influence of bicultural identity integration on socio-cognitive processes and outcomes. *The Oxford Handbook of Multicultural Identity*, 276–299. https://doi.org/10.1093/oxfordhb/9780199796694.013.025

Cheng, C. Y., Sanchez-Burks, J., & Lee, F. (2008). Connecting the dots within: Creative performance and identity integration. *Psychological Science, 19*(11), 1178–1184. https://doi.org/10.1111/j.1467-9280.2008.02220.x

Chung, R. H. (2001). Gender, ethnicity, and acculturation in intergenerational conflict of Asian American college students. *Cultural Diversity and Ethnic Minority Psychology, 7*(4), 376–386. https://doi.org/10.1037/1099-9809.7.4.376

Dasgupta, S. S. (1998). Gender roles and cultural continuity in the Asian Indian immigrant community in the U.S. *Sex Roles, 38*(11/12), 953–974. https://doi.org/10.1023/A:1018822525427

Deepak, A. C. (2005). Parenting and the process of migration: Possibilities within South Asian families. *Child Welfare, 84*(5), 585. https://doi.org/10.1111/0022-4537.00226

Dion, K. K., & Dion, K. L. (2001). Gender and cultural adaptation in immigrant families. *Journal of Social Issues, 57*(3), 511–521. https://doi.org/10.1111/0022-4537.00226

Dugsin, R. (2001). Conflict and healing in family experience of second-generation emigrants from India living in North America. *Family Process, 40*(2), 233–241. https://doi.org/10.1111/j.1545-5300.2001.4020100233.x

Erikson, E. (1968). *Youth: Identity and crisis*. Norton.

Farver, J. A. M., Narang, S. K., & Bhadha, B. R. (2002). East meets West: Ethnic identity, acculturation, and conflict in Asian Indian families. *Journal of Family Psychology, 16*(3), 338–350. https://doi.org/10.1037/0893-3200.16.3.338

Farver, J., Xu, Y., Bhadha, B., Narang, S., & Lieber, E. (2007). Ethnic identity, acculturation, parenting beliefs, and adolescent adjustment: a comparison of Asian Indian and European American families. *Merrill-Palmer Quarterly, 53*(2), 184–215. http://www.jstor.org/stable/23096105

Giguère, B., Lalonde, R., & Lou, E. (2010). Living at the crossroads of cultural worlds: The experience of normative conflicts by second generation immigrant youth. *Social and Personality Psychology Compass, 4*(1), 14–29. https://doi.org/10.1111/j.1751-9004.2009.00228.x

Gross, J. J. (1998). Antecedent-and response-focused emotion regulation: Divergent consequences for experience, expression, and physiology. *Journal of Personality and Social Psychology, 74*(1), 224–237.

Gross, J. J., & John, O. P. (2003). Individual differences in two emotion regulation processes: Implications for affect, relationships, and well-being. *Journal of Personality and Social Psychology, 85*(2), 348–362. https://doi.org/10.1037/0022-3514.85.2.348

Hegde, R. S. (1998). Swinging the trapeze: The negotiation of identity among Asian Indian immigrant women in the United States. In D. V. Tanno & A. González (Eds.), *International and intercultural communications annual. Communication and identity across cultures* (pp. 34–55). Sage.

Ho, J., & Birman, D. (2010). Acculturation gaps in Vietnamese immigrant families: Impact on family relationships. *International Journal of Intercultural Relations, 34*, 22–33. https://doi.org/10.1016/j.ijintrel.2009.10.002

Hong, Y. Y., Morris, M. W., Chiu, C. Y., & Benet-Martínez, V. (2000). Multicultural minds: A dynamic constructivist approach to culture and cognition. *American Psychologist, 55*(7), 709–720. https://doi.org/10.1037//0003-066X.55.7.709

Hsu, L., Woody, S. R., Lee, H. J., Peng, Y., Zhou, X., & Ryder, A. G. (2012). Social anxiety among East Asians in North America: East Asian socialization or the challenge of acculturation? *Cultural Diversity and Ethnic Minority Psychology, 18*, 181. https://doi.org/10.1037/a0027690

Huynh, Q. L., Nguyen, A. M. D., & Benet-Martínez, V. (2011). Bicultural identity integration. In S. J. Schwartz et al. (Eds.), *Handbook of identity theory and research* (pp. 827–842). Springer. https://doi.org/10.1007/978-1-4419-7988-9_35

Ibrahim, F., Ohnishi, H., & Sandhu, D. S. (1997). Asian American identity development: A culture specific model for South Asian Americans. *Journal of Multicultural Counseling and Development, 25*(1), 34–50. https://doi.org/10.1002/j.2161-1912.1997.tb00314.x

Inman, A. G., Howard, E. E., Beaumont, R. L., & Walker, J. A. (2007). Cultural transmission: Influence of contextual factors in Asian Indian immigrant parents' experiences. *Journal of Counseling Psychology, 54*(1), 93–100. https://doi.org/10.1037/0022-0167.54.1.93

Jambunathan, S., & Counselman, K. (2002). Parenting attitudes of Asian Indian mothers living in the United States and in India. *Early Child Development and Care, 172*(6), 657–662. https://doi.org/10.1080/03004430215102

Juang, L. P., Syed, M., Cookston, J. T., Wang, Y., & Kim, S. Y. (2012). Acculturation-based and everyday family conflict in Chinese American families. *New directions for child and adolescent development, 2012*(135), 13–34. https://doi.org/10.1002/cd.20002

Kalia, V. K., & Weatherall, A. (2009). "When mister right comes along": gender and ethnic identity in narratives from spontaneous Indian New Zealander mother-daughter conversations. *Qualitative Research in Psychology, 6*(3), 219–232. https://doi.org/10.1080/14780880802070542

Kallivayalil, D. (2004). Gender and cultural socialization in Indian immigrant families in the United States. *Feminism & Psychology, 14*(4), 535–559. https://doi.org/10.1177%2F0959353504046871

Kieu, T. (2013). *Why immigration is an Asian American issue*. http://www.americanprogress.org/

King, R. C. (2000). Racialization, recognition, and rights: Lumping and splitting multiracial Asian Americans in the 2000 Census. *Journal of Asian American Studies, 3*(2), 191–217. https://doi.org/10.1353/jaas.2000.0017

Koepke, S., & Denissen, J. J. (2012). Dynamics of identity development and separation–individuation in parent–child relationships during adolescence and emerging adulthood–A conceptual integration. *Developmental Review, 32*(1), 67–88. https://doi.org/10.1016/j.dr.2012.01.001

Lehtinen, V. (2002). America would lose its soul: The immigration restriction debate, 1920–1924 (Master's thesis). http://ethesis.helsinki.fi/julkaisut/hum/histo/pg/lehtinen/americaw.pdf

Li, G., & Wen, K. (2015). East Asian heritage language education for a plurilingual reality in the United States: Practices, potholes, and possibilities. *International Multilingual Research Journal, 9*(4), 274–290. https://doi.org/10.1080/19313152.2015.1086623

Lopez, G., Ruiz, N. G., & Patten, E. (2017, September). Key facts about Asian Americans, a diverse and growing population. *Pew Research Center.* https://www.pewresearch.org/fact-tank/2017/09/08/key-facts-about-asian-americans/

Lou, E., Lalonde, R. N., & Giguère, B. (2012). Making the decision to move out: Bicultural young adults and the negotiation of cultural demands and family relationships. *Journal of Cross-Cultural Psychology, 43*(5), 663–670. https://doi.org/10.1177/0022022112443414

Malik, S. (2015). Asian Americans in the United States today. *Center for American Progress*: http://www.americanprogress.org/

Marcia, J. E. (1966). Development and validation of ego-identity status. *Journal of Personality and Social Psychology, 3*, 551–558. https://doi.org/10.1037/h0023281

Markus, H. R., & Kitayama, S. (1991). Culture and the self: Implications for cognition, emotion, and motivation. *Psychological Review, 98*(2), 224–253. https://doi.org/10.1037/0033295X.98.2.224

Masood, N., Okazaki, S., & Takeuchi, D. T. (2009). Gender, family, and community correlates of mental health in South Asian Americans. *Cultural Diversity and Ethnic Minority Psychology, 15*(3), 265–274. https://doi.org/10.1037/a0014301

McAdams, D. P. (2001). The psychology of life stories. *Review of General Psychology, 5*, 100–122. https://doi.org/10.1037/1089-2680.5.2.100

Mendoza-Denton, R., Downey, G., Purdie, V. J., Davis, A., & Pietrzak, J. (2002). Sensitivity to status-based rejection: Implications for African American students' college experience. *Journal of Personality and Social Psychology, 83*(4), 896–918. https://doi.org/10.1037/0022-3514.83.4.896

Miller, M. J., Yang, M., Farrell, J. A., & Lin, L. L. (2011). Racial and cultural factors affecting the mental health of Asian Americans. *American Journal of Orthopsychiatry, 81*(4), 489–497. https://doi.org/10.1111/j.1939-0025.2011.01118.x

Mistry, J., Li, J., Yoshikawa, H., Tseng, V., Tirrell, J., Kiang, L., Mistry, R., & Wang, Y. (2016). An integrated conceptual framework for the development of Asian American children and youth. *Child Development, 87*(4), 1014–1032. https://doi.org/10.1111/cdev.12577

Mohanty, C. T. (1991). Cartographies of struggle: Third world women and the politics of feminism. In C. T. Mohanty, A. Russo, & L. Torres (Eds.), *Third world women and the politics of feminism* (pp. 2–47). Indiana University Press.

Nam, Y., Lee, E. J., Huang, J., & Kim, J. (2015). Financial capability, asset ownership, and later-age immigration: Evidence from a sample of low-income older Asian immigrants. *Journal of Gerontological Social Work, 58*(2), 114–127. https://doi.org/10.1080/01634372.2014.923085

Ngai, M. M. (1999). The architecture of race in American immigration law: A reexamination of the Immigration Act of 1924. *The Journal of American History, 86*(1), 67–92. https://doi.org/10.2307/2567407

Nguyen, A. M. D., & Benet-Martínez, V. (2013). Biculturalism and adjustment: A meta-analysis. *Journal of Cross-Cultural Psychology, 44*(1), 122–159. https://doi.org/10.1177/0022022111435097

Raj, S. P., Daga, S. S., & Raval, V. V. (2021). Cultural identity across the lifespan: Using an ecological framework to contextualize the experiences of Asian Indian immigrant mothers and their children. In H. Chu & B. Thelamour (Eds.), *Conceptual and. methodological approaches to navigating immigrant ecologies* (pp. 29–46). Springer.

Ramírez-Esparza, N., Gosling, S. D., Benet-Martínez, V., Potter, J. P., & Pennebaker, J. W. (2006). Do bilinguals have two personalities?: A special case of cultural frame switching. *Journal of Research in Personality, 40*(2), 99–120. https://doi.org/10.1016/j.jrp.2004.09.001

Richardson, J. T. (2003). Howard Andrew Knox and the origins of performance testing on Ellis Island, 1912–1916. *History of Psychology, 6*(2), 143–170. https://doi.org/10.1037/1093-4510.6.2.143

Roland, A. (1988). *In search of self in India and Japan*. Princeton University Press.

Saad, C. S., Damian, R. I., Benet-Martínez, V., Moons, W. G., & Robins, R. W. (2013). Multiculturalism and creativity: Effects of cultural context, bicultural identity, and ideational fluency. *Social Psychological and Personality Science, 4*(3), 369–375. https://doi.org/10.1177/1948550612456560

Sam, D. L., & Berry, J. W. (2010). Acculturation: When individuals and groups of different cultural backgrounds meet. *Perspectives on Psychological Science, 5*(4), 472–481. https://doi.org/10.1177/1745691610373075

Segal, U. A. (1991). Cultural variables in Asian Indian families. *Families in society, 72*(4), 233–242. https://doi.org/10.1177%2F104438949107200406

Shariff, A. (2009). Ethnic identity and parenting stress in South Asian families: Implications for culturally sensitive counselling. *Canadian Journal of Counselling and Psychotherapy, 43*(1), 35–46. https://cjc-rcc.ucalgary.ca/article/view/58908

Sirikantraporn, S. (2013). Biculturalism as a protective factor: An exploratory study on resilience and the bicultural level of acculturation among Southeast Asian American youth who have witnessed domestic violence. *Asian American Journal of Psychology, 4*(2), 109–115. https://doi.org/10.1037/a0030433

Steiger, J. H. (1980). Tests for comparing elements of a correlation matrix. *Psychological Bulletin, 87*(2), 245–251. https://doi.org/10.1037/0033-2909.87.2.245
Sue, D. W., Bucceri, J., Lin, A. I., Nadal, K. L., & Torino, G. C. (2009). Racial microaggressions and the Asian American experience. *Cultural Diversity and Ethnic Minority Psychology, 13*(1), 72–81. https://doi.org/10.1037/1099-9809.13.1.72
Sue, D. W., & Sue, D. (2003). *Counseling the culturally diverse: Theory and practice* (4th ed.). Wiley.
Syed, M., Azmitia, M., & Phinney, J. S. (2007). Stability and change in ethnic identity among Latino emerging adults in two contexts. *Identity: An International Journal of Theory and Research, 7*(2), 155–178. https://doi.org/10.1080/15283480701326117
Tardiff-Williams, S. Y., & Fisher, L. (2009). Clarifying the link between acculturation experiences and parent–child relationships among families in cultural transition: The promise of contemporary critiques of acculturation psychology. *International Journal of Intercultural Relations, 33*, 150–161. https://doi.org/10.1016/j.ijintrel.2009.01.001
Triandis, H. C. (1996). The psychological measurement of cultural syndromes. *American Psychologist, 51*(4), 407–415. https://doi.org/10.1037/0003-066X.51.4.407
Vygotsky, L. S. (1978). *Mind in society*. Harvard University Press.

Part II
Language, Language Role Identity, and Literacy

Chapter 7
Immigrant Identity Narratives: Youth Literacy Project

Stephen M. Quintana, Julia Z. Benjamin, and Numan Turan

Erik Erikson (1956) articulated a complex, multifaceted theory of identity development, which can be applied to immigrant identity (Schwartz et al., 2006). Erikson (1956) suggested, "the term identity ... connotes both a persistent sameness within oneself (self-sameness) and a persistent sharing of some kind of essential character with others" (p. 57). He identified three facets of adolescent development: personal, social, and ego identity. Personal identity refers to the youth's commitment to personal values, goals, social roles, and ideology, including religious values, vocational aspirations, and political affiliations. Personal identity evolves from childhood identifications with significant others and their values, goals, roles, and ideologies. As adolescents have increased awareness of their individuality, childhood identifications with others' values, goals, roles, and ideologies become integrated with adolescents' growing self-awareness. In the context of immigration, youth may form personal identity commitments based on the religious practice, cultural values, and socialized goals of their immigrant community. Additionally, as immigrant youth are exposed to host or receiving cultural contexts, their personal commitments may also represent their acculturation processes. Consequently, the personal identity development for immigrant youth results from the integration of (1) childhood identifications with family members and significant others, many of whom are from youth's country of origin, and (2) their own experiences acculturating to new cultural contexts. Often immigrant identity commitments represent complex integrations of cultures of origin with receiving or host cultures (Schwartz et al., 2006).

Analogous to personal identity, social identity develops from childhood identifications as a member of a social group. During adolescence, identifications with social groups become integrated with adolescents' growing awareness of how their

S. M. Quintana (✉)
University of Wisconsin-Madison, Madison, WI, USA
e-mail: Stephen.Quintana@wisc.edu

J. Z. Benjamin
American Family Children's Hospital, Madison, WI, USA
e-mail: JBenjamin@uwhealth.org

N. Turan
Istanbul Medeniyet University, Istanbul, Turkey
e-mail: numan.turan@medeniyet.edu.tr

D. J. Johnson et al. (eds.), *Re/Formation and Identity*, Advances in Immigrant Family Research, https://doi.org/10.1007/978-3-030-86426-2_7

own lived experiences reflect the experiences of larger social groups. Therefore, social identity—according to Erikson (1956)—refers to a sense of identification and solidarity with a larger social group with which one perceives a sense of shared history, life circumstances, and lifestyle. Quintana (1998) described this social identity, in the context of ethnic identity, as an "ethnic group consciousness" in which the identification with a social group is associated with a spirit of 'one for all, all for one' and esprit de corps and as illustrating a strong sense of connection to a larger social group. As the personal convictions of immigrant youth are often connected to their social identifications with immigrant communities, there may be a merging of personal and social identities, reinforcing the centrality and importance of immigrant identities.

Much of the extant research on immigrant youth identity is focused on the personal and social versions of identity. There is, however, relatively little research with immigrant youth focused on the ego identity component of Erikson's (1956) theory of identity development. From a developmental standpoint, ego identity processes are central to Eriksonian theory. Specifically, ego identity involves the processes by which personally relevant and socially relevant experiences are integrated into personal and social identity constructions, with these constructions referred to as "self-identities." The process by which relevant experiences are used to form personal and social self-identity constructions is considered to be ego identity formation, or how the ego integrates experiences into constructions of self. Ego identity formation, therefore, involves the processes, biases, and tendencies by which meaning is abstracted from relevant personal and social experiences. In the context of immigration, immigrant youth's ego identity functions to provide thematic heuristics regarding how the experiences of immigration and being an immigrant can be integrated into a coherent identity narrative or life story. For example, thematic heuristics of ego identity function could involve constructing a life narrative as a comedy, drama, or tragedy. In short, self-identity, like self-concept, represents social labels or roles with which the youth identifies. Ego identity functions, therefore, construct the identity narratives that describe the characteristic movement of the self through time, space, and context.

Much of the research on identity is focused on methodology that investigates, in Erikson's (1956) terms, self-identities, such as the social labels that represent individuals' identities. However, different methodologies are required to examine ego identity processes. For example, a narrative identity methodology has been applied to examine ego identity processes (Syed et al., 2007). According to narrative identity theorists (McAdams & Olson, 2010; Syed et al., 2007; Syed & Azmitia, 2008), individuals make meaning of experiences and life through stories. Thus, individuals arrange, synthesize, and integrate discrete autobiographical memories into narratives that provide explanations for how prior experiences have shaped who they are and where they are going in life. A narrative identity, therefore, involves reconstructing past events, incorporating contemporary experiences, and anticipating future directions, which are critical components of ego identity processes described by Erikson (1956; see also Syed & Azmitia, 2008).

Identity narratives serve introspective and social functions by being a way of understanding the self and introducing the self to others (Erikson, 1962). Erikson (1962, 1970) often relied on narratives to characterize ego and self-development.

Furthermore, narrative identity theorists maintain that "...individuals' life stories *are* [emphasis added] their identities, for stories—and identities—are constantly being renegotiated and reassessed as new experiences shed light on and help make meaning of the old ones, which contributes to a sense of integration and coherence" (Syed & Azmitia, 2008, p. 1014–1015). Thus, the narrative identity approach provides a compelling foundation for examining how adolescents and adults synthesize their life experiences into a singular, coherent sense of self (McAdams & McLean, 2013; Syed & Azmitia, 2008).

Although the narrative identity approach has been applied to adolescents and adults, it has been infrequently applied to immigrants and, more rarely, to immigrant youth. However, a narrative approach seems particularly relevant to immigrant youth who undergo dramatic life changes when immigrating to a different country, with different customs, lifestyles, cultures, values, and ways of defining self. Formulating a coherent identity narrative, despite the profound changes associated with immigration, may be particularly challenging and important for immigrant youth. Immigration poses many dilemmas for youth that need to be addressed in order to develop a coherent sense of self. To develop a coherent sense of self, immigrants need to reconcile their experiences of self in their home culture with the sense of self experienced in a foreign culture.

A coherent identity narrative will need to account for a shift in how the expected life trajectory originally shaped by the country of origin has been transformed into a new life trajectory influenced by the immigration process and the cultural norms within the adopted country. Because the immigration journey and subsequent adjustment to a new country and culture have a critical impact on individuals' lives, identity narratives need to be constructed that infer meaning and purpose for immigration and the life trajectory that is informed by experiences within a new country and culture. Furthermore, to maintain coherence across life experiences for immigrants, identity narratives need to include what vestiges of life, relationships, and the self that were established in the country of origin have been preserved, maintained, transformed, and lost. Identity narratives, for example, will address whether life circumstances are better for the immigrant, and whether the sacrifices involved in leaving home and culture of origin to immigrate to a new country are worth the gains. These issues are relevant for immigrants in order to formulate a sense of continuity and coherence in their identity narratives that spans the dramatic shift associated with immigration.

Researchers suggest that narrative identities can be evaluated based on their coherence (McAdams, 2006; Syed & Azmitia, 2008). As McAdams (2006, p. 109) suggested, "Like all stories, life stories [or identity narratives] exist to be told or performed in social contexts. Most criteria for identity coherence reflect the culture within which the story is told and the life is lived." Narrative coherence involves being able to sequence events into meaningful order such that the identity narrative is contextualized with personally relevant events. For immigrant youth, coherence would involve drawing out the personal relevance of the process of immigration—pre-migration preparation, immigration journey, and post-migration settlement. Coherence also involves being able to narrate a personal life story that is connected to and informed by the kinds of life stories common to a young person's culture of origin and their newly adopted cultural context (McAdams, 2006).

Canonical narratives about the U.S. for immigrants include that the U.S. is a "land of opportunity" where it is possible to pursue the "American Dream" in which anyone, with hard work, can succeed (Sowards & Pineda, 2013). Immigrant youth constructing their identity narratives contrast and reconcile the reality of their own and their families' circumstances with those canonical myths about the U.S. Furthermore, coherence involves narratives in which the self is depicted as being shaped by critical life experiences or sequences of events. Immigrant youth narratives are likely to cast immigration as a significant and influential life experience that may be perceived as enhancing and/or limiting life circumstances in certain ways. Finally, coherence involves narratives in which there is a clear theme or plot that meaningfully integrates life events across time to narrate transformations of self in personal and social domains. Immigrant narratives often involve some purpose or end goal of the immigration. A commonly stated end goal for immigrating to the U.S. is for "a better life," often involving parental sacrifices in order to provide improved prospects for their children, relative to the life the parents had in their country of origin.

Narrative identity scholarship for non-immigrants has identified several important themes, including agency or achievement motivation (e.g., de Fina, 2003; Todorova et al., 2008) and communion or belonging (e.g., Lam & Chan, 2004; McAdams & Olson, 2010). Narrative identity inquiry specific to immigrant youth has focused mostly on their post-migration adjustment, often in U.S. schools (e.g., Deng, 2017; Lam & Chan, 2004; Sheth, 2015; Todorova et al., 2008; Vall & Botella, 2015). Much of the research on immigrant narrative identities examined how expressions of self and identity were influenced by cultural practices from the country of origin—these cultural practices were expressions of identity which expressed the immigrants' connection to their cultures of origin and distance from host or receiving cultures (e.g., Sánchez, 2007). Narrative research has provided rich metaphors for immigrant identity including poignant quests to find a sense of 'home' in a new land (Sheth, 2015) and a sense of 'uprootedness' (Berman et al., 2009). Several researchers have examined the connection between maladjustment and narrative disruption for those immigrants whose life stories lack coherence as a result of the disruption the immigration process has created in developing a coherent sense of self (de Fina, 2003; Vall & Botella, 2015).

The purpose of the present research was to build on past research, using a narrative identity approach to take advantage of several innovations associated with using a large sample of recent immigrants. The sample of immigrants used for this study allowed us to code individual narratives for each of the immigrants as well as identify patterns across multiple narratives. As is standard in narrative identity research (Adler et al., 2017), we examined narrative components or features that were reflected across narratives, but we also coded the entire narratives into clusters associated with different narrative plots. The identification of complete narratives into clusters is a novel contribution of the present study and reflects a critical component of understanding narrative identity. Thus, the primary research focus and unit of analysis was the identity narratives generated by immigrant youth, with a focus on the different components within youth's immigrant narratives as well as identification of clusters across immigrant narratives.

Method

Participants and Selections of Narratives

A nonprofit organization worked with 266 immigrant youth from 30 countries of origin in GED classrooms at five New York City public schools (38.4% women; age range = 17–21 years). One of the five high schools served a predominantly immigrant community from the Dominican Republic. Because the current study involved a secondary data analysis on the stories that had been narrated for the purposes of the anthologies, very little demographic information about individual participants is available, limited to what was available in the printed anthologies. Nonetheless, the goal of the research was to identify common themes in immigrant narratives and to identify major canonical identity narratives for immigrant youth rather than attempting to connect specific components of the immigrant experience to demographic characteristics. Future quantitative research could then connect these themes with specific demographic characteristics for stochastic purposes.

Graduate students from Columbia University conducted life story interviews with youth lasting between 20 and 45 min, and then worked with youth to create written narratives, which were then compiled into published anthologies for each school. There were three main purposes of the project: foster literacy development by narrating and then expressing in written form life narratives of immigrant youth, promote self-affirmation for immigrant youth while they narrate their immigrant stories, and promote wider appreciation for the variety of stories about immigrant youth across immigrant and non-immigrant communities. The present study represents a systematic process for promoting a broad appreciation for the kinds of immigrant stories narrated by youth. To realize these three project goals, interviewers asked open-ended questions such as "Tell me your life story," in response to which youth described their immigration, events leading up to it, the journey from their country to the U.S., and experiences living in the U.S. Interviewers asked follow-up questions to clarify components of the immigrant's story. The goal of the interview was to allow youth to narrate their stories in an authentic manner, permitting them to choose which components to describe in their life narrative. This process is different from that of researchers with a set of interview questions which can impose researcher bias and preconceptions on the research methodology. From the 266 written stories, 45 were selected based on three coders being assigned 15 narratives each, which represented a relatively large sample for qualitative studies and allowed researchers to have each of the five high schools equally represented with nine narratives from each high school. Because one narrative was unrelated to immigration, the final sample analyzed was 44 narratives. We withheld a subsample of eight narratives from the 44 to determine if saturation had been established for the first 36 narratives that were analyzed. Saturation is established when the coding schema obtained for a set of data is evaluated against a second set of data to determine if the coding schema developed on the first set is sufficiently comprehensive, with no additions or changes to the coding schema indicated. Saturation had been reached in the 36 narratives as all of the data for the remaining eight narratives could be readily coded within the system developed on the original set of narratives. Consequently, for the 44 youth from 17 different countries of origin (i.e., Bangladesh, Burkina Faso, China, Dominican Republic, Ecuador, El Salvador, Gambia, Guinea, Haiti, Honduras, Liberia, Mali, Mexico, Morocco, Senegal, Tibet, and West Africa, and 40.5% women) that were selected, narratives ranged from 262 to 2203 words in length (*M* word = 947.32, *SD* = 470.94).

Narrative Methodology

The researchers based the narrative coding on Riessman's (2008) approach to thematic analysis and Lieblich, Tuval-Mashiach, and Zilber's (1998) categorical-content approach in exploring narratives because both Riessman's and Libelich et al.'s approaches focus on the content of the interviews rather than their expressive manner or style. The coding team consisted of the three authors, one bi-racial (Latinx-White) male faculty with a research focus on ethnic identity development, one male immigrant from Turkey of Turkish descent who was a doctoral student with research experience in cross-cultural differences, and one White, European American female doctoral student who had worked for 18 months as an assistant director and co-coordinator of the nonprofit program in NYC high schools which solicited these narratives. Each team member then used Atlas.ti software to record the coding of the 44 narratives.

The data analysis involved identifying meaning units, which represent the smallest unit of data that represents some meaning, usually one or a few words. Upon development of consensus for a convention of coding meaning units (e.g., labels for similar meaning units), the interview transcripts are coded into series of meaning units, which, in turn, are coded into themes that are abstracted from meaning units. An analogous process was followed for coding of the overall plot for each narrative, with each coder providing a brief summary of the plot for each individual narrative. The researchers found that the plots could be summarized across chronological order with the youth describing the immigrant journey and subsequent adjustment to a new lifestyle, comparing the new to old lifestyle and how the immigration experience influences social relationships, usually family relationships. By comparing youth's descriptions across these major components, the research team identified four main clusters from the 44 narrative plots.

Results

Story Clusters

The broad themes that emerged from the meaning units for this study included immigration journey (including the different phases of immigration described above), adjustment, family, and meaning making. Within each of the broad themes, four clusters of story patterns were identified: Smooth/Positive, Loss/Bitterness, Redemption, and Hard Work/Hard Life. Two stories did not fit the inclusion criteria because they focused on other individual's experiences instead of the narrator's. We conducted analyses across the four clusters and found no differences based on the available demographic and descriptive variables, which included gender, school, participation in the first or second year of the program, geographic region of the youth's country of origin, and number of words in the narrative. Below are descriptions of those four broad themes that were expressed in each of the four narrative clusters. As is typical for qualitative analyses, individual statements were coded into meaning units with a meaning label attached. After coding the same four narratives separately, members of the research team met to develop agreement about how to generate meaning units by comparing notes and sorting through the initial codes. After meaning units were generated, members of the team developed hierarchical themes (i.e., themes that captured a set of meaning labels) individually and then with one other team member. As one example, the meaning units "immigrating for a 'better life,'" "immigrating to flee violence," and "immigrating to escape poverty" were combined into a broader theme of "Motivations for Immigration." The larger theme of "Immigration Journey" included "Motivations for Immigration" along with "Immigration Choice," which was composed of meaning units associated with the youth indicating why they wanted or did not want to immigrate. The Family theme included subthemes associated with whether the youth immigrated with their nuclear families, with some family members (e.g., mother, father, sibling), or with no family members (e.g., alone or with friends or strangers). The four narrative clusters were formed by examining which narratives tended to cluster together across the major themes of Immigration Journey, Adjustment, Family, and Meaning Making. After discovering and further examining the four clusters, the team noted that the four clusters could be organized by whether immigration was narrated as involving a significant transition in their life story. The Smooth/Positive and Hard Work/Hard Life clusters had identity narratives characterizing continuity in their life narratives based on whether life opportunities before and after immigration were readily available or whether they had to work hard for opportunities all of their lives, respectively. Conversely, Redemption and Loss/Bitterness clusters constructed identity narratives in which immigration led to respectively positive or negative shifts in their life trajectories, with Redemption narratives involving youth who had overcome challenges. The Loss/Bitterness narratives characterized immigration as involving insurmountable setbacks and difficulties.

Smooth/Positive Cluster

Immigration Journey

Although all youth who provided narratives had immigrated to the U.S., their stories were told in markedly different ways, beginning with their motivation for immigration and the journey itself. For those youth who wrote Smooth/Positive narratives (12 youth), there was a smooth transition between their home to a new culture and a positive adjustment to their new cultural context. For these youth, the journey to the U.S. was made with other family members. These youth expressed how the decision to immigrate was motivated by expanding life opportunities either for the parents or the youth, such as securing a U.S. education or positioning themselves to secure influential or high-earning occupations. For example, one youth from China suggested: "…if you get your diploma [in the US], you can do whatever you want." None of these youth expressed having been pressured or forced to immigrate by others, with most of them describing their immigration as being voluntary.

Adjustment

Within this cluster, youth tended to have a carefree or casual tone and mentioned few concerns about life. These youth rarely described working to earn money and many expressed continuity in relationships and activities between their home country and the U.S. One youth wrote: "In Ecuador, the best memory I had was playing soccer with friends on the street… I still like to play soccer here, in Flushing Park." Although most youth across clusters expressed missing their home countries, Smooth/Positive narratives generally viewed circumstances in the U.S. favorably compared to their countries of origin, as illustrated as follows: "The first thing I remember seeing is the most beautiful plane in JFK, which is the most beautiful airport I have ever been in." Another youth wrote: "It is very different from schools in the Dominican Republic. The system is different, but there is also a lot of poverty there, so people cannot pay for school or books." Becoming more adjusted to life in the U.S. they expressed appreciation for their life here, as indicated by one youth from Sierra Leone: "Now I live with my dad. It is good because he is supporting me, so I go to school."

Family

These narratives focused on pleasant memories and simple pastimes with family. As one youth from the Dominican Republic stated, "My house was nice and I lived with my mom and all was well." Unlike the other clusters, many of the Smooth/Positive narratives indicated that the family as a whole immigrated, including nuclear and sometimes ($n = 4$) extended family, so that the process of immigration involved less family disruption than in the other clusters. To illustrate, one youth from Bangladesh commented: "I don't really miss my country, maybe a little bit. This is because everybody's here—I'm here with my parents, my uncle, auntie, sister and brother."

Meaning Making

Of the four clusters, narratives within this cluster had the fewest markers of self-reflection and meaning making. There were a few descriptions of personal change resulting from immigration, as in the following. "I will not forget my memories of my country, but I am in the United States now, and it is not Guinea… To start my new life, I have to change my ways and act differently, and do what I have to." However, these narratives implied that the life trajectory was unchanged, compared to other clusters, due to immigration. Additionally, these narratives included mostly concrete advice to readers about how to succeed as an immigrant, as a youth from Haiti advised: "I would tell youth who read my story to be serious. If you want to be a man or a woman, you have to go to school and obey your parents."

Loss/Bitterness Cluster

Immigration Journey

In stark contrast, narratives in this cluster ($n = 7$) were least likely to describe wanting to immigrate to the U.S. Several narratives included explicit descriptions about the youth having to immigrate against their will or about feeling deceived about the opportunities in the U.S. and wishing to return to their home countries. One youth wrote, "I did not want to come here. I miss my family and would like to be able to live with my mother and my sister in the Dominican Republic." As another youth stated: "I don't see New York the way others do. I think it's good, but it's not the fantasy people imagine." Another youth described this disillusionment as follows:

> I remember that people used to talk about America like it was heaven. I would sit with my friend Fatu, and it was the only place we would talk about. We would pray to come here together and actually see what we heard about every day. But if I told Fatu what it is really like, she would not believe me until she came and saw it for herself. They said there were no rats here, but I see rats here every day. They have very big rats here, so big they could swallow you.

Adjustment

The narratives within this cluster included descriptions of difficulties adjusting to life in the U.S. and having unresolved concerns in their home countries. For some youth, these difficulties were due to significant traumas or tragedies experienced in their home country, including witnessing death, loss of loved ones, sexual and physical assault, and pervasive neglect or mistreatment by caretakers. A youth who had recently arrived in New York with her two brothers stated, "Then we heard that our mother had passed away, but it was too late; we didn't even go to the funeral… It still hurts me that I didn't go to my mother's funeral." Some experienced harsh treatment from adults or parents/guardians, both in their native countries and upon arrival in the United States, such as described in one narrative constructed by a youth from the Dominican Republic who exhorted readers to support their children, saying, "Don't use harsh words that will scar their souls. Words like, 'I'd rather have a dead son than a gay son,' or 'all those people have AIDS.' My mom has always used those words on me." For others, immigration to the U.S. resulted in a sense of loss. Youth narratives described what they had lost due to immigration, including family, community, and living conditions. One youth wrote, "The hardest thing about this change has been getting used to the poverty, because although I'm not poor, I live as if I were because I had to get rid of all my belongings." Their descriptions of post-immigration settlement were generally more negative, conveying a greater sense of emptiness and hopelessness, relative to all the other clusters. For instance, a youth from China described his parents' divorce and father's immigration to the U.S. as "terrible" and described his current struggle to connect with his father, writing poignantly, "Every day, my father is busy. I'd like to talk with him more, but he has a wife, children, a family, so I can't. I just close my bedroom door and study English."

Family

These narratives reflected significant sacrifice and loss of family connections. Many youth described leaving behind close family members in a way that caused irreconcilable breaches and deep sadness. A youth from China expressed this sense of loss in his narrative: "When I go to school or to a shopping mall, I see children with both of their parents, and I envy them. They have a mother and a father." Not only have these youth lost family relationships through immigration, but they also felt abandoned or neglected by family members. One youth from the Dominican Republic wrote about living with his abusive aunt:

> Each time I say something, she says, 'Shut up! Shut up!' She does not hit me with a stick anymore, but she still wants to rule me. When I get nervous, I have trouble speaking. Since my aunt does not let me talk, everything gets stuck inside.

The overall sense of family relations within Loss/Bitterness narratives was that life circumstances, largely due to family circumstances, had been difficult and that there was little hope to establish better conditions in the U.S.

Meaning Making

Similar to the Smooth/Positive cluster, narratives in the Loss/Bitterness cluster had few descriptions of meaning making. None of these youth expressed any identity change, and the messages that they received from others were often negative, such as having learned by observing negative role models. One Haitian youth described problems living with her stepmother: "Because she [step-mother] was mean to me, I learned not to be mean the same way to her and to treat her better than she treated me." The principal meanings or 'moral of the story' in these narratives reflected pessimism, as illustrated by the following quote: "The fact that he was a human being reminded me that one day we are all going to die, one way or another." Despite loss and bitterness, youth still exhorted readers to fight for dreams and not give up hope. As one youth observed: "no pain, no gain."

Redemption Cluster

Immigration Journey

Youth in this cluster narrated their stories to focus on challenges, similar to youth in the Loss/Bitterness cluster, but the youth in the present cluster included some form of redemption in their identity narratives. Youth with narratives in this cluster often mentioned they had not made the immigrant journey voluntarily but were forced to immigrate, or described significant legal hardships along the way. The overall tone of this cluster's ($n = 11$) narratives with regard to the immigration journey was purposeful. Even when immigration had not been their choice, they expressed having learned about life as a result. For example, one youth from the Dominican Republic stated, "I only needed to finish the last year of high school, but then they sent me to the United States," and went on to write, "Coming to New York has made me grow as a person and my attitude has improved because I've realized what I really want for my present and my future."

Adjustment

Some narratives in this cluster also focused on loss of resources or relationships during the transition to the U.S. including financial stability as well as connections with family and community. Rather than emphasizing losses, however, these narratives focused on lessons learned and making a new start in life. One youth living in the Bronx narrated this process:

> Now, as I think back, I didn't realize that my friends were bad for me. We would go to the nightclub, and we'd smoke and drink. It was bad. My friends didn't like me for me. They liked me for my money, but my money wasn't mine. My mother had to work to give me that money. I had to leave those friends. Now, I want my life to be good. My past was crazy, so now I want to take more responsibility for my family.

The overall theme of this cluster was the youth's determination to grow as individuals and make more of their lives than they had in their home countries, as illustrated in statements like "I want to help my family to have a better quality of life, so I should be smarter and work harder." Education often played a central role in adjusting to life in the U.S. and achieving their goals. One youth from the Dominican Republic described the importance of finding social support through school:

> In school I found real friends… Two teachers have given me special support here as well… Along with my mom, these teachers have made me feel that I am as valuable as a precious stone. Although I understand that nothing in life is perfect, people accept me as I am now.

While acknowledging the academic disruptions and other hardships faced due to the immigration process, youth described persevering in order to work to better themselves and/or their families. Although some immediate hard work may have been required, the narratives in this cluster differed from the Hard Work/Hard Life cluster because the redemption narratives suggested youth would be in a better position to earn more money or not have to work as hard in order to attain a desired standard of living or reputation As a youth from Hong Kong wrote, "I want to get a better job, and I know that I need a diploma to go to college and find a better job." One youth described taking computer courses at a parochial school in Haiti, writing, "I started to progress, but then I moved away and all my plans fell apart, so I had to start all over again," but despite that setback, the youth maintained a strong dedication to her academic career, stating, "In college, I would like to learn civil engineering. That takes a lot of math, but I am good at math. Luckily, my education and going to church make me who I am today."

Narratives within this cluster emphasized significant differences between U.S. culture and their culture of origin, but identified strengths and limitations in each:

> Here, I have more freedom to go out. In the Dominican Republic, it was much more dangerous. I could only go from my house to school. If I went out, I would have to go with a friend or someone I knew, but I couldn't go out by myself. Here, people are kind of mean and they aren't very social. There, most people know one another.

Although a youth from Mali expressed being intimidated by cultural difference upon her arrival—"I felt different in my heart. I saw different things, like women and men doing the same work… I kept thinking, 'Will I live with these differences? How will I learn English?'"—ultimately she described adapting to life here so that "in the end, I enjoyed coming to the U.S."

Family

Similar to Loss/Bitterness, youth who wrote Redemption stories described family difficulties and separations; however, nearly every story mentioned feeling loved or having developmental needs met despite these difficulties. This social support may have fostered resilience in the youth who wrote Redemption stories. For example, a youth from Burkina Faso wrote, "Growing up, I never lived with my father, but I would talk to him sometimes. We talked about everything. He asked me questions and gave me advice about life, about what to do in different situations." Many of the youth who wrote Redemption stories reflected on having had few responsibilities growing up and their desire to now approach life in a more serious way to better provide for their families. As one youth from China commented:

> My new life began when I moved to America, because out here in a new environment, I changed myself completely. When I was in China, I was quite rebellious. I would do certain things just because my parents told me not to do them. After I came to America, I changed myself because I saw how much my family was struggling.

Meaning Making

Both Redemption and Hard Work/Hard Life stories had higher levels of self-reflection and meaning making than did the Smooth/Positive and Loss/Bitterness stories. Youth who wrote Redemption narratives felt that the immigration process had a positive formative effect on who they were becoming, for example "I learned that traveling around the world opens your eyes. You learn a lot, because people and countries are different," and "I had an important turning point when I came to America because of the things I experienced and saw. I did not think about my future before, but I have become someone with a goal." For these youth, immigration represented an opportunity to take charge of their futures. When giving advice to readers, they strongly encouraged everyone to make positive choices and create their own futures as well. One youth told her readers, "Never stop fighting for your dreams and your goals. You create your own present and future by knowing how much you are worth," while another wrote, "As long as you do not go too far down the wrong path, you can always turn back. During my turning point, I had some trouble, but I also found a goal and hope." In sum, the overall message these youth conveyed was optimistic: that "although life is not always the way people want it to be, I believe it will be okay."

Hard Work/Hard Life Cluster

Immigration Journey

Youth who wrote stories that were categorized in this cluster ($n = 12$) expressed a wide variety of reasons for immigrating, including improved academic and financial opportunities as well as escaping harsh living conditions and, sometimes, trauma in their country of origin. A youth who arrived in New York immediately following an earthquake stated, "I left Haiti after the earthquake because I was thinking of my future. I came here to find something different and to go to school." Many of these youth expressed greater agency in choosing to immigrate and some made the journey separately from their families. Indeed, a young woman from Tibet described her choice to leave her family by saying:

> One day, my friend asked me, 'Do you want to go to India?' I was surprised and answered, 'Yes.' We discussed it for a long time and decided to do it. Then we talked to our families. They were all surprised, and my mother cried. She said she was old and was not willing for me to go to another country, because she thought that if I went, then we would never see each other again. She cried a lot, and her face was full of sorrow, but I couldn't change it because I had already made my plans to leave.

For youth in this cluster, the journey itself was often dangerous, and presented significant threats to physical and emotional health. One youth from El Salvador described it as "an incredible journey that I'll never forget—surviving hunger, the cold, and the heat of the desert. I rode trains, buses, taxis, and climbed hills full of animals so dangerous that many people are killed just trying to get over them," while one from the Dominican Republic described the fear of family separation at the border, writing, "My mom arrived here first, and then we followed. The most difficult thing about moving was the fear that not all of us would be allowed in and that my mom would have to live here alone."

Adjustment

While youth who narrated stories within this cluster also noted cultural differences, the most significant difference was escaping poverty or other harsh living conditions in their country of origin. As a youth from Tibet described:

> We needed to take our lunch and walk to school and then back home to my family's village every day because we had no bus or train. Here, in the USA there are very good conditions, because we ride the train and bus to school every day. The school here gives us lunch and whatever we need… I think America is very good for everything—food, education, and opportunities. Anything we need is available; it is all very good here.

Despite appreciating an apparent improvement in living conditions, youth in this cluster also expressed sadness about having left their countries of origin, writing statements like "At first I was crying every day, but that's life," and "sometimes I don't know what to do because I miss our country so much." Many youth in this cluster also wrote about a strong desire to return to their home countries after 'making it' in the U.S., saying, "[in the future] I'd like to go back and forth every six months, to have two lives—one in my country and one here." Although youth who wrote Hard Work/Hard Life narratives experienced intense homesickness, they also accepted the challenges of being separated from their home countries with resolve and determination. One youth wrote, "Sometimes I'm very homesick, and I want to go back to Morocco and give up everything here. At these moments, I remember my dear teacher's words and decide to stay even though it's really tough."

As with youth who wrote Redemption stories, youth writing Hard Work/Hard Life stories were committed to accepting challenges in order to improve life for themselves and family members. Hard Work/Hard Life narratives differed from Redemption narratives because the former indicated life was hard before and life will continue to be hard and anticipated having to continue to work hard, whereas the latter described transformations, such as obtaining a good education that would allow for a well-paying job. Many of the Hard Work/Hard Life youth described education as the key to future stability and remaining out of poverty. Narratives in this cluster emphasized realistic rather than overly idealistic occupational and educational goals, sometimes reflective of Redemption narratives. For example, two youth from Latin America in the Hard Work/Hard Life cluster wrote, "I am studying to get my GED, learning English, and taking a course in Co-op Tech to get a good job and take care of my wife and daughter," and "My life goal is to go to college and to have a good job so my kids do not have to go through what I went through. I would like to be either a nurse or work in immigration."

Youth in this cluster also wrote about working to earn money to support their families while going to school. A youth from the Dominican Republic described having to divide the money he earns between family members, saying, "When I work, I give her ten pesos so she can buy her cigarettes. She wants me to give her the whole check, but I tell her, 'I have a mother in Santo Domingo who I need to send money to as well.'" The combination of homesickness and the pressures of work and school made adjustment to the U.S. difficult for these youth. These demands on their time limited opportunities to socialize with peers while language abilities and legal barriers made it difficult for some to find legitimate employment. A youth who arrived in the U.S. from El Salvador without legal documentation described this challenge:

> Once you are in this country, you face so many obstacles that are nearly impossible to beat, because you don't have official documents, or because your English isn't perfect. I don't have a social security card—a piece of paper that would make me as important as everyone else. Without that piece of paper, you're made to feel different from the people of this country. When you're not from here, you feel discriminated against—just because of a piece of paper.

However, the youth in this cluster accepted these challenges in order to move forward with their lives. The youth from El Salvador went on to write:

> The obstacles this country throws at you makes it hard to survive, but the U.S. also has its good side. There are so many opportunities for those who are willing to make the most of things and are willing to act right… Because I want to be someone, I want to tell everyone—both recent immigrants and people who were born here—that you should work hard for what you want because nothing's impossible.

This perseverance in the face of numerous obstacles was the defining feature of Hard Work/Hard Life narratives and how these youth approached the task of adjusting to life in the U.S.

Family

Narratives within this cluster often included recollections of positive, supportive relationships that helped sustain them through their current challenges. A youth from Haiti wrote about this feeling of support despite family separations, "I lived with my grandma and grandpa, and my mom came to see me every day. My brother and sister both loved me so much. And because I didn't know my father, my grandfather was like my father." Family circumstances in this cluster's narratives were often difficult due to socioeconomic conditions, rather than the neglectful and abusive conditions described within Loss/Bitterness narratives. As a youth from El Salvador described, "Our childhood wasn't that easy, but we shared everything and studied as hard as we could so that we could make it in life." Family sacrifice was a common theme within these stories. Youth often expressed a strong sense of obligation to fulfill family roles, and often took on more adult roles at earlier ages than youth in other clusters. As an example, one youth wrote, "I want to be someone, someone important in life, to help my brother [who has a disability] and give him everything that he deserves in life." Another youth wrote about growing up without his parents but is living in the U.S. with grandparents who are both sick, and he now strives "to work to help them pay for the apartment, the lights, food, everything."

Meaning Making

The meaning making within Hard Work/Hard Life stories resembled a combination of the Redemption and the Loss/Bitterness narratives. Youth repeatedly emphasized that life is difficult and that they hope to prevent their children from having to go through what they did. One youth wrote, "Today, I do not worry because sometimes I see things can get difficult, but I always trust that as long as I have a purpose, I can achieve anything. I will find whatever I need to help me reach my goal." There was a stronger emphasis on identity in these stories, relative to the first two clusters. Youth expressed knowing who they were despite the judgment of others, and in contrast to family members who had hurt them, as expressed by a youth who identified as gay: "There are people who point their fingers at me, judge me for who I am. But I know who I am, where I come from, and who my family is." Many in this category actively considered cultivating a new identity. One youth from Haiti illustrated this, stating, "When I moved to New York, I asked myself, 'Who do I want to become?'" These youth wanted to leave the negativity of their past behind and make progress to become better people. A youth who had participated in gang activity in his country of origin wrote, "I still have pain both physical and mental from what happened during these times, but I know that people can change."

Like in the Redemption cluster, the advice these youth gave readers included having faith in God, seeking support from others for problems, and avoiding dangerous activities, but also included exhortations to persevere and sacrifice to realize life goals. These narratives conveyed pragmatic acceptance of the realities of life, while also portraying the transition to the U.S. as a new beginning, "I'm a girl who has suffered so much in life, but I made a new start in my life even with the pain. Despite the struggle, I want the best."

Summary Across General Themes

In summary, the clusters varied regarding how the four general themes were integrated into their narratives. Specifically, they varied regarding whether youth wanted or volunteered to immigrate or were forced to by family circumstances or safety concerns in their countries of origin. There was also diversity of motivations to immigrate, depending on whether youth were escaping difficult circumstances or motivated by the promise of opportunity in the U.S. Youth narratives that referred to their adjustment to the U.S. indicated further variability, whether the youth remained somewhat idealistic, realistic, or pessimistic about their life circumstances in the U.S. Although youth in each of the clusters missed and longed to return to their country of origin, some were able to look forward to a brighter future because of immigration, while others seemed to dwell on what was left behind during immigration.

The Family theme also revealed important differences across the clusters. Those with a more positive outlook maintained important family relationships either by immigrating with family and sometimes with extended family or by experiencing nurturing familial relationships after immigrating. Some youth who had experienced greater difficulty adjusting to the U.S. reported mistreatment either by parents or guardians.

Finally, stories differed based on level of meaning making and the extent to which they reflected on how immigration had changed or was changing them and their identity. Those who experienced hardships but were able to either overcome or at least not be overwhelmed by the hardships narrated stories that were self-reflective and introspective. Those youth who either had very smooth transitions or were overwhelmed by hardships constructed somewhat superficial narratives or narratives that showed preoccupation with life challenges after and sometimes before immigration.

Discussion

The results of the present study demonstrate the value of examining complete narratives as the unit of analysis for immigrant narrative identities among youth. Within each of the narrative clusters, there was coherence across the four major themes (immigrant journey, adjustment, family, and meaning making) as these themes reflected the respective narrative motifs. Specifically, the Smooth/Positive cluster included narratives in which the immigrant journey, adjustment, and family connection were smooth and relatively unaffected by immigration: the stories revealed no fresh starts or major changes in the youth's identity. Conversely, authors in the Loss/Bitterness cluster constructed life stories in which their immigrant journey, adjustment, and family relations reflected themes of loss, difficulty, hardships, and sometimes traumas, and in which they had little control over life circumstances. Their pessimism was understandable given the overwhelming adversity described in their autobiographies. The immigrant narratives within the other two clusters, Redemption and Hard Work/Hard Life, reflected more identity work, including more self-reflection and making meaning out of the challenges they faced and felt they could overcome. Hard Work/Hard Life and Loss/Bitterness narratives included adversity and difficulties, but the Hard Work/Hard Life cluster included stories in which family connection and support made the hard work of overcoming challenges meaningful, while the Loss/Bitterness cluster included themes of being alone or isolated from positive sources of social support. Finally, Redemption and Hard Work/Hard Life clusters were differentiated because the former reflected optimistic orientations toward the future, while the latter cluster reflected what may be considered more realistic future orientations.

Recent research on immigrant youth (e.g., Lucić, 2013) suggests that their immersion in a new, foreign culture promotes meta-cognitive advantages over non-immigrant youth, with this meta-cognitive perspective allowing them to perceive and challenge some dominant socialization messages about life in the immigrant country. Two identity clusters—Hard Work/Hard Life and Loss/Bitterness—clearly demonstrated this meta-cognitive advantage narrative by describing how they received messages pre-journey about the 'American dream' that the U.S. is the land of opportunity for immigrants in the Hard Work/Hard Life and Loss/Bitterness clusters, but the disappointment described in the narratives reflects the hard realities of life in the U.S., for themselves and many other immigrants. The narratives in the Redemption cluster also demonstrated the meta-cognitive perspective, but these narratives differed from the previous two because they reflected the youth finding a niche in the new society that would allow them to return to the previously optimistic life story. Interestingly, our research suggests Smooth/Positive narratives did not mention skills associated with the meta-cognitive perspective, perhaps because the immigrant experience did not disrupt their previous identity narrative. Consequently, our findings caution against the field overgeneralizing this meta-cognitive advantage and suggest that the meta-cognitive advantage may be in response to disruptions and challenges to the identity narratives constructed by immigrant youth.

McCabe and Dinh (2016) examined the dimension of agency/ineffectiveness in the life narratives of immigrant youth, finding that the narratives reflected more ineffectiveness than sense of agency, but that agency or sense of success was strongly related to connection with others. Our study had some similar findings but was able to parse out four different variations or clusters associated with agency and avoid the tendency for literature to discuss immigrant youth in monolithic terms or as one population. In our study, there appeared to be strong concordance between access to social capital and expressions of optimism or pessimism and expressions of agency in the youth identity narratives. The two clusters that were most extreme regarding access to social capital—Smooth/Positive and Loss/Bitterness—were also most extreme regarding optimism or pessimism. For Smooth/Positive, there was a sense that a positive life was waiting for them; for Loss/Bitterness, there was more of a sense of tragedy waiting for them. The other two clusters, however, constructed life stories suggesting that positive rewards were possible if they cultivated themselves, worked hard, and persevered. Narratives that reflected the Redemption and Hard Work/Hard Life motifs described having some, albeit limited, social capital, and that if they worked hard and dedicated themselves their future would be brighter.

It is interesting that the narratives associated with the latter two clusters were highest in meaning making and identity work. For youth in these clusters, their identity work may represent some personological capital that they are attempting to cultivate, and which could be used to compensate for limited access to external sources of capital (i.e., social or material). It is also noteworthy that the narrative clusters in which there are high levels of meaning making and identity work correspond to the narratives that most resemble the societal narratives for immigration, including belief in the 'American dream,' success is possible with hard work, and 'a better life' is possible with sacrifice. The Loss/Bitterness cluster seems to be most inconsistent with the societal narratives associated with immigration, and corresponds to the cluster associated with the most disillusionment and disappointment. Lastly, the Smooth/Positive cluster seems to express an implicit belief that the 'American dream' is available for the taking rather than requiring hard work or sacrifice that is part of narratives regarding immigrants' strong work ethic—having access to greater social capital may help them realize their dreams without the immigrant work ethic. Consequently, the extent to which their own identity narratives reflect the societal narratives associated with immigrants is linked to the extent to which their future outlook is idealistic (Smooth/Positive), cautiously optimistic (Redemption), realistic (Hard Work/Hard Life), or pessimistic (Loss/Bitterness) and the extent to which they are buoyed or disillusioned by the American dream.

Our results can also be interpreted in light of Erikson (1956) theory. The more extensive meaning making and identity work characteristic of Redemption and Hard Work/Hard Life narratives seems consistent with the exploration of different identity alternatives that lead to identity formation and achievement (Marcia, 1966). Conversely, Loss/Bitterness narratives seem to resemble Erikson's (1980) description of identity diffusion in which the youth establish a sense of identity and social roles in society that are affirmed and validated by others. Finally, the Smooth/Positive narratives seem most consistent with identity foreclosure as described by Marcia (1966), in which there is no significant reworking of childhood identifications or exploration of identity alternatives. It is important to note that the high school youth were still relatively early in their identity formation and may be subject to change or further development. Transitions from high school to either advanced education or work may prompt further identity development. It is also interesting that the focus of these youth's ego identity development was more strongly connected to developing a personal identity rather than social identity, relative to their immigrant status. Thus, the youth's identity narratives focused most on their own personal identity and the implications of immigration for their own individual development. The social group most often included in the narratives across clusters seemed to be their family as there was no mention in these narratives of a sense of group consciousness (Quintana, 1998) or aspects of a social identification with a community of immigrants.

The study has important limitations. Because youth were recruited for a literacy project and not specifically for a research project, there was no attempt to structure the interviews to address specific research questions. We view this also as a strength given that the results were not simple artifacts of a research agenda. Results may be specific to immigrants in New York City: the fact that there are vibrant and visible immigrant communities may have influenced youth's construction of their narratives. The youth narratives were also collected prior to increased public expression of anti-immigrant sentiments and emphasis on stigmatizing immigrant narratives, including that many are rapists, terrorists, drug dealers, and gang members. Future research will need to examine the ways in which immigrant narratives are constructed in the context of dehumanizing conceptions and policy toward immigrants.

In sum, our research examined how immigrant youth narratives depict a sense of continuity as well as transformation during life journeys as immigrants. The immigration experience and subsequent settlement made it challenging for the youth to maintain a sense of continuity and sameness of self from their home culture to the newly adopted culture and nation. Our identification of four separate identity clusters challenges the depiction of immigrant groups as a monolithic population that only varies along different quantitative dimensions. Instead, our research found that across life satisfaction and formulation of meaning to the immigrant experiences, four narrative themes were apparent. We hope our research promotes greater attention to understanding the diversity of experiences of immigrant youth as they formulate an identity and sense of self across dramatically different cultural contexts and differences in status with respect to national contexts. Our research indicates that not all immigrant youth construct identity narratives based on disruption. Additionally, not all identity narratives of immigrant youth reflect growth or transformation in their sense of self, despite moving to a new culture. Finally and importantly, our study provides some insight into the identity narratives in which youth describe the meaning that the immigration journey has had for them and will have for their life stories, despite the challenges and disruptiveness of immigration for individuals' life courses.

References

Adler, J. M., Dunlop, W. L., Fivush, R., Lilgendahl, J. P., Lodi-Smith, J., McAdams, D. P., McLean, K. C., Pasupathi, M., & Syed, M. (2017). Research methods for studying narrative identity: A primer. *Social Psychological and Personality Science, 8*(5), 519–527. https://doi.org/10.1177/1948550617698202

Berman, H., Mulcahy, G. A., Forchuk, C., Edmunds, K. A., Haldenby, A., & Lopez, R. (2009). Uprooted and displaced: A critical narrative study of homeless, aboriginal, and newcomer girls in Canada. *Issues in Mental Health Nursing, 30*(7), 418–430. https://doi.org/10.1080/01612840802624475

de Fina, A. (2003). Crossing borders: Time, space, and disorientation in narrative. *Narrative Inquiry, 13*(2), 367–391. https://doi.org/10.1075/ni.13.2.08def

Deng, Y. (2017). A narrative inquiry of Chinese immigrant students' educational experiences in the United States: Language, culture, and identity. Theses and dissertations. Available from ProQuest, pp. 1–257.

Erikson, E. H. (1956). The problem of ego identity. *Journal of the American Psychoanalytic Association, 4*, 56–121. https://doi.org/10.1177/000306515600400104

Erikson, E. H. (Erik H. (1980). *Identity and the life cycle*. https://search.library.wisc.edu/catalog/999505300202121

Erikson, E. H. (1962). *Youth: Fidelity and diversity*. Daedalus, 5–27.

Erikson, E. H. (1970). *Reflections on the dissent of contemporary youth*. Daedalus, 154–176.

Lam, M. C., & Chan, T. S. (2004). Life themes in the narratives of young Chinese immigrants who have successfully adjusted to life in Hong Kong. *Journal of Youth Studies, 7*(4), 433–449. https://doi.org/10.1080/1367626042000315211

Lieblich, A., Tuval-Mashiach, R., & Zilber, T. (1998). *Narrative research: Reading, analysis, and interpretation*. Sage.

Lucić, L. (2013). Use of evaluative devices by youth for sense-making of culturally diverse interpersonal interactions. *International Journal of Intercultural Relations, 37*(4), 434–449..

Marcia, J. E. (1966). Development and validation of ego-identity status. *Journal of Personality and Social Psychology, 3*(5), 551–558. https://doi.org/10.1037/h0023281

McAdams, D. P. (2006). The problem of narrative coherence. *Journal of constructivist psychology, 19*(2), 109–125.

McAdams, D. P., & McLean, K. C. (2013). Narrative identity. *Current Directions in Psychological Science, 22*(3), 233–238. https://doi.org/10.1177/0963721413475622

McAdams, D. P., & Olson, B. D. (2010). Personality development: Continuity and change over the life course. *Annual Review of Psychology, 61*(1), 517–542. https://doi.org/10.1146/annurev.psych.093008.100507

McCabe, A., & Dinh, K. T. (2016). Agency and communion, ineffectiveness and alienation: Themes in the life stories of Latino and southeast Asian adolescents. *Imagination, Cognition and Personality, 36*(2), 150–171.

Quintana, S. M. (1998). Children's developmental understanding of ethnicity and race. *Applied and Preventive Psychology, 7*(1), 27–45.

Riessman, C. K. (2008). *Narrative methods for the human sciences*. Sage.

Sánchez, P. (2007). Cultural authenticity and transnational Latina youth: Constructing a meta-narrative across borders. *Linguistics and Education, 18*(3–4), 258–282. https://doi.org/10.1016/j.linged.2007.07.007

Schwartz, S. J., Montgomery, M. J., & Briones, E. (2006). The role of identity in acculturation among immigrant people: Theoretical propositions, empirical questions, and applied recommendations. *Human Development, 49*(1), 1–30. https://doi.org/10.1159/000090300

Sheth, R. (2015). Exploring meanings of identity, selfhood, and migration in the lives of first-generation Gujarati Asian Indian Americans: A narrative study. *Electronic Theses and Dissertations*. https://dsc.duq.edu/etd/1184

Sowards, S. K., & Pineda, R. D. (2013). Immigrant narratives and popular culture in the United States: Border spectacle, unmotivated sympathies, and individualized responsibilities. *Western Journal of Communication, 77*(1), 72–91. https://doi.org/10.1080/10570314.2012.693648

Syed, M., & Azmitia, M. (2008). A narrative approach to ethnic identity in emerging adulthood: Bringing life to the identity status model. *Developmental Psychology, 44*(4), 1012–1027. https://doi.org/10.1037/0012-1649.44.4.1012

Syed, M., Azmitia, M., & Phinney, J. S. (2007). Stability and change in ethnic identity among Latino emerging adults in two contexts. *Identity: An International Journal of Theory and Research, 7*(2), 155–178. https://doi.org/10.1080/15283480701326117

Todorova, I. L. G., Suárez-Orozco, C., & Suárez-Orozco, M. (2008). Changing stories: The evolving narratives of immigrant children. *Cognition, Brain, Behavior: An Interdisciplinary Journal, 12*(4), 345–367.

Vall, B., & Botella, L. (2015). Making sense of immigration processes: Overcoming narrative disruption [Text]. *Narrative Inquiry, 25*(2), 203–223. https://doi.org/10.1075/ni.25.2.01val

Chapter 8
Role-Based Identity Development in Latinx Adolescents from Immigrant Families: The Development of Language Broker Role Identity

Lester Sim, Su Yeong Kim, Melissa M. Vélez, Yishan Shen, and Minyu Zhang

Studies examining adolescents from immigrant families often focus predominantly on the development of their ethnic and racial identities, with a less pronounced emphasis on other forms of identity development (Kiang et al., 2008). Specifically, familial factors such as ethnic socialization are vital for shaping individuals' sense of ethnic identities (Kiang & Fuligni, 2009; Umaña-Taylor et al., 2004). However, it is likely that the influence of family may have far-reaching effects on other aspects of adolescents' social identities. One overlooked area is the study of role identities in ethnic minority adolescents. The social roles that children from immigrant families possess can promote the development of distinct role identities. For example, adolescents with English-limited immigrant parents were likely to provide linguistic support in the form of translation and interpretation on a regular basis (McQuillan & Tse, 1995). Over time, repeated exposure to such experiences among minority adolescents in immigrant families can manifest as the development of a unique role-based identity as a "language broker." Yet within the extant literature, little is understood about how these role-based identities are constructed and how they influence adolescents' overall identity development.

L. Sim (✉)
University of Michigan, Ann Arbor, MI, USA
e-mail: lsim@umich.edu

S. Y. Kim · M. Zhang
University of Texas at Austin, Austin, TX, USA
e-mail: su.yeong.kim@utexas.edu; zhangminyu@utexas.edu

M. M. Vélez
University of Minnesota, Minneapolis, MN, USA
e-mail: velez059@umn.edu

Y. Shen
Texas State University, Austin, TX, USA
e-mail: y_s77@txstate.edu

D. J. Johnson et al. (eds.), *Re/Formation and Identity*, Advances in Immigrant Family Research, https://doi.org/10.1007/978-3-030-86426-2_8

To understand the language broker role identity, we focus our examination among Mexican-origin adolescents for two important reasons. First, cultural contexts and embedded cultural values can meaningfully affect the uptake of social roles. For Latinx communities, communal values that heighten obligation to engage in translation are likely to be emphasized (Weisskirch et al., 2011); in turn, this may affect identity processes relevant to the language broker role identity. Second, we focus on Latinx adolescents because the largest ethnic minority immigrant population in the U.S. is of Mexican descent (United States Census Bureau, 2011).

This chapter provides three interrelated sections to substantiate the theoretical and empirical underpinnings of the construction of role identities and how the possession of multiple social identities can collectively influence identity development among ethnic minority language brokers in immigrant families. First, we build upon research on the intergenerational transmission of ethnic identities and illustrate how ethnic identities may be influenced by important social roles. In Study 1, we extend a social interactionist framework to substantiate the validity of this language broker role identity in Latinx adolescents with immigrant parents based on existing literatures and preliminary findings from our ongoing research. In so doing, we build a case for the language broker role identity and how parents can influence adolescents' identity development. Second, in Study 2, we use preliminary empirical evidence to investigate how various factors such as adolescents' families (e.g., parenting practices) via intrapersonal variables such as sense of life meaning can function as antecedents for the development of adolescents' language broker role identities. Finally, we discuss potential avenues for future research that can broaden our understanding of language broker role identities. We examine cultural values and their role in adolescent identity development. We also discuss the implications for the possession of multiple identities and how these identities inform identity integration (Benet-Martínez et al., 2002) for Latinx adolescents from immigrant families (see Fig. 8.1 for conceptual model).

The Case for Language Broker Role Identity

Identity Development and Identity Theory

Deconstructing identity development is complex because "identity" is multifaceted. Multiple pathways exist within the literature in an attempt to explain adolescent identity development processes. Some scholars espouse the role of psychosocial factors (i.e., life meaning) and relate it to identity development through promoting identity-seeking processes (Negru-Subtirica et al., 2016). Others suggest that identity is influenced by contextual and external environmental inputs. For example, social identity theory (Tajfel, 1974) may be suitable to explain the aspect of the adolescents' ethnic identity based on affiliation to their ethnic group. However, a unitary perspective of identity development (i.e., exploring only group membership

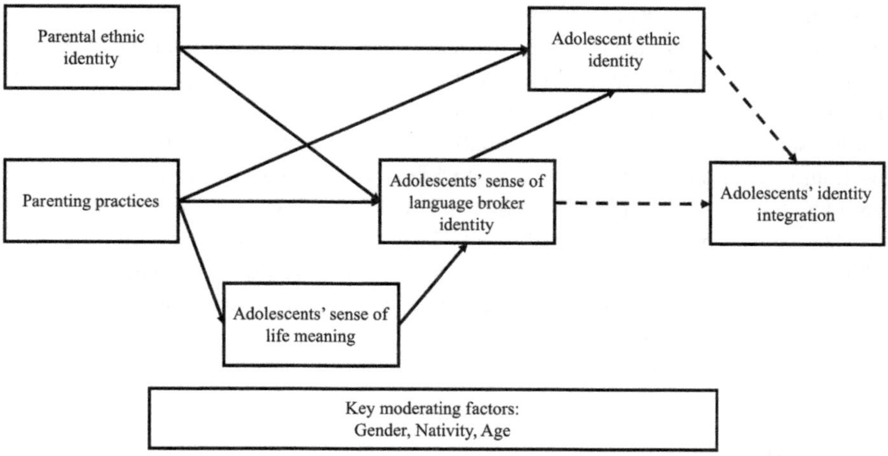

Fig. 8.1 Conceptual model of factors underlying formation of role-based identities and associated theoretical implications. Solid lines represent empirically tested pathways and dashed arrows represent hypothesized pathways

or examining only psychosocial factors) is insufficient. For example, identity theory (Stryker, 1980) not only considers the self-views that individuals possess toward identity but also takes into consideration how external actors (e.g., family members) facilitate the individuals' own sense of identity. Thus, we extend from work in identity theory and suggest a perspective that emphasizes both contextual (e.g., group membership) and intrapersonal (e.g., psychosocial) factors as important and necessary to provide a comprehensive elucidation to the identity development process.

Briefly, identity theory (Stryker, 1980) proposed that identity is the amalgam of the multiple roles that individuals possess in their daily lives. Individuals interact symbolically with other members in society (e.g., family members, peers) and collectively define shared identity meanings of their occupied statuses. Associated with identity meanings are roles, which are culturally defined expectations of a person's multiple role identities; identity then is the internalization of roles into individuals' self-concepts (Stryker, 1980; Stryker & Burke, 2000). All individuals seek to verify self-perceptions and culturally defined identity meanings by behaving in ways that maintain congruence and consistency (Burke, 1991). When self-perceptions are in sync, identity theory suggested that verification has taken place and this contributes positively to the individuals' identity development (Cast & Burke, 2002).

While studies have examined both individual and contextual factors to substantiate how identity theory is relevant to different forms of social identity (e.g., student identity, spousal identity; Burke & Reitzes, 1991; Stets & Burke, 2014), an extension of identity theory to language broker role identity is absent in the literature. We suggest that the development of the language broker identity parallels the role identity development processes. Chao (2006) revealed that approximately 60–80% of all Mexican-origin youths engage in some form of translation for their immigrant parents. Through their cumulative translation interactions, ethnic minority adolescents from immigrant families assume the status of a language broker. Identity meanings are attached to the language broker role and confer expectations that necessitate the demand to translate for parents. The combined influence of language broker identity meanings and familial demands for translation, in turn, lead adolescents to work toward successful verification. Especially for adolescents who actively engage in the language broker role, they are more likely to facilitate the integration of the language broker role meanings into their self-concepts. Previous work demonstrates that language brokering is related to adolescents' social identities, including their ethnic identities (Weisskirch, 2005). Other scholars also postulate that adolescent translators possess social identities that capture their role as language brokers (Del Torto, 2008; Weisskirch, 2017).

Ethnic Identity

Similar to the broader construct of identity, ethnic identity has been revealed to be multifaceted. Umaña-Taylor et al. (2004) conceptualized ethnic identity through three important components: affirmation, exploration, and resolution. Affirmation examines the positive or negative feelings toward individuals' ethnicity; exploration indicates the extent to which individuals explore their ethnicity; while resolution is an achieved understanding of what ethnicity means to individuals (Umaña-Taylor et al., 2004). Additionally, Sellers et al. (1998) extended the work on identity theory and suggested important determinants (salience, centrality, ideology, and regard) as integral in shaping the individuals' sense of racial identity. Specifically, salience assesses how individuals construe events and behaviors related to their racial identity (Sellers et al., 1997). Centrality underscores the extent to which individuals define themselves with regard to race/ethnicity and whether race/ethnicity is a part of their self-concepts (Sellers et al., 1998). Ideology refers to individuals' beliefs, opinions, and attitudes toward the way members of the particular race should act, while regard refers to the affective and evaluative judgment of the individuals' own race. In our research, we adapt Sellers' model within an ethnic identity framework because (1) ethnic identity captures individuals' social and cultural experiences and (2) language brokering is a culturally relevant experience for Mexican-origin adolescents that may affect adolescents' ethnic identities (Weisskirch et al., 2011).

Intergenerational Transmission of Parent Ethnic Identity to Adolescent Ethnic Identity

Parents play an integral part in adolescents' sense of ethnic identity by teaching ethnic customs and traditions (Umaña-Taylor & Guimond, 2010). As parental ethnic identities reflect parents' personal beliefs about the culture, adolescents' attitudes toward the cultural practices that they engage in (e.g., language brokering) may be transformed based on the culturally relevant messages that parents impress upon their children. For example, adolescents may perceive translating as a way of helping parents, which can lead to closer identification with being a translator. These feelings about language brokering may increase individuals' closeness with the home culture (Weisskirch, 2005) and relate to their own sense of ethnic identities. In our work, we explore whether parents' ethnic identities can facilitate the transmission of their ethnic beliefs to children by enforcing cultural practices, such as language brokering, and examine how various elements of parental ethnic identities affect this relation differently.

A potential complication to the intergenerational transmission of ethnic identities may manifest through the distinct components of ethnic affirmation, exploration, and resolution. In fact, parental ethnic affirmation, exploration, and resolution have, through similar or contrasting pathways, demonstrated varying influence on adolescents' own dimensions of ethnic identities. From a practical standpoint, investigating individual components of ethnic identity may provide distinct value as each component implicates identity development differently (Umaña-Taylor et al., 2004; Umaña-Taylor et al., 2009). For example, only exploration and resolution, but not affirmation, were significantly related to familial ethnic socialization, which was in turn associated positively with adolescents' cultural orientations and ethnic identities (Umaña-Taylor et al., 2009; Umaña-Taylor & Guimond, 2010). As familial ethnic socialization is a concept that closely parallels the current study's notion of intergenerational transmission of ethnic identities, our research focuses only on these two components of ethnic identity: exploration and resolution.

We postulate that parents who have explored and are resolved about their ethnic identities also impress stronger cultural beliefs and expectations onto their adolescents. This indirectly creates an environment in which adolescents discern a greater need to translate for parents, which can consequentially increase the adolescents' own identification with being a language broker. Engaging in language brokering may therefore promote the adolescents' own ethnic identities as they begin to engage with their ethnic language more and perceive being a translator as important to their own sense of ethnic identities. Specifically, we hypothesize that parents' sense of exploration and resolution yield a positive relation with adolescents' reports of identification as a translator; in turn, the latter is positively associated with the adolescents' own sense of ethnic identities.

Language Broker Role Identity Centrality

To capture the notion of identifying as a translator, we utilize the multidimensional model of racial identity (Sellers et al., 1998). Extending from the model, language broker salience measures individuals' identification with the role of a translator. Given the high prevalence of language brokering (Chao, 2006), adolescents' language broker salience is expected to be elevated as well. Corresponding to the high degree of salience, adolescents are more likely to identify themselves or internalize this role identity of a language broker into their self-concepts (language broker centrality). Language broker ideology in the current context corresponds to adolescents' beliefs regarding how language brokers, in general, should behave. Conversely, language broker regard refers to individuals' subjective feelings toward how they view themselves as a translator and how others view their role identity as a translator for parents (positive versus negative). In sum, the four components (salience, centrality, ideology, and regard) provide a cohesive theoretical framework for language broker role identities.

In Study 1, we suggest language broker centrality to be particularly relevant to the intergenerational transmission of ethnic identities for two reasons. First, centrality reflects the individuals' normative perceptions of who they are across multiple social identities (Sellers et al., 1998). Examining centrality provides a bridge that links the two variants of social identities (ethnic and language broker) together. Second, identity centrality inherently conveys the notion that some social identities are hierarchically superordinate (vs. subordinate) (Sellers et al., 1998). For example, language brokers may perceive their ethnic identities as more central over their role identities, but this may manifest as a result of individuals' recognition that their language broker role identities are a subcomponent of their ethnic identities (i.e., being a language broker for my parents reflects who I am ethnically). Especially for individuals whose parents possess high levels of ethnic identity centrality, we suggest that these individuals also acknowledge their language broker role identities as complementary to their own ethnic identities. Thus, we hypothesize that parents' ethnic identity centrality is positively related to adolescents' language broker centrality, which is in turn positively related to adolescents' ethnic identity centrality.

Gender and Nativity as Moderators of Ethnic and Language Broker Identity

Adolescents' feelings toward language brokering may be influenced by important demographic characteristics such as gender. Villanueva and Buriel (2010) found that adolescent girls translated the most for their mothers. Due to mother-daughter discrepancies in expectations of autonomy during adolescence (Bámaca-Colbert et al., 2012), tensions rose between mothers and adolescent daughters in Mexican families. When coupled with the intensity of language brokering, it is probable that female adolescents' positive emotions toward their mothers would be diminished, which would negatively impact the influence of the mothers' ethnic identities on female adolescents' feelings of identification as a translator. However, an inverse pattern is observed for young male adolescents. Young male adolescents in Mexican-origin families enjoyed lower amounts of parent-child conflict as they adhered to the roles and expectations set by parents (Updegraff et al., 2014). This decreased conflict resulted in stronger relationships between younger male adolescents and their parents, potentially leading to higher levels of identification as a language broker in male adolescent language brokers.

Adolescent nativity may also influence this process of intergenerational transmission of ethnic identities by affecting adolescents' feelings about language brokering. Phinney et al. (2000) found higher levels of parent-child conflict in families with U.S. born adolescents than in families with foreign-born adolescents. When compared to their Mexico-born counterparts, U.S. born individuals revealed a lower sense of familial obligation (Updegraff et al., 2014), which can negatively affect adolescents feelings toward language brokering. This finding suggests that the potential association between the adolescent's own perceptions of being a language broker (identification as a translator) and the intergenerational transmission of ethnic identities may depend on the individual's nativity.

Testing Language Broker Identity as a Mechanism for the Intergenerational Transmission of Ethnic Identities

We conducted analyses to examine the intergenerational transmission of ethnic identities through the potential mediator of adolescents' language broker centrality. Specifically, we sought to determine the relation between parents' ethnic identities to adolescents' identification as a language broker, and in turn, adolescents' identification as a language broker to their own sense of ethnic identities, while taking into account adolescent gender and nativity (refer to Fig. 8.1).

Participants

Using data from 557 Mexican-origin families recruited in a large metropolitan city in central Texas, we assessed a sample of adolescents between 11 and 15 years of age ($M = 12.96$ years, $SD = 0.94$, 54% girls, 75.4% America born). Both mothers' ($N = 455$) and fathers' ($N = 254$) reports were incorporated to fully understand the role of parents in transmitting parental ethnic identities to adolescents' ethnic identities. Median family income was $10,000–$20,000, while mean education level for parents was "finished middle school." Only one adolescent in the study reported that parents were separated at the time of study. Results were distinguished based on adolescent gender and nativity.

Measures

Ethnic identity exploration (3 items; α's = 0.81–0.84) and resolution (3 items; α's = 0.85–0.91) were assessed by administering the Ethnic Identity Scale (Umaña-Taylor et al., 2004) to both parents and adolescents. A sample item for exploration was "I have spent time trying to find out more about being Mexican, such as its history, traditions, and customs." A sample resolution item was "I have a clear sense of what being Mexican means to me."

In addition, parents filled out an ethnic identity centrality measure while adolescents reported on their ethnic identity and language broker centrality through three items which were "A part of who I am is being a Mexican (translator for my mother/father)," "Being a Mexican (translator for my mother/father) is important to who I am," and "A key part of how I see myself is being a Mexican (translator for my mother/father)" (Kim et al., 2017). Cronbach alphas for all ethnic identity and language broker centrality scales, across both parents and adolescents, were found to be high (α's = 0.81–0.91).

Analysis Plan

We attempted to ascertain if adolescents' language broker centrality facilitated the intergenerational transmission of ethnic identities from parents to adolescents. To reduce model complexity, our models were tested separately for mother-adolescent and father-adolescent dyads to identify how the gender of parents may affect adolescents' ethnic identities. Accordingly, we included either mothers' or fathers' self-reported ethnic identity exploration, resolution, and centrality and adolescents' self-reported language broker centrality, ethnic identity exploration, resolution, and centrality and tested the model based on adolescent nativity and gender.

Indirect effects were tested through the mediating pathways from the exogenous variables (mother/father ethnic identity centrality, exploration, and resolution) and the mediator (adolescents' own report of language broker centrality) to the outcomes (adolescent's ethnic identity centrality, exploration, and resolution). We included mothers' education and income as covariates for the mother-adolescent model, while fathers' education and income were covariates for the father-adolescent model.

Maternal and Paternal Discrepancies in the Intergenerational Transmission of Ethnic Identity

Maternal Effects for the Intergenerational Transmission of Ethnic Identities

The results of our analyses revealed that adolescents' feelings toward language brokering underlie the intergenerational transmission of ethnic identities, albeit through contrasting pathways (see Tables 8.1 and 8.2). Specifically, indirect paths for mothers' resolution to adolescents' language broker centrality to adolescents' ethnic identity centrality, exploration, and resolution were all found to be significant, but only for U.S. born female adolescents (see Table 8.1). Mothers' resolution was

Table 8.1 Indirect effects from mother's ethnic identity, to adolescent language broker centrality, to adolescent's ethnic identity by adolescent gender and nativity

Indirect Paths	U.S. Male	MEX Male	U.S. Female	MEX Female
Mother EthCent→ Adolescent LBCent→ Adolescent EthCent				
Mother EthCent → Adolescent LBCent → Adolescent Exploration				
Mother EthCent → Adolescent LBCent → Adolescent Resolution				
Mother Exploration→ Adolescent LBCent → Adolescent EthCent				
Mother Exploration→ Adolescent LBCent → Adolescent Exploration				
Mother Exploration→ Adolescent LBCent → Adolescent Resolution				
Mother Resolution→ Adolescent LBCent → Adolescent EthCent			0.064 ($p = 0.029$)	
Mother Resolution→ Adolescent LBCent → Adolescent Exploration			0.098 ($p = 0.021$)	
Mother Resolution→ Adolescent LBCent → Adolescent Resolution			0.086 ($p = 0.022$)	

Note. All indirect effects are standardized coefficients. Only significant indirect effects are shown in the table. *EthCent* Ethnic Identity Centrality, *Adolescent LBCent* Language Broker Centrality for Mother, *U.S.* U.S born, *MEX* Mexico-born. Covariates included mother's educational level and mother-reported family income

related to higher levels of adolescents' language broker centrality ($\beta = 0.26$, $p < 0.05$), while adolescents' language broker centrality was associated with adolescents' ethnic identity centrality ($\beta = 0.25$, $p < 0.001$), exploration ($\beta = 0.49$, $p < 0.001$), and resolution ($\beta = 0.37$, $p < 0.001$).

Paternal Effects for the Intergenerational Transmission of Ethnic Identities

Indirect paths for fathers' resolution to adolescents' language broker centrality to adolescents' ethnic identity centrality and exploration, but not resolution, were also found to be significant but only in Mexico-born male adolescents (see Table 8.2). Fathers' resolution was positively related to adolescent language broker centrality ($\beta = 0.44$, $p < 0.01$), which was associated with adolescents' ethnic identity centrality ($\beta = 0.36$, $p < 0.01$) and exploration ($\beta = 0.42$, $p < 0.01$). Similarly, the results also revealed additional significant indirect paths from fathers' ethnic identity centrality to adolescents' language broker centrality to adolescents' ethnic identity centrality and exploration, but not resolution. Fathers' ethnic centrality was positively related to adolescents' language broker centrality ($\beta = 0.58$, $p < 0.05$), which was associated with adolescents' ethnic identity centrality ($\beta = 0.51$, $p = 0.001$) and exploration ($\beta = 0.42$, $p < 0.01$) for only Mexico-born male adolescents.

Table 8.2 Indirect effects from father's ethnic identity, to adolescent language broker centrality, to adolescent's ethnic identity by adolescent gender and nativity

Indirect paths	U.S. Male	MEX Male	U.S. Female	MEX Female
Father EthCent → Adolescent LBCent → Adolescent EthCent		0.184 ($p = .042$)		
Father EthCent → Adolescent LBCent → Adolescent Exploration		0.212 ($p = 0.023$)		
Father EthCent → Adolescent LBCent → Adolescent Resolution				
Father Exploration→ Adolescent LBCent → Adolescent EthCent				
Father Exploration→ Adolescent LBCent → Adolescent Exploration				
Father Exploration→ Adolescent LBCent → Adolescent Resolution				
Father Resolution→ Adolescent LBCent → Adolescent EthCent		0.157 ($p = 0.044$)		
Father Resolution→ Adolescent LBCent → Adolescent Exploration		0.181 ($p = 0.024$)		
Father Resolution→ Adolescent LBCent → Adolescent Resolution				

Note. All indirect effects are standardized coefficients. Only significant indifferent effects are shown in the table. *EthCent* Ethnic Identity Centrality, *Adolescent LBCent* Language Broker Centrality for Father, *U.S.* U.S born, *MEX* Mexico-born. Covariates included father's educational level and father-reported family income

We investigated whether different dimensions of parental ethnic identities were related to adolescents' perception of their role as a language broker and, in turn, adolescents' own dimensions of ethnic identities. The results suggested that language brokering enhanced adolescents' sense of ethnic identities through identifying as a language broker. Although we found disparate effects from mothers' and fathers' ethnic identities to daughters' and sons' ethnic identities, respectively, our findings may be explained by the differential cultural roles that Latinx parents play in the language brokers' experiences. Male adolescents are expected to take over the fathers' role when the fathers are not present, and this expectation may create closer relationships between Mexican-origin fathers and their sons (Updegraff et al., 2014). Despite our earlier discussion on mother-daughter conflict, we found that mothers' ethnic identities were positively associated with daughters' ethnic identities through language broker centrality. Perhaps, language brokering is perceived as a normative cultural script for Mexican-origin adolescents (Kam & Lazarevic, 2014) and therefore does not jeopardize Latina adolescent identity processes. In sum, our research emphasized that the intergenerational transmission of ethnic identities is a family-driven process with parents' ethnic identity having a unique influence depending on whether analyses are focused on mothers or fathers.

Based on the results, we found evidence for the intergenerational transmission of ethnic identities through adolescents' language broker centrality. To provide a clearer appreciation of the language broker role identity however, we underscore the importance of investigating factors that may influence language broker centrality. For example, familial processes such as parenting practices, including parental warmth and hostility, can contribute/attenuate adolescents' subjective feelings toward the translation experience (Tilghman-Osborne et al., 2015; Weisskirch, 2007). Additionally, innate psychosocial factors may vary adolescents' language broker centrality. Arguably, adolescents who possess a greater desire to discover their own identities are more likely to internalize their language broker role identities into their self-concepts through various identity verification means. In the next section, we provide preliminary empirical evidence through integrating factors such as parenting and a possible identity verification mechanism, adolescents' life meaning, to further substantiate the language broker role identity in Latinx adolescents from immigrant families.

Factors Influencing Language Broker Role Identity

Parenting, Life Meaning, and Their Effects on Language Broker Role-Based Identity

Past research indicates that identity verification relies on important contextual and psychological processes that can affect the uptake of individuals' role identities (Cast & Burke, 2002; Stryker & Burke, 2000). For example, it is likely that the combined influence of individuals' and close others' perception of identity meanings impacts how identities are construed and developed. One important contextual factor that is often implicated alongside Latinx adolescents' identity development is the familial environment. Parents help in the maintenance and continued transmission of culture through engaging in parenting practices informed by their own culture (Grau et al., 2015). Especially for Latinx communities, specific cultural values such as familismo (i.e., strong identification with one's family including extended family ties) encourage warm parenting practices in Latinx households (Gonzales et al., 2011). An emphasis on the need to develop supportive interdependent relationships with family members and the promotion of prosocial behaviors toward in-group members (Boer & Fischer, 2013) can, in turn, heighten adolescents' responsibility for translation, affecting their language broker identities. Indeed, a number of studies surrounding ethnic identities have found that positive parenting practices such as parental warmth were associated with identity processes, including higher ethnic belonging and ethnic identity exploration (Davey et al., 2003). Furthermore, warm environments have been found to be more conducive for the transmission of parent-child values (Knafo & Schwartz, 2001), whereas in negative environments, pressure from parents can culminate in identity development that contradicts parents' wishes (Cheng & Kuo, 2000).

To expand the research on role identities further, we posit that specific psychosocial factors may simultaneously vary adolescents' role identity development through parental warmth and hostility. Specifically, we borrow from past studies that have demonstrated life meaning to be positively associated with identity exploration and development (Kiang & Fuligni, 2010) and suggest that life meaning discerned from the parent-child relationship potentially serves as a resource that fosters role identity development. Baumeister (1991) suggested that life meaning is attained when individuals' need for value, purpose, efficacy, and self-worth are met. For adolescents who report high levels of parental warmth, supportive acts from parents and a warm environment suggest an abundance of resources (e.g., psychological or emotional), which can positively contribute to identity development. On the other hand, higher levels of parental hostility may underline a less beneficial environment and fewer resources. Accordingly, adolescents who report higher levels of parental hostility may be less successful in deriving a sense of purpose and thus negatively relate to their levels of role identity development.

Life meaning is critical to identity development as it can encourage relevant identity exploration processes. Healthy identity development in adolescence is related to the search and acquisition of meaning in one's life. For example, Negru-Subtirica and colleagues (2016) revealed the interdependent nature of identity and meaning, highlighting the presence of meaning in life as an impetus for youth to search for and discover their identities. Therefore, the quest for meaning promotes identity exploration and, by extension, affects adolescents' identity development. When adolescents are able to experience caring relationships, and achieve a sense of belonging, this manifests in positive identities and meaning (Noble-Carr et al., 2014). Moreover, past research showed that interpersonal relationships, especially those associated with family and close others, are often cited as a source for meaning in life (O'Donnell et al., 2014). Latinx cultural values' emphasis on solidarity with family members can further enhance Mexican-origin adolescents' life meaning through advocating for stronger familial relationships. As life meaning may be channeled to identity verification of adolescents' language broker role identities, cultural values can therefore indirectly enhance Latinx adolescents' uptake of language broker role identities.

Parenting Practices and Language Broker Identity Development

In Study 2, we were interested in unraveling potential antecedents of the language broker identity by examining the role of parenting practices (i.e., parental warmth and parental hostility) and adolescents' perceived life meaning. Past findings have reported that mothers and fathers relate to their children's identity development in differing ways (Grotevant & Cooper, 1985; Koh et al., 2009). For example, most research examining warmth and nurturing acts within the parent-child dyad reveal mothers, more so than fathers, to possess profound influence on adolescents' psychological development (Davidov & Grusec, 2006; Shearer et al., 2005; Sun et al.,

2017). Considering the exploratory nature of our work, we therefore focus our attention on mother-adolescent dyads in the current study to understand the effects of mothers' parenting practices on adolescents' role-based identity development. We specifically examined language broker centrality because it allows for assessment of adolescents' internalization of the language broker role identity into their self-concepts. As with Study 1, we examined if these findings differed when taking into consideration the adolescents' gender.

Parenting Practices Influence Language Broker Role Identity Development

Participants

In Study 2, participants included adolescents from all families in Study 1 and an additional group of adolescents that were subsequently recruited for the study, bringing the total count of participants to 604 Mexican-origin adolescents ($M = 12.41$ years, $SD = 0.97$, 54.3% female). Self-report data was collected to investigate the hypothesized effects of perceived maternal warmth and perceived maternal hostility to adolescents' life meaning to adolescents' language broker centrality. Mothers' mean education level was "finished middle school." Covariates included adolescent gender, age, and mothers' education level and reported household income.

Measures

Perceived maternal warmth and perceived maternal hostility were assessed based on seven and six items, respectively, from the Iowa Youth and Families Project (Ge et al., 1996). Sample items for maternal warmth included whether mothers "acted supportive and understanding" and whether mothers "listened carefully." Sample items for maternal hostility examined hostile behaviors such as "get into a fight or argument" and "shout or yell." Cronbach alpha was high for maternal warmth ($\alpha = 0.91$) and maternal hostility ($\alpha = 0.84$).

Life meaning ($\alpha = 0.87$) was assessed through the Meaning in Life Questionnaire (Steger et al., 2006). A sample item is "My life has a clear sense of purpose." Language broker centrality ($\alpha = 0.86$) included the same three items as previously stated in Study 1.

Analysis Plan

In Study 2, we tested the indirect effects of adolescents' life meaning on the relationship from adolescents' perceived maternal parenting practices and adolescents' language broker centrality. Accordingly, exogenous variables included adolescents' perceived maternal warmth and perceived maternal hostility, the mediator (adolescents' self-reported life meaning), and the outcome was adolescents' language broker centrality. Separate models for maternal warmth and maternal hostility were conducted before assessing both predictors together to understand how the two contrasting parenting practices individually and simultaneously influenced adolescents' role identity development, while including adolescent gender, age, mothers' education and household income as covariates.

Maternal Parenting Practices and Language Broker Role Identity Development

Maternal Warmth, Maternal Hostility, and Language Broker Centrality

When we included only the predictor, maternal warmth, the path from maternal warmth to life meaning ($\beta = 0.40, p < 0.001$), the path from life meaning to language broker centrality ($\beta = 0.19, p < 0.001$), and the positive indirect effect from maternal warmth to adolescents' language broker centrality (see Table 8.3) were all found to be statistically significant.

When we substituted the predictor with maternal hostility, we found a similar pattern, albeit in the opposite direction. The path from maternal hostility to life meaning ($\beta = -0.19, p < 0.001$), life meaning to language broker centrality ($\beta = 0.30$, $p < 0.001$), and the negative indirect effect from maternal hostility adolescents' language broker centrality (see Table 8.3) were all found to be statistically significant.

When we included the predictors, maternal warmth and maternal hostility together, results revealed that only the path from maternal warmth to life meaning was significant ($\beta = 0.38$, $p < 0.001$), while the path from maternal hostility to life meaning was found to be marginally significant ($\beta = -0.08$, $p = 0.05$). Though the path from life meaning to language broker centrality was also statistically significant ($\beta = 0.19$, $p < 0.001$), we found only significant indirect effects of maternal warmth, and not maternal hostility, to language broker centrality (see Table 8.3).

Table 8.3 Test of indirect effects for maternal warmth and maternal hostility for language broker centrality

Path	Indirect	Direct	Total
Perceived maternal warmth			
Warmth → life meaning → centrality	0.08*	0.32*	0.39
Perceived maternal hostility			
Hostility → life meaning → centrality	−0.11*	−0.06*	−0.17
Perceived maternal warmth and hostility			
Warmth → life meaning → centrality	0.07*	0.31*	0.38
Hostility → life meaning → centrality	−0.02	−0.04	−0.06

Note: Standardized coefficients and significant indirect paths are shown above. *$p < 0.001$.

The results suggested that life meaning was a mediator that bridged the association between maternal warmth and language broker centrality.

Additionally, we conducted gender invariance tests on (a) maternal warmth only, (b) maternal hostility only, and (c) maternal warmth and maternal hostility together model. Results show that none of the Satorra-Bentler chi-square difference tests between unconstrained models and constrained models (constrained all paths to be equal between male and female adolescent participants) were significant: (a) $X^2(3) = 0.468$, $p = 0.926$; (b) $X^2(3) = 1.090$, $p = 0.779$; (c) $X^2(5) = 4.210$, $p = 0.520$. These results indicate that effects of maternal warmth and/or hostility on language brokering centrality through life meaning are not different between male and female adolescents.

Influence of Parenting Practices on Language Broker Role-based Identities

Our findings suggested that maternal parenting practices were important factors that related to language broker centrality through life meaning. Building on identity theory, we posit the derived sense of purpose from mother-child interactions to function as a resource that is channeled to the identity verification process (Burke, 1991). Adolescent language brokers infer the need to translate for their mothers based on both their internalized language broker role identity meanings and also their ability to derive important resources from the positive/negative parent-child relationships. For example, adolescents who are successful in obtaining a sense of purpose because of their close relationships with their mothers are likely to feel that their language broker roles are important and purposeful for the further development of the mother-child relationship. This is one of the first studies to examine the construct of language broker role identity and the first to underscore the contextual and psychological antecedents that lead to language broker role identity formulation. Considering the exploratory nature of our study, more work needs to be undertaken to provide a comprehensive understanding of language broker role identities.

Notably, our results showed that parental practices, maternal warmth and hostility, had significant indirect effects only in models with one predictor (either maternal warmth or maternal hostility only), yet the effects of maternal warmth remained significant in the combined model. Our results suggested that maternal warmth confers effects on adolescents' role identity formation and development above and beyond that of maternal hostility. It may be important to investigate other parenting-related factors that may contribute to individuals' role identity development. Additionally, we assessed only mothers' influence in this preliminary analysis, but it is vital (as we have suggested at different points in this chapter) to analyze both mothers' and fathers' influence on their children's identity formation. Because parents can play different roles or exact influence in contrasting manners, incorporating fathers' warmth and hostility in future studies can therefore shed light on how parents' parenting practices contribute to identity formulation differentially.

All in, there is evidence that the language broker role identity is shaped by both contextual parental inputs and individual psychological resources. In the final sections of this chapter, we continue to discuss the importance of language broker role identities, and the role of culture, with special attention to how specific cultural values may influence the antecedents of adolescents' role identities. Finally, we discuss the possible integration of multiple identities in minority adolescents from immigrant families through identity integration work in the extant literature.

Future Directions

Mexican-origin adolescents from immigrant households possess unique sociocultural experiences that enable the development of the language broker role identity. Across two studies, we provided phenomenological evidence for the language broker role identity and identified potential antecedents for the role identity in Mexican-origin adolescents. In the final portion of this chapter, we discuss possible future directions by offering two related domains of research that can deepen our knowledge regarding the language broker identity: (1) cultural values, and (2) adolescents' identity integration.

The Effect of Cultural Values on Language Broker Role Identity

As language brokering occurs among ethnic minority households, ethnic traditions, teachings, and customs, encapsulated by various cultural values, can also impart influence in affecting Latinx adolescents' uptake of the language broker role identity. As we have examined briefly throughout the chapter, cultural values such as familismo and respeto (i.e., filial obligation to yield to older member of the family) can shape Latinx parenting, promoting warmer parenting practices, and directly affect language broker identity centrality. Notwithstanding the effects of cultural values on parenting however, the emphasis of various Latinx cultural values may also directly enhance adolescents' commitment to the family and foster greater desire to provide assistance with familial chores and duties (Arcia et al., 2000; Fuligni et al., 1999). A possible manifestation may be evident in increasing translation frequency that ultimately affects language broker identities. Past research has demonstrated that children with immigrant parents emphasized their cultural identities more than children of non-immigrant parents (Urdan & Munoz, 2012). Thus, greater endorsement of cultural values in home cultures may be observed among Mexican-origin adolescents from immigrant households. Borrowing from both identity theory and the multidimensional model of racial identity, cultural values (e.g., familismo and respeto) possibly serve as enhancers for language broker centrality, salience, ideology, and regard. Over time, this can manifest in greater self-identification as language brokers and crystallize language broker role identities.

Furthermore, we argue that it is important to consider how language broker role identities may vary across cultures as a function of the priorities and the degree of endorsement different cultures place on their cultural values (Cheung et al., 2016). Investigating communities with seemingly similar values, for example, Latinx culture and their value of respeto and Chinese culture and their value of xiao (filial piety), can reveal diverging motivations and meanings associated with language brokering. On the one hand, xiao may encourage Chinese adolescent language brokers to engage in translation because filial piety is considered an important virtue by Confucius teachings (Nichols, 2013). On the other hand, familismo and respeto can together convey repeated importance of the prominence of family for Latinx adolescents. Cultures and their values therefore foster different levels of need for adolescents to engage in translation, indirectly affecting adolescents' language broker role identities. Additionally, while xiao is often understood with direct reference to individuals' own parents in Chinese cultures (Yeh & Bedford, 2003), respeto and familismo in Latinx cultures not only include deference to parents but often include extended members in the household (Schwartz et al., 2010). The degree of familial obligation experienced by Latinx adolescents thus differs, and it is likely that Latinx cultural values enhance the expectation for adolescents to language broker. These different underlying motivations may illuminate distinct identity verification pathways that propel language brokers to translate. Perhaps the language broker role is perceived as instrumental in particular cultural groups where the inherent cultural values stress a need for adolescents to tangibly help out at home. In the current chapter, we highlight only the language broker role identity because of its relevance in Latinx ethnic immigrant households. Future research should pay close attention to both the emphasis of cultural values and the motivational differences regarding specific role identities to bring greater clarity to the work on language broker role identities.

Navigating Adolescents' Multiple Identities

Beyond the role of cultural values, it is equally important to understand how Latinx adolescents manage their multiple social identities. Already in the literature, an important challenge lies in understanding how different social identities complement and interact with one another. For example, the literature on identity integration provides vital information about how individuals manage their varied identities through (1) conflict: the degree of endorsement between individuals' multiple identities, and (2) distance: their ability to integrate their multiple selves and how they come to make sense of them as a whole (Benet-Martínez & Haritatos, 2005; Cheng & Lee, 2009).

Past research demonstrates that perceiving identities as compatible and inherently similar provides individuals with a coherent understanding of who they are, while low identity integration is costly in a number of important domains such as lowered social support, lowered performance in professional settings, and less innovation (Cheng et al., 2008). Enhancing Latinx adolescents' competencies in managing their multiple social identities together with language broker role identities (e.g., through reductions in perceived conflict and distance) can help language brokers manage their complex identities. For example, Cheng and Lee (2013) showed that thinking about positive biculturalism experiences led to short-term increases in identity integration. Creating positive language brokering experiences can therefore situate Latinx adolescent language brokers in a more advantageous position to capitalize on positive translation experiences. For instance, awarding praises that relate to both language broker and familial identities (e.g., engaging in language brokering can serve as a form of filial service, language broker is praised for being a good son/daughter for translating for parents) or cultural identities (e.g., language brokering is a normative cultural activity) may frame adolescents' language brokering interactions as favorable and positive, improving overall identity integration levels. Additionally, positive relationships with others who possess similar identities have been associated with higher levels of identity integration (Mok & Morris, 2012). Thus, securing a peer group who also translates regularly may serve as an additional means to foster identity integration for adolescent language brokers.

Helping adolescents make sense of their identities as language brokers is therefore an important goal for healthy adolescent identity development. Coupled with our earlier discussion on the role of life meaning and language broker role identity development and research elsewhere demonstrating that individuals' meaning making abilities are critical for helping the self manage multiple dimensions of identity (e.g., social identities) (Abes et al., 2007), we posit that helping Latinx adolescents understand that the different aspects of their lives are not individual fragmented pieces but are instead parts of a summed whole can encourage positive appraisals of the language broker role identity. This may facilitate the integration of language broker role identities into their overall self-concepts. In fact, this constant process of negotiating and renegotiating what it means to be a language broker may further allow adolescents to appreciate the function that they play within their households. Indeed, previous research has typically focused only on social identities that are related to one's ascribed space (e.g., ethnic identity); less attention has been devoted to identities that originate from adolescents', especially immigrant adolescents', lived experiences. Acknowledging adolescents' role identities, a product of their lived experiences, can help with consolidating, reforming, and perpetuating Latinx adolescents' identities. As few researchers have incorporated the study of identity integration and language broker role identities, future studies may examine the possibility of an enhanced life meaning in adolescents with high levels of identity integration.

In sum, helping adolescent language brokers become comfortable with their role identities is a crucial step toward successful identity integration. Future work examining new identities, especially role-based identities in immigrant adolescents, should emphasize positive aspects of how their role identities can complement the other social roles that these adolescents fulfill. When Mexican-origin adolescents are able to integrate their language broker role identities with their other social identities, they are more likely to perceive a wholesome understanding of their selves.

Conclusion

Our research suggests that it is important to investigate role-based identity development. Specifically, studying role identities confer information about a dimension of adolescent identity that has been less examined in the literature. Our work sheds light on two important findings related to the language broker role identity. Language broker role identities are important to (1) bridging parents' social identities (e.g., ethnic identity) to adolescents' own sense of identity development, and (2) the interrelatedness between contextual familial factors like parenting practices and psychosocial determinants such as life meaning. In particular, we found that life meaning mediates the construction of the language broker role identity in Latinx adolescents from specific parenting practices. In addition, we discussed how cultural values influence adolescents' identity development, paying specific attention to their language broker role identities. Finally, we stressed the importance of understanding how role identities are integrated alongside other social identities to aid in the elucidation of identity development in immigrant ethnic minority adolescents. Our work provides a more nuanced understanding of the language broker role-based identity in immigrant adolescents. More work is needed to explore how role identities are established and managed, as well as their associated implications in ethnic minority adolescents from immigrant families.

References

Abes, E. S., Jones, S. R., & McEwen, M. K. (2007). Reconceptualizing the model of multiple dimensions of identity: The role of meaning-making capacity in the construction of multiple identities. *Journal of College Student Development, 48*(1), 1–22. https://doi.org/10.1353/csd.2007.0000

Arcia, E., Reyes-Blanes, M. E., & Vazquez-Montilla, E. (2000). Constructions and reconstructions: Latino parents' values for children. *Journal of Child and Family Studies, 9*(3), 333–350. https://doi.org/10.1023/a:1026444507343

Bámaca-Colbert, M. Y., Umaña-Taylor, A. J., & Gayles, J. G. (2012). A developmental-contextual model of depressive symptoms in Mexican-origin female adolescents. *Developmental Psychology, 48*(2), 406–421. https://doi.org/10.1037/a0025666

Baumeister, R. F. (1991). *Meanings of life*. Guilford Press.

Benet-Martínez, V., & Haritatos, J. (2005). Bicultural identity integration (BII): Components and psychosocial antecedents. *Journal of Personality, 73*(4), 1015–1050. https://doi.org/10.1111/j.1467-6494.2005.00337.x

Benet-Martínez, V., Leu, J., Lee, F., & Morris, M. W. (2002). Negotiating biculturalism: Cultural frame switching in biculturals with oppositional versus compatible cultural identities. *Journal of Cross-Cultural Psychology, 33*(5), 492–516. https://doi.org/10.1177/0022022102033005005

Boer, D., & Fischer, R. (2013). How and when do personal values guide our attitudes and sociality? Explaining cross-cultural variability in attitude–value linkages. *Psychological Bulletin, 139*(5), 1113–1147. https://doi.org/10.1037/a0031347

Bureau, U. S. C. (2011). *The Hispanic population: 2010*. United States Census Bureau.

Burke, P. J. (1991). Identity processes and social stress. *American Sociological Review, 56*(6), 836–849. https://doi.org/10.2307/2096259

Burke, P. J., & Reitzes, D. C. (1991). An identity theory approach to commitment. *Social Psychology Quarterly, 54*(3), 239–251. https://doi.org/10.2307/2786653

Cast, A. D., & Burke, P. J. (2002). A theory of self-esteem. *Social Forces, 80*(3), 1041–1068. https://doi.org/10.1353/sof.2002.0003

Chao, R. K. (2006). The prevalence and consequences of adolescents' language brokering for their immigrant parents. In M. H. Bornstein & L. R. Cote (Eds.), *Acculturation and parent-child relationships: Measurement and development* (pp. 271–296). Erlbaum.

Cheng, C.-Y., & Lee, F. (2009). Multiracial identity integration: Perceptions of conflict and distance among multiracial individuals. *Journal of Social Issues, 65*(1), 51–68. https://doi.org/10.1111/j.1540-4560.2008.01587.x

Cheng, C.-Y., & Lee, F. (2013). The malleability of bicultural identity integration (BII). *Journal of Cross-Cultural Psychology, 44*(8), 1235–1240. https://doi.org/10.1177/0022022113490071

Cheng, C. Y., Sanders, M., Sanchez-Burks, J., Molina, K., Lee, F., Darling, E., & Zhao, Y. (2008). Reaping the rewards of diversity: The role of identity integration. *Social and Personality Psychology Compass, 2*(3), 1182–1198. https://doi.org/10.1111/j.1751-9004.2008.00103.x

Cheng, S. H., & Kuo, W. H. (2000). Family socialization of ethnic identity among Chinese American pre-adolescents. *Journal of Comparative Family Studies, 31*(4), 463–484.

Cheung, W.-Y., Maio, G. R., Rees, K. J., Kamble, S., & Mane, S. (2016). Cultural differences in values as self-guides. *Personality and Social Psychology Bulletin, 42*(6), 769–781.

Davey, M., Fish, L. S., Askew, J., & Robila, M. (2003). Parenting practices and the transmission of ethnic identity. *Journal of Marital and Family Therapy, 29*(2), 195–208. https://doi.org/10.1111/j.1752-0606.2003.tb01200.x

Davidov, M., & Grusec, J. E. (2006). Untangling the links of parental responsiveness to distress and warmth to child outcomes. *Child Development, 77*(1), 44–58. https://doi.org/10.1111/j.1467-8624.2006.00855.x

Del Torto, L. M. (2008). Once a broker, always a broker: Non-professional interpreting as identity accomplishment in multigenerational Italian–English bilingual family interaction. *Multilingua - Journal of Cross-Cultural and Interlanguage Communication, 27*, 77.

Fuligni, A. J., Tseng, V., & Lam, M. (1999). Attitudes toward family obligations among American adolescents with Asian, Latin American, and European backgrounds. *Child Development, 70*(4), 1030–1044. https://doi.org/10.1111/1467-8624.00075

Ge, X., Best, K. M., Conger, R. D., & Simons, R. L. (1996). Parenting behaviors and the occurrence and co-occurrence of adolescent depressive symptoms and conduct problems. *Developmental Psychology, 32*, 717–731. https://doi.org/10.1037/0012-1649.32.4.717

Gonzales, N. A., Coxe, S., Roosa, M. W., White, R. M. B., Knight, G. P., Zeiders, K. H., & Saenz, D. (2011). Economic hardship, neighborhood context, and parenting: Prospective effects on Mexican–American adolescent's mental health. *American Journal of Community Psychology, 47*(1–2), 98–113. https://doi.org/10.1007/s10464-010-9366-1

Grau, J. M., Duran, P. A., Castellanos, P., Smith, E. N., Silberman, S. G., & Wood, L. E. (2015). Developmental outcomes of toddlers of young Latina mothers: Cultural, family, and parenting factors. *Infant Behavior & Development, 41*, 113–126. https://doi.org/10.1016/j.infbeh.2015.08.001

Grotevant, H. D., & Cooper, C. R. (1985). Patterns of interaction in family relationships and the development of identity exploration in adolescence. *Child Development, 56*(2), 415–428. https://doi.org/10.2307/1129730

Kam, J. A., & Lazarevic, V. (2014). The stressful (and not so stressful) nature of language brokering: Identifying when brokering functions as a cultural stressor for Latino immigrant children in early adolescence. *Journal of Youth and Adolescence, 43*(12), 1994–2011. https://doi.org/10.1007/s10964-013-0061-z

Kiang, L., & Fuligni, A. J. (2009). Ethnic identity and family processes among adolescents from Latin American, Asian, and European backgrounds. *Journal of Youth and Adolescence, 38*(2), 228–241. https://doi.org/10.1007/s10964-008-9353-0

Kiang, L., & Fuligni, A. J. (2010). Meaning in life as a mediator of ethnic identity and adjustment among adolescents from Latin, Asian, and European American backgrounds. *Journal of Youth and Adolescence, 39*(11), 1253–1264. https://doi.org/10.1007/s10964-009-9475-z

Kiang, L., Yip, T., & Fuligni, A. J. (2008). Multiple social identities and adjustment in young adults from ethnically diverse backgrounds. *Journal of Research on Adolescence, 18*(4), 643–670. https://doi.org/10.1111/j.1532-7795.2008.00575.x

Kim, S. Y., Hou, Y., Shen, Y., & Zhang, M. (2017). Longitudinal measurement equivalence of subjective language brokering experiences scale in Mexican American adolescents. *Cultural Diversity and Ethnic Minority Psychology, 23*(2), 230–243. https://doi.org/10.1037/cdp0000117

Knafo, A., & Schwartz, S. H. (2001). Value socialization in families of Israeli-born and soviet-born adolescents in Israel. *Journal of Cross-Cultural Psychology, 32*(2), 213–228. https://doi.org/10.1177/0022022101032002008

Koh, J. B. K., Shao, Y., & Wang, Q. (2009). Father, mother and me: Parental value orientations and child self-identity in Asian American immigrants. *Sex Roles, 60*(7–8), 600–610. https://doi.org/10.1007/s11199-008-9550-z

McQuillan, J., & Tse, L. (1995). Child language brokering in linguistic minority communities: Effects on cultural interaction, cognition, and literacy. *Language and Education, 9*(3), 195–215. https://doi.org/10.1080/09500789509541413

Mok, A., & Morris, M. W. (2012). Managing two cultural identities: The malleability of bicultural identity integration as a function of induced global or local processing. *Personality and Social Psychology Bulletin, 38*(2), 233–246. https://doi.org/10.1177/0146167211426438

Negru-Subtirica, O., Pop, E. I., Luyckx, K., Dezutter, J., & Steger, M. F. (2016). The meaningful identity: A longitudinal look at the interplay between identity and meaning in life in adolescence. *Developmental Psychology, 52*(11), 1926–1936. https://doi.org/10.1037/dev0000176

Nichols, R. (2013). The origins and effects of filial piety (xiao): How culture solves an evolutionary problem for parents. *Journal of Cognition and Culture, 13*(3–4), 201–230. https://doi.org/10.1163/15685373-12342092

Noble-Carr, D., Barker, J., McArthur, M., & Woodman, E. (2014). Improving practice: The importance of connections in establishing positive identity and meaning in the lives of vulnerable young people. *Children and Youth Services Review, 47*(Part 3), 389–396. https://doi.org/10.1016/j.childyouth.2014.10.017

O'Donnell, M. B., Bentele, C. N., Grossman, H. B., Le, Y., Jang, H., & Steger, M. F. (2014). You, me, and meaning: An integrative review of connections between relationships and meaning in life. *Journal of Psychology in Africa, 24*(1), 44–50.

Phinney, J. S., Ong, A., & Madden, T. (2000). Cultural values and intergenerational value discrepancies in immigrant and non-immigrant families. *Child Development, 71*(2), 528–539. https://doi.org/10.1111/1467-8624.00162

Schwartz, S. J., Weisskirch, R. S., Hurley, E. A., Zamboanga, B. L., Park, I. J. K., Kim, S. Y., Umaña-Taylor, A., Castillo, L. G., Brown, E., & Greene, A. D. (2010). Communalism, familism, and filial piety: Are they birds of a collectivist feather? *Cultural Diversity and Ethnic Minority Psychology, 16*(4), 548–560. https://doi.org/10.1037/a0021370

Sellers, R. M., Rowley, S. A. J., Chavous, T. M., Shelton, J. N., & Smith, M. A. (1997). Multidimensional inventory of black identity: A preliminary investigation of reliability and construct validity. *Journal of Personality and Social Psychology, 73*(4), 805–815. https://doi.org/10.1037/0022-3514.73.4.805

Sellers, R. M., Smith, M. A., Shelton, J. N., Rowley, S. A. J., & Chavous, T. M. (1998). Multidimensional model of racial identity: A reconceptualization of African American racial identity. *Personality and Social Psychology Review, 2*(1), 18–39. https://doi.org/10.1207/s15327957pspr0201_2

Shearer, C. L., Crouter, A. C., & McHale, S. M. (2005). Parents' perceptions of changes in mother-child and father-child relationships during adolescence. *Journal of Adolescent Research, 20*(6), 662–684. https://doi.org/10.1177/0743558405275086

Steger, M. F., Frazier, P., Oishi, S., & Kaler, M. (2006). The meaning in life questionnaire: Assessing the presence of and search for meaning in life. *Journal of Counseling Psychology, 53*(1), 80–93. https://doi.org/10.1037/0022-0167.53.1.80

Stets, J. E., & Burke, P. J. (2014). The development of identity theory. *Advances in Group Processes, 31,* 57–97. Emerald Group Publishing Limited. https://doi.org/10.1108/S0882-614520140000031002

Stryker, S. (1980). *Symbolic interactionism: A social structural version.* Benjamin/Cummings Pub Co.

Stryker, S., & Burke, P. J. (2000). The past, present, and future of an identity theory. *Social Psychology Quarterly, 63*(4), 284–297. https://doi.org/10.2307/2695840

Sun, X., McHale, S. M., & Updegraff, K. A. (2017). Maternal and paternal resources across childhood and adolescence as predictors of young adult achievement. *Journal of Vocational Behavior, 100*, 111–123. https://doi.org/10.1016/j.jvb.2017.03.003

Tajfel, H. (1974). Social identity and intergroup behaviour. *Information (International Social Science Council), 13*(2), 65–93. https://doi.org/10.1177/053901847401300204

Tilghman-Osborne, E. M., Bámaca-Colbert, M., Witherspoon, D., Wadsworth, M. E., & Hecht, M. L. (2015). Longitudinal associations of language brokering and parent-adolescent closeness in immigrant Latino families. *The Journal of Early Adolescence*. https://doi.org/10.1177/0272431614566944

Umaña-Taylor, A. J., Alfaro, E. C., Bámaca, M. Y., & Guimond, A. B. (2009). The central role of familial ethnic socialization in Latino adolescents' cultural orientation. *Journal of Marriage and Family, 71*(1), 46–60. https://doi.org/10.1111/j.1741-3737.2008.00579.x

Umaña-Taylor, A. J., & Guimond, A. B. (2010). A longitudinal examination of parenting behaviors and perceived discrimination predicting Latino adolescents' ethnic identity. *Developmental Psychology, 46*(3), 636–650. https://doi.org/10.1037/a0019376

Umaña-Taylor, A. J., Yazedjian, A., & Bámaca-Gómez, M. (2004). Developing the ethnic identity scale using Eriksonian and social identity perspectives. *Identity: An International Journal of Theory and Research, 4*(1), 9–38. https://doi.org/10.1207/s1532706xid0401_2

Updegraff, K. A., McHale, S. M., Zeiders, K. H., Umaña-Taylor, A. J., Perez-Brena, N. J., Wheeler, L. A., & Rodríguez De Jesús, S. A. (2014). Mexican-American adolescents' gender role attitude development: The role of adolescents' gender and nativity and parents' gender role attitudes. *Journal of Youth and Adolescence, 43*(12), 2041–2053. https://doi.org/10.1007/s10964-014-0128-5

Urdan, T., & Munoz, C. (2012). Multiple contexts, multiple methods: A study of academic and cultural identity among children of immigrant parents. *European Journal of Psychology of Education, 27*(2), 247–265. https://doi.org/10.1007/s10212-011-0085-2

Villanueva, C. M., & Buriel, R. (2010). Speaking on behalf of others: A qualitative study of the perceptions and feelings of adolescent Latina language brokers. *Journal of Social Issues, 66*(1), 197–210. https://doi.org/10.1111/j.1540-4560.2009.01640.x

Weisskirch, R. S. (2005). The relationship of language brokering to ethnic identity for Latino early adolescents. *Hispanic Journal of Behavioral Sciences, 27*(3), 286–299. https://doi.org/10.1177/0739986305277931

Weisskirch, R. S. (2007). Feelings about language brokering and family relations among Mexican American early adolescents. *The Journal of Early Adolescence, 27*(4), 545–561. https://doi.org/10.1177/0272431607302935

Weisskirch, R. S. (2017). A developmental perspective on language brokering. In R. S. Weisskirch & R. S. Weisskirch (Eds.), *Language brokering in immigrant families: Theories and contexts* (pp. 7–25). Routledge/Taylor & Francis Group.

Weisskirch, R. S., Kim, S. Y., Zamboanga, B. L., Schwartz, S. J., Bersamin, M., & Umaña-Taylor, A. J. (2011). Cultural influences for college student language brokers. *Cultural Diversity and Ethnic Minority Psychology, 17*(1), 43–51. https://doi.org/10.1037/a0021665

Yeh, K.-H., & Bedford, O. (2003). A test of the dual filial piety model. *Asian Journal of Social Psychology, 6*(3), 215–228. https://doi.org/10.1046/j.1467-839X.2003.00122.x

Chapter 9
"Talking Black": Examining Black Immigrant Youth's Cultural Identity Through Peer Culture and Black American English

Barbara Thelamour

The process of adopting Black culture as manifested in the United States is one that many Black immigrants undergo. This chapter focuses on Black immigrant adolescents' cultural identity development in the United States. Like immigrants from other ethnic groups, these first- and second-generation adolescents of African or Caribbean descent learn to navigate and balance their cultures of origin with the destination (host) culture (Berry, 2005). For many, the differences between the two cultural frames can cause some conflict for the individual immigrant and their families (Benet-Martinez & Haritatos, 2005). In this chapter, the host culture under consideration is the Black American (sub)culture: As racially Black, there is an implicit expectation that Black immigrant youth learn about, adopt, and grow to identify with Black American culture, despite inherent ethnic differences. For Black immigrant youth especially, there are several socializing spaces that can promote a connection to Black American culture, including schools, communities, and peer groups.

This chapter examines the peer ecology as a critical context for cultural identity development. The developmental literature cites peer relationships as a critical influence on adolescents and their evolving senses of self (Phinney et al., 2001; Rageliene, 2016). Through peer socialization, immigrant adolescents can also internalize aspects of culture and language. This chapter uses qualitative data to explore how Black immigrant youth, through their peers, learn about and grow fluent in Black American English[1] (BAE) or the dialect that has been spoken by generations of Black Americans (Smitherman, 1977). The responses from a small focus group study provide evidence for this chapter's organizing framework where I argue that Black American youth peer culture is a catalyst for language exploration which then

[1] This language has also been known elsewhere as African American Vernacular English, Ebonics, African American Language, and Black English Vernacular, among other names (Long, 2003).

B. Thelamour (✉)
Swarthmore College, Swarthmore, PA, USA
e-mail: bthelam1@swarthmore.edu

D. J. Johnson et al. (eds.), *Re/Formation and Identity*, Advances in Immigrant Family Research, https://doi.org/10.1007/978-3-030-86426-2_9

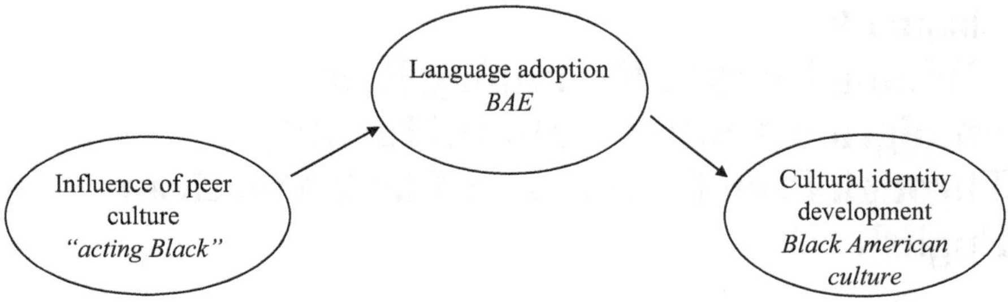

Fig. 9.1 Conceptual model of Black immigrant cultural identity development

supports Black immigrants' cultural identity development. Specifically, this chapter examines a model where adolescent peer culture influences immigrant cultural identity exploration through language (Fig. 9.1). In the existing research, the relationship between language and identity has been researched heavily; however, there is a dearth of studies focusing on all three—peers, language, and identity. Further, much of the work has focused on ethnic identity and the ways that learning a new language might take away from one's identification with their heritage culture(s) (e.g., Laroche et al., 1998; Vedder & Virta, 2005). In this chapter, a developing cultural identity is seen as an additional self-definitional frame for the immigrant participants.

In this chapter, I first present acculturation as the overarching theoretical framework for this study. Next comes the method section detailing the procedure used to collect and analyze the focus group data. Then, in order to outline this conceptual model of cultural identity development, I will present a review of the existing literature in three parts: First, I review the research pertaining to Black American youth culture with regard to race-acting (i.e., acting Black). Then, I explore the adoption of language, namely, Black American English, as one way adolescents act Black. Last, I show how speaking BAE is demonstrative of a developing Black American cultural identity. The relevant findings from the focus group study is incorporated with the overview of the literature in order to support the proposed pathway from Black American youth culture to language adoption to cultural identity development. This approach is advantageous for displaying the interplay between the literature and the experiences of the study's participants.

Theoretical Framework

To a large extent, Black immigrants remain an understudied group in the United States. Demographic trends indicate that they are a rapidly growing segment of the overall foreign-born population, comprising almost 9% of immigrants. Black immigrants also represent 10% of the Black population in the United States, and the numbers are projected to almost double in the next decades (Morgan-Trostle et al., 2016). Though categorized under the umbrella term "Black immigrants," they emigrate from numerous countries from the Caribbean region and the African continent with their respective histories, cultures, and languages. These immigrants, like so many other groups, grapple with reconciling or rejecting aspects of their respective heritage cultures and the host culture, a process described by acculturation theory (Berry et al., 2006) which provides the theoretical backdrop for this chapter.

The bidimensional framework of immigrant acculturation has conceptualized most of the literature on immigrant adjustment. Here, acculturation is the navigation and negotiation of heritage and host cultures (Berry et al., 2006). However, for Black immigrants, in addition to adjusting to mainstream (European American) culture and maintaining their cultural backgrounds, they also experience racism due to their new racial categorization as Black. This also means this population of immigrants also learn about and adjust to Black American culture and its expectations (Ferguson & Bornstein, 2014; Thelamour, 2017). The grouping of Black immigrants with Black Americans often means there is an assumption of the same culture. Research has indicated that Black immigrants engage with Black American culture to varying degrees (Thelamour, 2021); however, as with the mainstream culture, Black immigrants determine how integrated they will be with Black American culture.

The simultaneous study of African and Caribbean immigrants is facilitated by their shared experience of being racialized upon arrival to the United States. Despite the various countries of origin, Black immigrants are categorized as "Black or African American" under the current demographic labels (www.census.gov), effectively dampening their ethnic backgrounds. This racialization is often quite distinct from their countries of origin, where other socioeconomic and social class markers are more salient than racial labels (Phinney & Onwughalu, 1996). Relatedly, Black immigrants stand to experience discrimination on the basis of their race, as do their Black American counterparts who have been in the United States for generations (Sanchez & Awad, 2016; Tormala & Deaux, 2006). Black immigrants can also experience racial identity invalidations from their Black American counterparts, where their Blackness is called to question (Durkee et al., 2019). Contextualizing Black immigrant acculturation in their racial and cultural experiences in the United States provides an entry way to examine the ways Black American youth culture can influence identity in Black immigrant youth.

Focus Group Study Method

Supporting data for this chapter come from a focus group project conducted at a small liberal arts college in the Midwestern region of the United States. Convenience sampling on the campus yielded a sample of nine undergraduate students who were interviewed in two focus groups divided by gender (Hughes & DuMont, 1993). Students who were first-generation immigrants (i.e., came to the United States from an African or Caribbean country) or second-generation immigrants (i.e., born in the United States to at least one immigrant parent) and who attended at least two years of high school in the United States were selected to participate. Although the number of participants in the sample was small, the focus group method allowed students to interact with each other and explore the topics under consideration with depth. The synergy among the participants was especially helpful when discussing these sometimes isolating events in their lives (Kress & Shoffner, 2007).

Demographic characteristics of the sample are summarized in Table 9.1. The four male and five female participants' ages ranged from 18 to 22 years ($M = 19.33$ years). Five of the nine participants either migrated ($N = 4$) to a predominately White city or were born and raised in one ($N = 1$). Four of these five lived in rural environments in the Midwestern region of the United States, and one was raised in an affluent suburb in the mid-Atlantic region. The remaining four participants in the study resided in large, ethnically diverse urban communities in major East Coast U.S. cities.

Open-ended interview questions were developed to investigate the role of race in the participants' adolescent experiences, how peers facilitated their acculturation to Black American culture, and their identities as (Black) American or as a member of their native ethnic group. Examples of questions included "Did the notion of acting Black or acting White come up for you as you were growing up?" "What did those mean?" and "Do you think you fit in with your Black American peers? What about with your White peers?" "Was BAE something you think you had to learn? What was your relationship to BAE?" The interview protocol was piloted with a second-generation Black immigrant graduate student. Both sets of interviews lasted approximately 90 min each.

Each participant also completed a short demographic questionnaire at the end of the interviews. Questions determined the participants' ethnic backgrounds, parents'

Table 9.1 Study participant demographics

Name[a]	Age	Generational status	Gender	Country of origin	Racial composition of community
Adofo	19	2.5	Female	Ghana (father) United States (mother)	Predominately Black
Antoinette	19	2nd	Female	Haiti (both parents)	Predominately Black and Latino
Arthur	20	2nd	Male	Honduras (mother) Jamaica (mother)	Predominately Black and Latino
Baba Tundai	19	1st	Male	Liberia (mother) South Africa (father)	Racially diverse/ multiracial
Erika	19	1st	Female	Madagascar (both parents)	Predominately White
Iman	19	2nd	Female	Sierra Leone (both parents)	Predominately White
Monee	18	1st	Female	Cameroon (both parents)	Predominately White
Sherlock	22	2nd	Male	Mauritania (both parents)	Predominately White
Sly	19	2nd	Male	Kenya (both parents)	Predominately White

aNames were changed to pseudonyms that the participants chose occupations, number of siblings, and the cultural make-up of their home neighborhoods. Participants also answered a one-question cultural identification measure, where they rated themselves as one of the following: "definitely Caribbean/African/Afro-Latino," "more Caribbean/African/Afro-Latino than Black American," "equally Caribbean/African/Afro-Latino and Black American," "More Black American than Caribbean/African/Afro-Latino," or "definitely Black American" (Deaux et al., 2007). At the end of each focus group, participants entered a drawing to win a $25 gift card.

Both focus interviews were transcribed by undergraduate research assistants and checked by the author. After transcription, participants were contacted via email with any remaining clarification questions. Thematic analysis (Braun & Clarke, 2006) was the approach used to highlight meaningful themes across the responses and to glean a rich understanding from the interviews. After transcription, the author analyzed the interview data first using open coding (breaking down the interview responses into chunks according to commonalities in responses) and then axial coding (relating the chunks to each other; Corbin & Strauss, 2008). These initial codes highlighted themes in the participants' use of BAE (their definitions and usage) and experiences with race-acting, specifically acting Black. These themes were then compared to the participants' racial identities to illustrate the study's conceptual model, specifically whether race-acting and BAE were related to participants' cultural identities.

Researcher Reflexivity

As an integral part of this study, it is important to consider my positionality in the context of this work. I am an educational psychologist who takes a mixed-methods approach to topics of culture, identity, and academic achievement for Black immigrant students. I am a second-generation Haitian immigrant woman who grew up in a predominately White city in the Southeastern United States. I have strong racial identity in that being Black is an important part of my self-concept, and I balance my Haitian (Caribbean) ethnic and Black American cultural identities.

Black Immigrant Adolescents and Black American Culture

Black Immigrant Adolescents and Peer Culture

Black immigrant adolescents often experience the effect of Black American cultural expectations via relationships with their Black American peers. The peer ecology is an understudied one in the literature on Black immigrant youth adjustment, a limitation that overlooks the role non-family members have on immigrant adolescents' social and emotional development. The emphasis on peer relationships during the adolescent stage is consistent with existing developmental theory and research: Same-age peers have significant influence on adolescents as the multiple facets of their identities develop (Azmitia et al., 2013). Compliance to peer group norms can mean acceptance by peer groups and a positive social status. On the other hand, rejection of peer group socialization can make for a difficult time for adolescents and can lead to ostracism (Oyserman et al., 2006).

For Black adolescents, one cultural norm that peer groups enforce is that of "acting Black" or behaviors, attitudes, and an esthetic that is in alignment with Black American culture (Cousins, 2008). Acting Black is often characterized by an attitude of cool and is seen as being in opposition to mainstream white American culture and ways or acting White (Durkee & Williams, 2015). Elements of acting Black include clothing, interests, musical tastes, and language (Burrell et al., 2013). Acting Black can have negative connotations and be attached to stereotypes about Black youth behavior and interests (e.g., "acting disrespectful" or "ignorant") (Peterson-Lewis & Bratton, 2004); it can also be associated with pride and respectfulness (Burrell et al., 2013). Similar to the peer norms of other groups, conformity to the expectations of American Blackness can provide Black adolescents with status and access to peers.

The research on Black immigrant adolescents' understanding of and alignment with acting Black is currently limited to one study on second-generation Caribbean and African youth (i.e., those born in the United States to immigrant parents) (Thelamour & Johnson, 2017). Findings from the longitudinal mixed-method study indicated that Black immigrant adolescents had less nuance in their perceptions of acting Black than their Black American peers. Over time, however, their definitions of acting Black reflected a deeper understanding that included the Black population's societal position and their seeking equality in the face of discrimination. This chapter adds to this line of research by examining the ways a sample of first- and second-generation immigrants both understood and encountered the notion of acting Black. Here, their peers' feedback on their behavior as shaping their understanding of acting Black or White is described.

Focus Group Findings

From the interviews, and consistent with existing literature (Thelamour & Johnson, 2017), most of the participants had encountered race-acting and acting Black during their adolescence. From their experiences, acting Black included being "ghetto" or loud and boisterous. Being held to "inner city standards" was another aspect. For women, how they wore their hair (i.e., not using chemical straightener) was under the acting Black umbrella. The element of cool or knowing what was current was especially salient: Knowing the current dances, music, and pop culture was the way these participants perceived "acting Black." For the men, at times, acting Black was associated with violence and aggression. High academic achievement (e.g., advanced classes) was described as not being Black enough. Based on these responses, and similar to previous studies, acting Black was multifaceted. The characteristics were varied, with some negative aspects.

The participants' conceptualizations of acting Black likely seemed to have developed from feedback from peers, especially when they were accused of "acting White." The male participants in the sample received pushback for the latter: As adolescents, they were often told by their Black and White peers that they did not act Black enough, and they felt pressure to conform to those standards. Sherlock, a second-generation Mauritanian immigrant, stated,

> A bunch of Black kids in high school who were like 'We don't fuck witchu because you ain't nigga enough to hang out with us.'...Like, people expected me to know how to dance, people expected me to like know what was up and coming with like rap music.

Similarly, Baba Tundai (first generation) found himself confused when confronted with "acting Black." He described how his Black peers would be surprised at his interests and activities:

> It's like 'No, no you're doing it differently. I haven't seen a Black person sound like you. No Black person takes AP calculus in high school. No, you're the only Black person on the soccer team. You're the captain of the soccer team. You're number one in the school. No Black person likes math.'

Sly, who attended a predominately White high school, would get feedback from his White peers about his not "acting Black," noting the negative, stereotypical definitions from his peers.

> There weren't a lot of Black people, so whenever I heard that, it was from a White person... So they were telling me what Black was and it didn't make sense because a lot of the times, when they said like 'Oh, that's so Black,' it would be kind of negative. So, it was kind of confusing to me because I never knew when they said 'Oh Sly, you got really Black,' I was like 'I don't even know how to take that now.'

The female participants had slightly more varied experiences with acting Black. Unlike the male group, where all described not acting or being Black enough, one woman was described as sometimes acting *too* Black and participating in the labeling. Antoinette (second generation) stated, "Yeah, I'm a victim of calling African American girls who act White 'oreos.[2]' In certain places I'm called 'acting too Black' or 'ratchet' or 'ghetto.' I get really offended when people are like 'Antoinette, you're ghetto,' cause I'm not, it's just who I am." Here, Antoinette's description of herself as a "victim" of making fun of her Black girl peers who do not act Black perhaps refers to her internalization of the stereotypes pertaining to Black youth. The use of victim connotes her regret of imposing those standards on others, especially in the context of others sharing their experiences.

Besides Antoinette, the other women were on the other side of acting Black. Adofo, a second-generation immigrant, stated, "There's always been this pressure,

[2] "Oreo" describes a Black person who is seen as acting White. Like the cookie, they are Black on the outside, White on the inside.

and definitely throughout high school, people were ostracized if they didn't fit these, you know, inner city standards of acting or performing." Iman stated that she "used to get called an 'oreo' all the time. And it really used to bother me because I can't help it, that's the kind of people I grew up around."

Across both focus groups, the use of Black American English was considered to be one more aspect of acting Black. As with other indicators of acting Black, the participants' peers' feedback about their behaviors contributed to their understanding. "Ebonics" was invoked in the women's focus group as one way that individuals were seen as acting Black. From their responses, those who did not speak BAE were ostracized from the peer group. Adofo stated, "When people would speak, you know, properly or when people would not use Ebonics, that's when 'Oh, you're trying to be white.'" Monee chimed in with her experiences that expectations about American Blackness were even forced upon her by her white peers, though she was a first-generation Cameroonian immigrant.

> They just looked at me like African American. So one time, my friend was like, 'Monee, you're so not ratchet.[3] You're different." And I was like 'What does ratchet mean?' And then she said, describing 'Oh, you're supposed to like talk like that…'

In the men's focus group, the participants also described BAE as an aspect of acting Black within the peer group. Again, contemporary slang as an aspect of BAE was used to describe some of the young men in the group. Baba Tundai responded, "Some kid came and asked me…'what does turnt up[4] mean?' I don't know." As a result, he notices a cultural distinction in what acting Black is, despite his being racially Black. Their responses focused on BAE as an aspect of pop culture that kids of immigrants might not acquire. This becomes particularly complicated with white peers. Arthur reflected, "I just feel like people are looking for a cultural informant, like, 'yo, tell us what being Black is about.'" BAE as a facet of acting Black that Black peers evaluated and to which even White peers wanted access.

The participants' statements demonstrated their encounters with race-acting through their interactions with their peers. Others misidentifying them as Black American lead to accusations of not acting Black enough based on longstanding stereotypes about academic and sports interests, hair, and, most relevant for this study, language use. These youth were especially criticized by their Black American peers for not understanding particular slang terms or speaking Mainstream White English (Baker-Bell, 2020). Situating BAE in the context of race-acting in Black youth culture sheds light on one factor explaining why these adolescents would adopt Black American language.

---

[3] "Ratchet" is a contemporary slang term often used by Black American youth to describe unrefined behavior that is in contrast to social conventions.

[4] "Turnt (turned) up" often refers to partying behavior, demonstrating great enthusiasm and little restraint.

Learning Black American English

As demonstrated in the participants' responses, language was an important aspect of their Black peers' cultures, as it was one way to act Black. The focus on language is expected given its importance to group functioning and identity. From a sociocultural perspective, language functions as a symbolic or psychological tool that mediates individual thinking and behavior (Vygotsky, 1978). Theoretically, cognitive and socioemotional development occurs through the internalization of one's group's language (John-Steiner & Mahn, 1996). At the group level, language is an important "marker" that can help delineate membership (Clément & Noels, 1992). Communities' identities are "established and maintained through language," and its explicit public function is to facilitate communication and relationships among the members of a given group (Gumperz & Cook-Gumperz, 1982, p. 7). Additionally, language is political, particularly for minority groups (Smitherman, 1977), and the use of one language over another foregrounds specific histories and perspectives that are important to the lives of the speakers (Makoni et al., 2003).

For immigrants, or any newcomers to a community who might seek to gain entry, language can be a useful mechanism for inclusion (McNamara, 2005; Perez, 2011). Immigrants can learn by paying attention to linguistic cues and observing dialogue (Hansell & Ajirotutu, 1982). The internalization of language for newcomers can facilitate relationships with and attachment to the target groups (Cheng et al., 2014). When learning a new language, immigrants are not only concerned with linguistic fluency or competence (e.g., grammar and vocabulary), they must also understand the ways language is context-dependent (Edwards, 1986). Certain situations or dynamics will require individuals to codeswitch or move between speech forms (Linnes, 1998).

Even with exposure, however, those who are not native to a language might always have some difficulty with the nuances of acquired language use and interpretation (Edwards, 2009). As such, acquiring a new language can be laden with stress (Clément & Noels, 1992), especially if the differences between one's own native language and the acquired language(s) are significant (Cheng et al., 2014). Another added layer of stress can be "shibboleths" or formal and informal language tests to determine whether an individual is a member of a group (McNamara, 2005). At the national level, these tests can come in the form of language policy, but interpersonal interactions can include shibboleths, such as the reactions when an individual does not know a word or phrase. Consequently, for immigrants, connections that language acquisition might afford can also be accompanied by some anxiety as they navigate the complexities of the new language.

Black American English

Black American English is "Euro-American speech with an Afro-American meaning, nuance, tone, and gesture" (Smitherman, 1981, p. 411). Its roots can be traced to a creolized pidgin developed from the multiple language backgrounds of enslaved Africans in the United States (Smitherman, 1999). Combined with English, BAE in its original form allowed for communication among enslaved peoples. Over time, it has also grown to include contemporary slang terminology that is frequently connected to hip-hop and rap music (Alim, 2003). The language style involves the active participation of the speakers, including call and response and the Dozens (Smitherman, 1977). Additionally, the Great Migration of Black Americans from the southern region of the United States to the Northeast, Midwest, and Western regions of the country during the mid-twentieth century has contributed to regional differences in BAE (Jones, 2015).

Across generations, the language has evolved to have its own grammar and vocabulary that runs counter to the rules of Mainstream White English. Similar to other languages and dialects, BAE is political in that it is a form of resistance and liberation in the face of racial and linguistic discrimination (Alim, 2003). The politics of BAE became part of a national educational conversation in the late 1990s (Rickford, 1999). Black American students who entered U.S. schools speaking mostly BAE were viewed as less capable than their mainstream English-speaking or codeswitching peers (Duncan, 2004). Additionally, school assessments often failed to consider the linguistic backgrounds of students, making for poor scores (Hilliard III, 2002). In the face of this institutional discrimination, the maintenance of BAE reinforced Black American resistance to marginalization (Duncan, 2004). Over time, pedagogies have developed that enforce a critical linguistic consciousness (Baker-Bell, 2020) empowering students to examine the ways Mainstream White English is privileged over BAE.

Within the Black American community, there has existed some tension with the use of BAE, and the divide has been present along generational lines. Historically, BAE was ascribed to young people (Smitherman & Cunningham, 1997) and was viewed as an impediment to young Black Americans joining the elite classes (Long, 2003). Even when the language was spoken in the home, speaking it in public was discouraged because it was thought to reinforce negative stereotypes about Black people's intelligence. Additionally, even though the language is associated with the Black American cultural group, not every member is fluent (Smitherman, 1981), either from a lack of exposure or a conscious decision to forego its use. Regardless of the precursor, lack of facility with BAE can lead to ostracism from the community (Smitherman, 1981). Relatedly, Black individuals who speak BAE have been seen as more likable by other Black Americans than those who do not (Billings, 2005). The importance of the language to Black culture and its opposition to Whiteness make it a mechanism through which Black individuals can feel connected to the culture. Using the race-acting concept, lack of proficiency in BAE can be seen as acting White, which can lead to negative social experiences.

As with other languages, outsiders to the Black American cultural group can learn to adopt BAE. Over time, the language has become a fixture in mainstream American popular culture, through music, movies, and social media (Jones, 2015). Because Blackness has been associated with the element of cool and resistance, those who seek access to a certain social status might choose to internalize the language, even if they are not racially Black. This has been shown in the corporate appropriation of BAE that has been related to capital gains for certain industries (Baker-Bell, 2020). Research has shown that even Asian American youth (Chun, 2001; Igoudin, 2011; Reyes, 2005) and White adolescents (Cutler, 2003) have often adopted BAE. As with other language learners, these outsiders to the community tend to internalize the surface characteristics of the language (e.g., contemporary slang) with less understanding of linguistic nuances.

Black Immigrant Adolescents and Black American English

The participants in this study described BAE as one facet of acting Black in their peer culture(s), which is consistent with the limited research on this topic (Ibrahim, 2003a, 2003b). In these studies, African immigrant adolescents in Canada internalized BAE through Black American popular culture, in particular rap and hip-hop. For the participants, investing in BAE was a form of connectedness with Blackness as manifest in a racialized North America (Ibrahim, 2003b). In both studies, becoming Black offered the African immigrant youth a form of expression that mainstream society failed to provide.

Where Ibrahim's focuses on cultural and societal influences on Black immigrants' language adoption, this chapter presents peers as a more localized factor. The cultural expectations of Blackness as socialized by peers have a direct impact on adolescents' behaviors. Specifically, peers' feedback about acting Black and using BAE can go far in shaping immigrants' perspectives on the language. The focus group data presented here delineate the how and why the participants learned BAE and why they chose not to.

Focus Group Findings

In each focus group, participants were asked about their relationship to BAE, and if they felt compelled to learn the language. Not surprisingly, their responses revealed diverse perspectives on the language.

Embracing BAE

Several participants declared their appreciation of BAE as well as their desire to be fluent in it. For instance, Adofo stated, "there's nothing really wrong with Ebonics. I love Ebonics. I just think um, that, it's all about codeswitching and being able to speak in a way that people can understand you." Sly described the language as "an addition to my vocabulary," something that "comes out of my mouth and I don't really think about it honestly."

Two participants described their acquisition of BAE over time. For them, the language was a form of access to peers. Antoinette responded,

> Once I hit high school and I was around more people of my- of like African American descent…I kind of put myself into that 'Well, I need to, I'm trying to assimilate I'm trying to make, make friends so I need to talk like them.'

Sherlock reflected on his high school experience, where his peers did not accept that he did not codeswitch from Mainstream White English to BAE. He took advantage of popular media (e.g., TV shows) to learn.

> In high school…like teaching myself BAE like straight up watching *Fresh Prince* DVDs and like watching *The Fresh Prince of Bel Air,* watching like, I don't know, *Martin,* watching *The Jamie Foxx Show,* like sitting there religiously watching BET. That was BAE.

These television shows, sitcoms that portrayed various aspects of Black American life in the 1990s, functioned as teaching tools for Sherlock. Over time, he internalized the characters' language models. He further stated that being a child of immigrants from Mauritania, codeswitching was a part of his upbringing, "At my home we speak different language, we had to codeswitch all the time. I've become really good at that with people too."

Tension with the Language

Although there were participants who sought to learn BAE in order to communicate in a more accepted way with some peers, not everyone felt a connection to BAE. Arthur stated, "Whatever Black vernacular is, I have it. I can do it. I don't cuz I feel like it's illegitimate. I just do my own thing with my own words." He expressed a contentment with being able to use language as he wanted to and not how Black peer culture dictated.

There also existed some generational differences with the use of BAE. Baba Tundai, a first-generation male, described his struggle between the use of Black language and his Black identity:

> I can't say I've gotten BAE or Black language, but I do understand certain slangs now. I do understand certain things that are thrown at me, but at my core, I am still African. I am still… Black. I'm Black, I am Black. And I wouldn't really say that there's a distinction between language, but it's culture, I think. And our perception of culture.

Baba Tundai, through his exploration of and ultimate releasing of BAE as an expression of his Blackness, evolved in his understanding that not only is his Black identity still legitimate, Black American culture is not that different from his native cultural background.

Another first-generation participant recounted her own experiences with not speaking BAE in the United States. Monee, who grew up in a French and English bilingual household, said, "Coming here, I speak English. So, I had people who were like, 'Why do you speak English so good?' And I was like, 'Well, that's the only language I know.' But only, it wasn't...the language. It was more my accent." Here, the notion of acting Black was imposed on Monee, who did not speak in a way her other Black peers recognized. For these three participants, Blackness was not something to act but it simply was, so embracing BAE was not a priority.

Language and Cultural Identity

The importance of language for individuals within groups is well documented, and the above findings highlight the ways this group of Black immigrants acquired or rejected BAE as a way to conform to their peers' cultures pertaining to acting Black. Aside from group affiliation, language acquisition can also have implications for immigrants' sense of self, as "language is the foundation of individual and group identity construction" (Smitherman & Cunningham, 1997, p. 230). Here, the connection between peer relationships, language, and identity, specifically cultural identity, is reviewed.

In this chapter, cultural identity is a construct that is distinct from ethnic identity, where the two are frequently used interchangeably in the research literature. Ethnic identity refers to the extent to which individuals self-categorize as members of particular ethnic groups and the meaning derived from that membership (Phinney & Ong, 2007). As a multifaceted construct, ethnic identity can also encompass one's cultural identity or the ways people incorporate and internalize the behaviors, traditions, roles, and language of a cultural group (McBride Murry et al., 2001). Making a distinction between these two types of identity is necessary for the purposes of this study. Ethnic identity focuses on membership to those groups to which individuals inherently belong, or in the case of immigrants, their native culture(s) (Umaña-Taylor et al., 2014). The use of the cultural identity construct allows for the examination of how individuals can grow to identify with a cultural group that is not their own (Nguyen & Brown, 2010). In pluralistic societies like the United States, where multiple cultures reside, individuals can gravitate to several cultural groups.

For immigrants to a new country, identifying with a host cultural group can be an asset in the adjustment process. While for first- and second-generation children of immigrants, internalizing host culture(s) can be a source of conflict with parents (Telzer, 2010), there are benefits for adolescents. Those who do not incorporate aspects of a host culture, especially when embedded in that culture, are likely to experience isolation and discrimination. Those who are able to maintain a connection with their native cultures and develop links with receiving cultures also show higher self-esteem and lower anxiety than those who struggle with being bicultural (Nguyen & Benet-Martínez, 2013).

The study of cultural identity is especially relevant for adolescence because theoretically the exploration and commitment to identity is a critical process for this developmental stage (Erikson, 1950), and much of the literature on identity development has focused on this stage (e.g., French et al., 2006; Tanti et al., 2011). Adolescents' experiences, relationships, and positionalities all contribute to their multiple social identities (e.g., racial/ethnic, gender, sexual orientation, and cultural) (Yoon et al., 2017). Peers and friends are relationships that can function as identity agents or supports for identity development (Azmitia et al., 2013), and As adolescents become more autonomous, they have been shown to rely more heavily on their peer relationships, especially friendships, than in previous years (Erikson, 1950). Within these peer relationships, the norms set and enforced by peers, particularly same-race peers, can become group inclusion or exclusion criteria, further influencing identity development (Hitti & Killen, 2015). For immigrants, peers can have similar roles: Setting norms for behavior and socializing cultures in ways that can shape identities.

Enforcing and socializing language use is one way that peers can have influence on others' identities. Research indicates that language ability and fluency can be inclusion criteria for peer groups (Beißert et al., 2019). Those who speak the language might also be identified by others as members of the group; in this case, Black immigrants who speak BAE might be seen as their Black American peers as part of the group (Ochs, 1993). For immigrants, adopting markers of culture that are not one's own can be associated with loss (Bhugra & Becker, 2005); however, language acquisition can be a sign that immigrant adolescents are developing an identity that enhances their positive self-image (Bashir-Ali, 2006; Noels et al., 1996).

This chapter extends the limited existing research examining peers, language adoption, and cultural identity. One relevant qualitative study of Hmong adolescents found that bilingualism designated a connection to mainstream American culture and peers (Nguyen & Brown, 2010). In that study, a cultural identity rooted in mainstream American culture was demonstrated by the use of English and speaking loudly when among peers. Unlike the Nguyen and Brown study, this chapter focuses on peer groups as setting the standards to which Black immigrants can align. For Black immigrants, there is evidence that this population of immigrant adolescents are bicultural in that they balance their Black American and heritage cultures (Allen et al., 2012; Ferguson et al., 2012). Culturally, many Black immigrant youth have to grapple with what it means to "become Black" in the United States, and research suggests a multifinality to a Black American cultural identity. Speaking BAE is one way that a cultural identity can be enhanced (Ibrahim, 2003a). In this section, the relationship between language and identity for the focus group participants is presented.

Focus Group Findings

As the interview responses indicated, the participants had different relationships to BAE. Some embraced BAE and chose to speak it frequently, while others opted not to speak it. Here, these first- and second-generation immigrants' BAE language use is considered in the context of their cultural identities. Their language use is positioned alongside the participants' responses to the cultural identity measure to explore the use of BAE is a mechanism through which Black immigrant adolescents grow to identify with Black American culture.

There were five participants who identified as "definitely African/Afro-Latino" or strongly identified with their cultures of origin. Three of the five were first-generation immigrants. Iman was frequently accused of "talking White" (i.e., not codeswitching to BAE), Erika did not describe any use of BAE when asked, and Baba Tundai tried to imitate BAE but decided to forego its use. The remaining two participants in this category were second generation. One of them, Arthur, thought that speaking BAE was illegitimate and did not want to ascribe to a singular way of talking because of his race. Sherlock was the only participant in this group who embraced BAE and studied it through television media.

Two participants identified as being "more African than Black American," one first generation and second generation each. According to their responses, their use of BAE was limited. Monee described her experiences speaking English and was accused of not talking Black by her peers. As an adolescent, Sly had a "gradual progression" into speaking BAE. He stated it is something he uses "from time to time."

Antoinette, a second-generation immigrant, identified as "equally Caribbean and Black American." She described her relationship with BAE as something she had to learn in order to fit in with her Black American peers. Adofo (2.5 generation) also embraced BAE ("Ebonics" as she put it) and identifies as "definitely Black American." These two women, who eagerly acquired BAE, were also highly identified with Black American culture, either balanced with their own native culture or entirely Black American.

The small number of participants and the qualitative design preclude any causal inferences. Nonetheless, there is a trend in the data. Generally, little use of BAE was associated with a stronger native culture identification, whereas those who adopted the language were more likely to at least balance their native cultures with Black American culture. Sherlock was the clear exception to this trend. He, like Antoinette, actively learned BAE and sought tools to facilitate this language acquisition, and he perceived BAE as a means to achieving acceptance from his peers. However, he remained strongly identified as African.

Discussion

For immigrants, cultural adjustment can take many forms, including how their identities change as a result of adopting a new culture. This chapter presented a framework wherein peers and peer culture influence immigrants' cultural identity through language. Data from a small focus group study were used to support the conceptual framework. The focus on Black immigrant adolescents allowed for an investigation of how first- and second-generation immigrants grow to feel a sense of connection with a culture that is not inherently their own. Considering the generational status of these youth is consistent with previous research that has shown variability in the ways Black immigrant youth identify (Waters, 1994). Focusing on their Black American cultural identities offers a unique perspective on identifying with a subculture in the United States, one to which Black immigrants may be pushed to identify with given a shared race.

The data in this study show a link between language use and identity; however, there was more evidence of participants choosing to forego the use of BAE and showing a stronger heritage cultural/ethnic identity. Two of the three participants who embraced BAE and made an effort to learn the language showed a stronger Black American cultural identity than their peers. However, there was variability in their responses: The one participant who described his process of immersing himself in Black American English indicated that he was very much identified with his native culture and not at all with Black American culture. In some ways, the conceptual model was supported with these qualitative data; however, more research is needed to investigate the mediators of this relationship between language and identity (Ochs, 1993; Pablé et al., 2010). Direct socialization, parental rearing, and immigrants' own perspectives on BAE as a legitimate language could be such mediators that could further explain this relationship.

Critical to this relationship is the role of peers, especially adolescent peers' imposing of race-acting on these Black immigrant youth. The ways that the participants behaved and spoke that fit or failed to fit peer groups' perception of acting Black often exposed them to negative feedback. In this model, their peers' assessment of their Black-aligned behaviors set the context for these Black immigrants to engage with the language. The peer ecology is a critical one for its position as helping to expose elements of Black American culture in a more immediate way than media. The peer "micro-group" (Pablé et al., 2010) offers insiders' perspectives on talking Black in a way that fits the local peer culture.

This small qualitative study provides an exploration into the relationship between the peer ecology, language, and identity for Black immigrant adolescent youth. While the data presented here suggest a relationship, future research using this framework would benefit from the use of larger-scale quantitative methods that test the predictive relationships among these three concepts. Quantitative studies could also add to the current model by considering person-level and environmental factors that can further moderate these relationships. For example, at the individual level, gender should be taken into account when thinking about the ways peer culture pertaining to race-acting is socialized. The statements here suggested gender differences in the messages received about acting Black. Generational status can also influence the extent to which immigrants identify with Black American culture given their more recent connection to their native environments. Contextually, the racial make-up of the neighborhoods can impact the content and frequency with which Black immigrants receive messages about acting Black (Lee et al., 2014).

Ongoing research should also consider how well Black immigrants learn the nuances of BAE. Theory and research suggest that newcomers to a culture might have some difficulty understanding the deeper and symbolic meanings of a new language (Edwards, 2009), and even with a command of BAE, there might be linguistic errors or idiosyncrasies on the part of the immigrants. Given its roots in the language traditions of enslaved Africans, the intonation, grammatical structure, and active participation inherent in the BAE language might not be unfamiliar to African and Caribbean immigrants, perhaps further promoting cultural identity development. On the other hand, because the language has changed over generations, there might still be a significant amount of learning on the part of these immigrants. Pursuits in this direction expand not only our understanding of the connection between language and identity, but also our knowledge of Black immigrants' participation in (and perhaps expansion of) another linguistic group besides the mainstream.

Finally, future research should investigate the ways Black immigrant adolescents, when presented with alternative groups and ways of belonging, negotiate their native cultures and Black American culture. Due to the way race is conceptualized in the United States, this population of immigrant youth is positioned to be bicultural within one racial category (Ferguson et al., 2012). While there may be similarities between Black American and African or Caribbean culture(s), the former is laden with negative stereotypes to which the latter groups are not immune (Thelamour, 2017). As such, for Black immigrant adolescents, identifying with either Black American or their native cultures can bring stress and conflict with their families or with their Black American peers. Biculturalism research indicates that those cultures that are compatible are more easily fused than those that are perceived as incompatible (Benet-Martinez & Haritatos, 2005). Ongoing questions about the degree of perceived compatibility of the cultures and the ways Black immigrant adolescents maneuver the demands of each remain and should be addressed.

This chapter provided evidence for a framework to understand the way peer groups, language, and identity are intertwined for Black immigrant adolescents. This focus underscores the way that immigrant recreate their identities in the United States; however, in lieu of focusing on culture at the national level, emphasis was placed on the way that Black American culture and the Black racial group are spaces for identity formation. While much more work to support and expand this framework is needed, this chapter highlighted social and individual factors that can promote Black immigrant identity development in ways that help the field better comprehend the journeys for this understudied population of immigrants.

References

Alim, H. S. (2003). "We are the streets": African American language and the strategic construction of a street conscious identity. In S. Makoni, G. Smitherman, A. F. Ball, & A. K. Spears (Eds.), *Black linguistics: Language, society, and politics in Africa and the Americas* (pp. 40–59). Routledge.

Allen, K. M., Jackson, I., & Knight, M. G. (2012). Complicating culturally relevant pedagogy: Unpacking west African immigrants' cultural identities. *International Journal of Multicultural Education, 14*(2), 1–28. https://doi.org/10.18251/ijme.v14i2.506

Azmitia, M., Syed, M., & Radmacher, K. (2013). Finding your niche: Identity and emotional support in emerging adults' adjustment to the transition to college. *Journal of Research on Adolescence, 23*(4), 744–761. https://doi.org/10.1111/jora.12037

Baker-Bell, A. (2020). *Linguistic justice: Black language, literacy, identity, and pedagogy*. Routledge.

Bashir-Ali, K. (2006). Language learning and the definition of one's social, cultural, and racial identity. *TESOL Quarterly, 40*, 628–639. https://doi.org/10.2307/40264549

Beißert, H., Gönültas, S., & Mulvey, K. L. (2019). Social inclusion of refugee and native peers among adolescents: It is the language that matters! *Journal of Research on Adolescents, 30*(1), 219–233. https://doi.org/10.1111/jora.12518

Benet-Martinez, V., & Haritatos, J. (2005). Bicultural identity integration (BII): Components and psychosocial antecedents. *Journal of Personality, 73*, 1015–1050. https://doi.org/10.1111/j.1467-6494.2005.00337.x

Berry, J. W. (2005). Acculturation: Living successfully in two cultures. *International Journal of Intercultural Relations, 29*(6), 697–712. https://doi.org/10.1016/j.ijintrel.2005.07.013

Berry, J. W., Phinney, J. S., Sam, D. L., & Vedder, P. (2006). Immigrant youth: Acculturation, identity, and adaptation. *Applied Psychology: An International Review, 55*(3), 303–332. https://doi.org/10.1111/j.1464-0597.2006.00256.x

Bhugra, D., & Becker, M. A. (2005). Migration, cultural bereavement, and cultural identity. *World Psychiatry, 4*(1), 18–24.

Braune, V., & Clarke, V. (2006). Using thematic analysis in psychology. *Qualitative Research in Psychology, 3*(2), 77–101. https://dx.doi.org/10.1191/1478088706qp063oa

Burrell, J. O., Winston, C. E., & Freeman, K. E. (2013). Race-acting: The varied and complex affirmative meaning of "acting Black" for African-American adolescents. *Culture & Psychology, 19*(1), 95–116. https://doi.org/10.1177/135407X12464981

Cheng, C., Lee, F., Benet-Martínez, V., & Huynh, Q. (2014). Variations in multicultural experience: Influence of bicultural identity integration on socio-cognitive process and outcomes. In V. Benet-Martínez & Y. Hong (Eds.), *The Oxford handbook of multicultural identity* (pp. 276–299). Oxford University Press.

Chun, E. W. (2001). The construction of white, black, and Korean American identities through African American vernacular English. *Journal of Linguistic Anthropology, 11*(1), 52–64. https://doi.org/10.1525/jlin.2001.11.1.52

Clément, R., & Noels, K. A. (1992). Towards a situated approach to ethnolinguistic identity: The effects of status on individuals and groups. *Journal of Language and Social Psychology, 11*(4), 203–232. https://doi.org/10.1177/0261927x92114002

Corbin, J., & Strauss, A. (2008). *Basics of qualitative research* (3rd ed.). Sage.

Cousins, L. (2008). Black students' identity and acting White and Black. In J. U. Ogbu (Ed.), *Minority status, oppositional culture and schooling* (pp. 167–189). Routledge.

Cutler, C. A. (2003). Hip-hop, White immigrant youth, and African American vernaclar English: Accommodation as an identity choice. *Journal of English Linguistics, 38*(3), 248–269. https://doi.org/10.1177/0075424210374551

Deaux, K., Bikmen, N., Gilkes, A., Ventumeac, A., Joseph, Y., Payne, Y., & Steele, C. (2007). Becoming American: Stereotype threat effects in afro-Caribbean immigrant groups. *Social Psychology Quarterly, 70*(4), 384–404. https://doi.org/10.1177/019027250707000408

Duncan, G. A. (2004). Ebonics and education: A critical appraisal of the post-1996 research literature. *African American Research Perspectives, 10*, 188–198.

Durkee, M. I., Gazley, E. R., Hope, E. C., & Keels, M. (2019). Cultural invalidations: Deconstructing the "acting white" phenomenon among black and Latinx college students. *Cultural Diversity and Ethnic Minority Psychology, 25*(4), 451–460. https://doi.org/10.1037/cdp0000288

Durkee, M. I., & Williams, J. L. (2015). Accusations of acting white: Links to black students' racial identity and mental health. *Journal of Black Psychology, 41*(1), 26–48. https://doi.org/10.1177/009578413505323

Edwards, V. (1986). *Language in a Black community*. College Hill Press.

Edwards, J. (2009). *Language and identity: An introduction*. Cambridge University Press.

Erikson, E. (1950). *Childhood and society*. Norton.

Ferguson, G. M., & Bornstein, M. H. (2014). Tridimensional (3D) acculturation: Culture and adaptation of black Caribbean immigrants in the USA. In R. Dimitrova, M. Bender, & F. J. R. van de Vijver (Eds.), *Global perspectives on Well-being in immigrant families* (pp. 31–51). Springer.

Ferguson, G. M., Bornstein, M. H., & Pottinger, A. M. (2012). Tridimensional acculturation and adaptation among Jamaican adolescent-mother dyads in the United States. *Child Development, 83*(5), 1486–1493. https://doi.org/10.1111/j-1467-8624.2012.01787.x

French, S. E., Seidman, E., & Allen, L. (2006). The development of ethnic identity during adolescence. *Developmental Psychology, 42*(1), 1–10. https://doi.org/10.1037/0012-1649.42.1.1

Gumperz, J. J., & Cook-Gumperz, J. (1982). Introduction: Language and the communication of social identity. In J. J. Gumperz (Ed.), *Language and social identity* (pp. 1–21). Cambridge University Press.

Hansell, M., & Ajirotutu, C. S. (1982). Negotiating interpretations in interethnic settings. In J. J. Gumperz (Ed.), *Language and social identity* (pp. 85–94). Cambridge University Press.

Hilliard, A. G., III. (2002). Language, culture, and the assessment of African American children. In L. Delpit & J. K. Dowdy (Eds.), *The skin that we speak: Thoughts on language and culture in the classroom* (pp. 87–105). The New Press.

Hitti, A., & Killen, M. (2015). Expectations about ethnic peer group inclusivity: The role of shared interests, group norms, and stereotypes. *Child Development, 86*(5), 1522–1537. https://doi.org/10.1111/cdev.12993

Hughes, D., & DuMont, K. (1993). Using focus groups to facilitate culturally anchored research. *American Journal of Community Psychology, 21*(6), 775–806. https://doi.org/10.1007/bf00942247

Ibrahim, A. (2003a). Marking the unmarked: Hip-hop, the gaze, and the African body in North America. *Critical Arts, 17*(1), 52–70. https://doi.org/10.1080/02560240385310051

Ibrahim, A. (2003b). Whassup, homeboy? Joining the African diaspora: BAE as a symbolic site of identification and language learning. In S. Makoni, G. Smitherman, A. F. Ball, & A. K. Spears (Eds.), *Black linguistics: Language, society, and politics in Africa and the Americas* (pp. 169–185). Routledge.

Igoudin, A. L. (2011). Asian American girls who speak African American English: A subcultural language study. In I. Du Bois & N. Baumgarten (Eds.), *Multilingual identities: New global perspectives* (pp. 51–66). Peter Lang.

John-Steiner, V., & Mahn, H. (1996). Sociocultural approaches to learning and development: A Vygotskian framework. *Educational Psychologist, 31*(3), 191–206. https://doi.org/10.1207/s15326985ep3103&4_4

Jones, T. (2015). Toward a description of African American Vernacular English dialect regions using "Black Twitter." *American Speech, 90*(4), 403–440. https://doi.org/10.1215/00031283-3442117

Kress, V. E., & Shoffner, M. F. (2007). Focus groups: A practical and applied research approach for counselors. *Journal of Counseling and Development, 85*, 189–195. https://doi.org/10.1002/j.1556-6678.2007.tb00462.x

Laroche, M., Kim, C., Hui, M. K., & Tomiuk, M. A. (1998). Test of a nonlinear relationship between linguistic acculturation and ethnic identification. *Journal of Cross-Cultural Psychology, 29*(3), 418–433. https://doi.org/10.1177/0022022198293003

Lee, M., Madyun, N., Lam, B. O., & Jumale, M. (2014). School contexts and "acting white": Peer networks of Somali immigrant youths in an Afrocentric charter school. *Schools: Studies in Education, 11*(1), 122–155. https://doi.org/10.1086/675752

Linnes, K. (1998). Middle-class AAVE versus middle-class bilingualism: Contrasting speech communities. *American Speech, 73*(4), 339–367.

Long, M. H. (2003). Ebonics, language, and power. In R. K. Blot (Ed.), *Language and social identity* (pp. 147–170). Praeger Books.

Makoni, S., Smitherman, G., Ball, A. F., & Spears, A. K. (2003). Introduction: Toward Black linguistics. In S. Makoni, G. Smitherman, A. F. Ball, & A. K. Spears (Eds.), *Black linguistics: Language, society, and politics in Africa and the Americas* (pp. 1–18). Routledge.

McBride Murry, V., Phillips Smith, E., & Hill, N. E. (2001). Introduction to the special section: Race, ethnicity, and culture in studies of families in context. *Journal of Marriage and Family, 63*(4), 911–914. https://doi.org/10.1111/j.1741-3737.2001.00911.x

McNamara, T. (2005). 21st century shibboleth: Language tests, identity, and intergroup conflict. *Language Policy, 4*, 351–370. https://doi.org/10.1007/s10993-005-2886-0

Morgan-Trostle, J., Zheng, K., & Lipscombe, C. (2016). *The state of Black immigrants, Part 1: A statistical portrayal of Black immigrants in the United States* (Research Report). http://www.stateofblackimmigrants.com/

Nguyen, A. D., & Benet-Martínez, V. (2013). Biculturalism and adjustment: A meta-analysis. *Journal of Cross-Cultural Psychology, 44*(1), 122–159. https://doi.org/10.1177/0022022111435097

Nguyen, J., & Brown, B. B. (2010). Making meanings, meaning identity: Hmong adolescent perceptions and use of language and style as identity symbols. *Journal of Research on Adolescence, 20*(4), 849–868. https://doi.org/10.1111/j.1532-7795.2010.00666.x

Noels, K. A., Pon, G., & Clement, R. (1996). Language, identity, and adjustment: The role of linguistic selfconfidence in the acculturation process. *Journal of Language and Social Psychology, 15*(3), 246–264.

Ochs, E. (1993). Constructing social identity: A language socialization perspective. *Research on Language and Social Interaction, 26*(3), 287–306. https://doi.org/10.1207/s15327973rlsi2603_3

Oyserman, D., Brickman, D., Bybee, D., & Celious, A. (2006). Fitting in matters: Markers of in-group belonging and academic outcomes. *Psychological Science, 17*(10), 854–861. https://doi.org/10.1111/j.1467-9280.2006.01794.x

Pablé, A., Haas, M., & Christe, N. (2010). Language and social identity: An integrationist critique. *Language Sciences, 32*, 671–676. https://doi.org/10.1016/j.langsci.2010.08.004

Perez, R. M. (2011). Linguistic acculturation and context on self-esteem: Hispanic youth between cultures. *Child and Adolescent Social Work Journal, 28*, 203–228. https://doi.org/10.1007/s10560-011-0228-y

Peterson-Lewis, S., & Bratton, L. M. (2004). Perceptions of "acting Black" among African American teens: Implications of racial dramaturgy for academic and social achievement. *The Urban Review, 36*(2), 81–100. https://doi.org/10.1007/s11256-004-0614-2

Phinney, J. S., & Ong, A. D. (2007). Conceptualization and measurement of ethnic identity: Current status and future directions. *Journal of Counseling Psychology, 54*(3), 271–281. https://doi.org/10.1037/0022-0167.54.3.271

Phinney, J. S., & Onwughalu, M. (1996). Racial identity and perception of American ideals among African American and African students in the United States. *International Journal of Intercultural Relations, 20*(2), 127–140.

Phinney, J. S., Romero, I., Nava, M., & Huang, D. (2001). The role of language, parents, and peers in ethnic identity among adolescents in immigrant families. *Journal of Youth and Adolescence, 30*(2), 135–153. https://doi.org/10.1023/a:1010389607319

Rageliene, T. (2016). Links of adolescents identity development and relationship with peers: A systematic literature review. *Journal of the Canadian Academy of Child and Adolescent Psychiatry, 25*(2), 97–105.

Reyes, A. (2005). Appropriation of African American slang by Asian American youth. *Journal of Sociolinguistics, 9*(4), 509–532. https://doi.org/10.1111/j.1360-6441.2005.00304.x
Rickford, J. R. (1999). The Ebonics controversy in my backyard: A sociolinguist's experiences and reflections. *Journal of SocioLinguistics, 3*(2), 267–275. https://doi.org/10.1111/1467-9481.00076
Sanchez, D., & Awad, G. H. (2016). Ethnic group differences in racial identity attitudes, perceived discrimination, and mental health outcomes in African American, Black Caribbean and Latino Caribbean college students. *International Journal of Culture and Mental Health, 9*(1), 31–43. https://doi.org/10.1080/17542863.2015.1081955
Smitherman, G. (1977). *Talkin and testifyin: The language of Black America*. Wayne State University Press.
Smitherman, G. (1981). From Africa to the New World and into the Space Age. In G. Smitherman (Ed.), *Black English and the Education of Black Children and Youth* (pp. 409–423). Detroit University Press.
Smitherman, G. (1999). *Talkin that talk: Language, culture, and education in African America*. Routledge.
Smitherman, G., & Cunningham, S. (1997). Moving beyond resistance: Ebonics and African American youth. *Journal of Black Psychology, 23*(3), 227–232. https://doi.org/10.1177/00957984970233004
Tanti, C., Stukas, A. A., Halloran, M. J., & Foddy, M. (2011). Social identity change: Shifts in social identity during adolescence. *Journal of Adolescence, 34*, 555–567. https://doi.org/10.1016/j.adolescence.2010.05.012
Telzer, E. H. (2010). Expanding the acculturation gap-distress model: An integrative review of research. *Human Development, 53*, 313–340. https://doi.org/10.1159/000322476
Thelamour, B. (2017). Applying the relative acculturation extended model to examine Black Americans' perspectives on African immigrant acculturation. *Journal of Cross-Cultural Psychology, 48*(9), 1457–1471. https://doi.org/10.1177/0022022117730614
Thelamour, B. (2021). A mixed-methods examination of acculturation and African immigrants' perceptions of black American culture. In H. Chu & B. Thelamour (Eds.), *Conceptual and methodological approaches to navigating immigrant ecologies* (pp. 107–127). Springer.
Thelamour, B., & Johnson, D. J. (2017). Exploring Black immigrants' and nonimmigrants' understanding of "acting Black" and "acting White." *Journal of Black Psychology, 43*(3), 280–304. https://doi.org/10.1177/0095798416641863
Tormala, T. T., & Deaux, K. (2006). Black immigrants to the United States: Confronting and constructing ethnicity and race. In R. Mahalingam (Ed.), *Cultural psychology of immigrants* (pp. 131–150). Erlbaum.
Umaña-Taylor, A. J., Quintana, S. M., Lee, R. M., Cross, W. E., Rivas-Drake, D., Schwartz, S. J., Syed, M., Yip, T., & Seaton, E. (2014). Ethnic and racial identity during adolescence and into young adulthood: An integrated conceptualization. *Child Development, 85*(1), 21–39. https://doi.org/10.1111/cdev.12196
Vedder, P., & Virta, E. (2005). Language, ethnic identity, and the adaptation of Turkish immigrant youth in the Netherlands and Sweden. *International Journal of Intercultural Relations, 29*, 317–337. https://doi.org/10.1016/j.ijintrel.2005.05.006
Vygotsky, L. (1978). *Mind in society: The development of higher psychological processes*. Harvard University Press.
Waters, M. (1994). Ethnic and racial identities of second-generation black immigrants in new York City. *International Migration Review, 28*(4), 795–820. https://doi.org/10.2307/2547158
Yoon, E., Adams, K., Clawson, A., Chang, H., Surya, S., & Jérémie-Brink, G. (2017). East Asian adolescents' ethnic identity development and cultural integration: A qualitative investigation. *Journal of Counseling Psychology, 64*(1), 65–79. https://doi.org/10.1037/cou0000181

Part III
Transnational Place and Identity: Resistant Re/Formation

Chapter 10
Transnational Community and Identity Among Mexican Migrants: A Third Space

Alfredo Mirandé

For generations, residents of Jalostotitlán ("Jalos") have been successfully migrating from the Mexican state of Jalisco to Turlock, California. Initially, most of this movement was what sociologists termed "circular migration" (Newland, 2009) because migrants would travel to the United States during the harvest and return to Jalos at the end of the season. In the 1960s, this pattern changed dramatically as migrants began to stay and to settle in the United States. Today, there is a vibrant Mexican community in Turlock, California, made up largely of people who originated in Jalos (Mirandé, 2014).[1] Although their adjustment to their adopted community has been remarkable, migrants and their descendants have retained a very strong sense of identification with Jalos. The retention of a Jalos identity has been so pervasive, in fact, that local residents often affectionately refer to this California Central Valley community as "Jalos USA" (Mirandé, 2014).

This chapter critically examines this complicated pattern of migration and identity formation which challenges the traditional "uprooted, Anglo conformity" sociological model of migration (Agius, 2012). The traditional model predicted that by the second or third generation, immigrants would assimilate into the dominant culture, and lose their language and allegiance to their native land (Park et al., 1925; Warner et al., 1945; Handlin, 1951; Gordon, 1964).

This chapter looks at how and why Jalos migrants challenged the Anglo conformity model and were able to adapt to life in the United States, while remaining connected physically, culturally, linguistically, and emotionally to their native land. It also addresses how this strong identification with Jalos, rather than impeding their progress, has facilitated their material and economic success in the United States. I propose that the master identity of "being from Jalos," regardless of one's place of birth, residence, citizenship, or immigration status, is facilitated by biannual religious fiestas that reinforce membership within a global transnational, fictive community—a hybrid third space between Mexico and the United States.

[1] The research was conducted prior to the outbreak of the COVID-19 pandemic.

A. Mirandé (✉)
University of California, Riverside, CA, USA
e-mail: alfredo.mirande@ucr.edu

D. J. Johnson et al. (eds.), *Re/Formation and Identity*, Advances in Immigrant Family Research, https://doi.org/10.1007/978-3-030-86426-2_10

More specifically, the chapter focuses on courtship, family, gender, and culture in this agricultural community, and points to the existence of an emergent and enduring Jalos identity that transcends both the border and traditional national boundaries. It also discusses an ideology termed "*El que quiere, puede*" [Those who want, can] that Jalos migrants readily subscribe to. I conclude that this ideology inspires Jalos residents to pursue a Mexican version of the American Dream by adherence to the belief which asserts that those who are highly motivated and hard-working can and do succeed in the United States.

Extant Immigration Theories

For most of the past century, the predominant sociological model for understanding immigration in the United States was an assimilationist paradigm that predicted the eventual acculturation, incorporation, and integration of immigrant groups (Berry, 2019; Briggs, 1978; Gordon, 1964). According to this model, the first generation of immigrants were driven to assimilate and quickly learn their new language, customs, and culture, and they were readily incorporated into the dominant society. The model posited that after a generation or two, the immigrant group would become an ethnic group, eventually seamlessly losing their culture and assimilating into American society. The ultimate fate of all immigrants, according to the assimilationist, Anglo conformity model, was incorporation into the dominant society, culture, economy, and polity. The model predicted that by the third generation, immigrant children would lose their language and culture, and meld into the great American melting pot, eventually losing their allegiance to their native land and becoming simply Americans (Handlin, 1951).

One of the problems with the assimilationist–uprooted, Anglo conformity model is that it was based largely on the experience of White European immigrants, rather than those of racial minorities like African Americans, Mexicans, Native Americans, and Asians, whose entrance into the United States historically was by force or coercion resulting from slavery, war, or conquest (Blauner, 1972). The assimilationist model was based on the experiences of groups that entered the United States as volunteer immigrant groups and quickly integrated into the primary labor force and the economy. It also assumed that, unlike Mexicans, who come from a country that shares a border with the United States, European immigrants had to travel long distances to escape poverty or religious persecution and were "uprooted" and unlikely to ever return to their native land (Handlin, 1951).

There is an emergent scholarship that has been extremely critical of the assimilationist, uprooted Anglo conformity model of immigration and offered alternative models. Portes and Rumbaut (2006) discussed several popular models of migration, including "push-pull" and "structural misbalancing" of core and peripheral nations. The push-pull model focuses on factors of expulsion or "push" (economic, social, and political hardships) in the sending countries and contrasts it with positive "pull" factors in receiving countries, such as economic and political advantages and demand for cheap immigrant labor. The model predicts that immigration will be greatest when there are factors simultaneously pushing people out of the sending nation and pulling them toward the host country (Portes & Rumbaut, 2006). Structural misbalancing also shifts the focus of analysis away from individualistic theories of immigration and examines the impact of macro forces such as colonization and dependence on migration. For example, the three largest Latino immigrant groups in the United States came from areas that have a long colonial or neo-colonial history with the United States (Mexico, Puerto Rico, and Cuba).

Social network analysis, on the other hand, proposes that once certain members of a community migrate, dissemination of information about economic opportunities and the migration process may become self-sustaining, as successful migrants bring back information that facilitates the eventual migration of others (Portes & Rumbaut, 2006; Vecchio, 2007). The fact that international migration is a network-creating and network-dependent process has been established by sociological studies that date back to Thomas and Znaniecki's (1996) classic study of Polish peasants in the United States at the beginning of the twentieth century. Other researchers have also explored the social networking of immigrants. For example, Portes and Bach's (1985) study on 822 adult Mexican immigrants arriving in two Texas ports of entry reported that over 90 percent of the respondents had obtained legal residence through family and/or employee connections.

Massey et al., (2002) also found that social networks largely accounted for the migration patterns of 19 Mexican communities, which were grouped into distinct categories. In the first category, migration was unusual or exceptional, whereas in the final category it was normative. The critical factor was the timing of the first pioneering trips, so that among the recruited adult male laborers, those who traveled abroad on their own, their multiple returns created cross-national networks (Massey & Goldring, 1994). Thus, Jalos is a community where immigration is normative and where social networks are critical for explaining migration patterns.

Emergent Transnational Theories of Migration

In response to the failure of traditional immigration theories to explain the persistence of close ties between migrants and their home countries (Smith, 2006), scholars began to propose and develop new transnational theories of migration. Glick Schiller (1999, 2003) and her colleagues were among the first to articulate a theory of transnationalism that had several essential elements. First, it refuted the uprooted view of immigration that predicted the inevitable rupture of ties to the old country. Instead, it proposed that immigrants today are able to continue to be linked to their home country long after they migrate. Second, it posited that "capitalism has created a system of global markets and processes that increased migration and extend beyond the nation-state, thereby creating a kind of civil global society that threatens the state's monopoly on politics" (Smith, 2006, p. 6). Thus, transnationalism is the process through which immigrants are able to develop and maintain familial, economic, and social relations that link together their countries of origin and settlement. Finally, the theory argued that transnationalism has created what David

Gutierrez (1996, 1998) terms a "third space" between the grasp of the nation state of both the host and the home society.

Glick Schiller and her associates distinguished between the global and the transnational, noting that the term "global" is reserved for processes that are not located in a single state or space but are carried out throughout the world, whereas international or binational migration refers to a pattern of migration in which persons move across international borders but maintain social relations with their home country (Glick Schiller, 1999, 2003). Transnational migrants, then, literally live their lives across international borders (Glick Schiller, 1999). Such populations live in different geographic locales but feel related to one another and retain social relations with each other (Gowricharn, 2006). Transnational communities are not societies in the sense that they have a state, a legal system, or a labor market, but they come close to being communities. Like Jalos, they are social entities that "are unified by religion, language, customs, rituals, tastes, and the like, and are also integrated at the micro level by kinship systems and primordial loyalties" (Gowricharn, 2006, p. 6).

This new transnationalist approach has challenged the traditional view of immigration and immigrants, which formerly focused on assimilation and incorporation into the dominant society and was based on a static view of the immigration process that assumed that immigrants play a passive role in their integration into the dominant society. The transnational model also impacts our conceptions of identity, as people develop a fictive community identity that transcends political and geographic borders.

The Present Study

The study was conducted in a transnational, fictive community that transcends national boundaries and includes persons "from Jalos" who reside either in Mexico or in Turlock, California. Jalos is located in the Los Altos ("Highlands") region of Jalisco.

Like Jalos, Turlock is an agricultural community. It is located in Stanislaus County, California, in the middle of the San Joaquin Valley, the second largest city in the county, and part of the Modesto Metropolitan Statistical area.

Founded in 1899 by a prominent cattle rancher, Turlock became known as the "Heart of the Valley" because of its agricultural production (Mirandé, 2014). By the early part of the twentieth century, intensive agricultural development surrounded most of the city. Agriculture and the dairy industry remain major economic forces in the region. The major products are milk, almonds, cattle and calves, chickens, walnuts, silage, peaches, alfalfa, deciduous fruit and nut (nursery), turkeys, and tomatoes (cityofturlock.org).

Jalos residents have a long history of migration to the United States, which dates.

back to the early twentieth century (Rojas, 2008). Initially, most men and a large number of women would make the journey to the United States in search of employment—some migrated as early as 14 or 15 years of age. While people from Jalos migrated to a number of cities in California, including Anaheim, Oakland,

Sacramento, and Los Angeles, Turlock was selected for study not only because of the extensive migration by people from Jalos but because it is an agriculture community that is roughly the same size as Jalos.

Thus, the present study explored the migration patterns of people from Jalos and how they were able to develop their sense of community as "being from Jalos" across national boundaries. The idea of being from Jalos was reinforced by the establishment of a vibrant Jalos community in Turlock, and the celebration of the traditional town fiestas, or festivals, on both sides of the U.S.–Mexico border.

Method

The findings of this study are based on a wide range of qualitative research methods, including participation observation, ethnographic field research, archival research, and 85 in-depth open-ended personal interviews conducted on both sides of the border (40 in Turlock and 45 in Jalos). The sample includes not only ordinary citizens but also priests, school principals, psychologists, and social service providers. Interviews were conducted in Spanish and translated by the author.

The open-ended interviews included background questions such as where the person was born and raised, family composition, and demographic information. There were also questions about how parents got along and who exercised the most control in the family. Other topics included *noviazgo* (courtship) and other dating rituals, decision-making in the family, conceptions of masculinity and machismo, gender roles, attitudes toward child-rearing, perceptions of the major problems facing the community and/or youth, and relations between *ausentes* (the "departed" or "absent") and *presentes* (current Jalos residents). With the Turlock sample, a great deal of attention was devoted to the migration process, social networks, and how life changes in the United States.

Finally, focus group interviews were conducted with youth (15–17 years of age) in the respective communities: two in Jalos and one in Turlock. One of the Jalos focus groups involved students at CONALEP (Colegio Nacional de Educación Profesional Técnica), a technical high school that prepares its largely working-class students for work in trades such as auto mechanics, electrical, and welding. The second consisted of students at the *Prepa*, or college preparatory school in Jalos, who came largely from middle- and upper-class families. The third was an hoc Turlock youth focus group. A final focus group was held with a class of kindergarten children in Jalos.

Results and Discussion

While this is one of an increasing number of studies of transnational migration, an important difference is that the present study is not on focused on transnational migrants per se but on transnational migration, community, social networks, and identity. Almost all families in Jalos have one or more family members who had migrated to the United States (*ausentes*), so that people who do not migrate are the exception, rather than the rule.

A second distinguishing feature of the study is that migration between Jalos and the United States has historically been circular, as people traveled back and forth between Jalos and the United States, with many families spending periods of time alternately living in Jalos and Turlock. Finally, the study is unique in focusing on the creation and maintenance of a transnational identity and community among persons who identify as being "from Jalos," regardless of their place of birth, citizenship, or immigration status.

The study seeks to analyze, evaluate, and hopefully gain a greater understanding of the persistence of Jalos identity and to interrogate what it means to say that one "is from Jalos." I argue that Jalos is a community that originated in the town of Jalostotitlán, Jalisco, but that has become a binational community without borders. It is a fictive, hybrid third space united by religion, language, rituals, and culture, and integrated at the micro level by kinship systems and "primordial loyalties" (Gowricharn, 2006).

The Jalos Fiestas

Jalos is located within a strongly Catholic region in Los Altos ("Highlands") of Jalisco, which has a rich religious history, including the *Cristero War*.[2] Two religious events in particular work to engender a strong village identity among both current (*presentes*) and former (*ausentes*) residents of Jalos (Mirandé, 2014). One is the *Fiesta Para La Virgen de la Asunción* Festival celebrating the Ascension of Mary. It should be Virgin of the Ascension of Mary [into heaven], the patron saint of the city, which is celebrated on August 15 but lasts for approximately two weeks (August 1–August 17). While primarily a religious festival, it is also a time for eating and drinking, and for an opportunity for *ausentes* to return and reunite with family and friends (Mirandé, 1984).

The second event, *Carnaval*, takes place a week before Lent in late February or early March, is less family oriented and generally caters more to young people, and is characterized by excessive dancing, drinking, and partying (Mirandé, 2014). *Carnaval* includes numerous activities like the *palenque*, or a music venue that attracts popular national signers and entertainers. There is also bull fighting, a rodeo, baseball games, mariachi music, a big-rig truck parade, fireworks display and entertainment, a parade with floats, and the crowning of the Queen of Jalostotitlán. Although the celebration for *La Virgen* is a more serious event that caters mostly to families, and *Carnaval* appeals more to the youth, "both fiestas draw thousands of

[2] A Holy War and counter-revolution from 1926 to 1929 which came in response to the strict enforcement of the anti-clerical provisions of the Mexican Constitution of 1917 and the expansion and brutal enforcement of these laws in the 1920s (Mirandé, 2014).

people, as *los ausentes* return in large numbers from the United States to visit family and friends" (Mirandé, 2014, 20).

During the Jalos fiestas, it is not uncommon to see vehicles, especially pick-up trucks, with bumper stickers that say "*Yo amo a Jalos*" [I love Jalos], as well as baseball caps, shirts, coffee mugs, and other paraphernalia with "Jalos" proudly displayed on them. In fact, in the *mercado* one can even find silver-and-black Raider ponchos with "Jalos" printed on the front. These are clearly visible symbols of people's strong identification with their native home.

Even more impressive is the fact that thousands of *ausentes*, absent community members living in the United States, make yearly pilgrimages back to Jalos to celebrate the fiestas. During an interview in Jalos, the assistant to the *presidente municipal*, or Assistant Mayor, commented that the Jalos population more than doubles during the fiestas, swelling from about 75,000 to more than 200,000 when the *ausentes* return. Although most people stay with relatives, it is virtually impossible to find a hotel room during the fiestas.

Several years ago, Turlock adopted the custom of honoring Jalos' patron saint, the Virgen de la Asunción [Virgin of the Assumption], by establishing its own fiesta in honor of *La Virgen.* A committee was established in Turlock to celebrate the annual fiesta, and after the first year, members commissioned the artist who made the original statute of the Jalos Virgin to produce a replica for use in the Turlock celebration. The fiesta for the Virgen de la Asunción is held each year on August 15 and has been so successful that some people believe that the Turlock celebration is as good or better than the Jalos event (Mirandé, 2014).

Elvia, a young mother in her twenties, commented on the importance of the August fiestas for her family, noting that for an *ausente* like herself, *las fiestas*

> are a great way to remain connected with our roots and ancestors. It is the only way to understand where we come from and why our families migrated to the U.S. Also, it is a time to revisit family from México and those who live in the U.S. that we do not visit for geographic reasons or simply because time did not permit during the rest of the year. But overall it's a family tradition that began when I was a child and that I hope to continue with my children.

Elvia was not born in Jalos but in Berkeley and moved to Oakland at a young age, but she considers herself an *ausente* because her parents moved back to San Gaspar in the municipal district of Jalostotitlán when she was a toddler, and she sees Jalos as home. She attended kindergarten in San Gaspar, returned to Oakland for second grade, and is currently an elementary school teacher in Turlock.

Third-generation twin 17-year-old sisters, part of the Turlock focus group, were similarly born and raised in Turlock but take great pride in and identify with "being from Jalos" because their family is originally from there, although they have never lived in Jalos. During the focus group, they smiled and reported that they tell people "they are from Jalos." They love Jalos and are very proud to be from there. To be from Jalos, then, is clearly not about a person's place of birth or residence but their place of origin. It is more about one's primal roots than where you were born.

Los Ausentes y Las Fiestas

The interviews revealed that people who lived in the United States, *ausentes*, generally had a very favorable view of Jalos and *las fiestas*, particularly the August fiestas, even if they were not able to attend. The primary factors preventing people from attending were being undocumented, having young children, not having enough money, or being elderly or infirm.

In years past, individuals and families would migrate and go back and forth freely between Jalos and the United States, spending part of the year in Turlock during the harvest season before returning to Jalos. This caused problems for children in school because they would spend part of the school year in Turlock and part in Jalos. With increased border enforcement, this has become increasingly more difficult, and many people simply believe that it is too risky to return to Mexico for the fiestas, and they remain in Turlock.

Cheno González was born in Jalos, came to Turlock in 1964, and is now a very successful farmer. During the interview, Cheno shared that his children love Jalos and that the family goes back almost every year, especially for the August fiestas or Christmas. Some years the family didn't go because he didn't like the children to miss school, but now that they are older and out of school, they go as a family.

Lola Olmedo, a community leader, and her husband, Beto, also love to return to the fiestas. Lola shared how she came up with the idea of organizing the fiesta for the *Virgen* in Turlock when she realized that many people, for one reason or another, could not go back to Jalos for the celebration. They were depressed that they could not visit the *Virgen*, so the idea was to bring the celebration and the *Virgen* to the people in Turlock. A remarkable, strong, energetic, and charismatic woman, she is credited with almost single-handedly organizing the fiesta to the *Virgen* in Turlock. She worked with the committee to bring the fiesta to Turlock, but everyone, including the local priest, shared that she was the one who made it all happen. After a year, they were also able to raise enough funds to purchase a replica of the Jalos statute of the *Virgen*. Each year they also invite one or more priests from Jalos to celebrate the mass and join in the procession.

Lola talked about the importance of the fiestas for *los ausentes* who return to Jalos. She also commented on the large number of new trucks and SUVs that you see at the Jalos fiestas. She believes that a lot of the young men want to impress when they go back. At times, the ones who go to Jalos and even here in the United States can't really afford to purchase a new vehicle, but they do so anyway. She smiled and said: "They buy the car, they buy it on time, they take it there [Jalos] and then a lot of times they have to return it because they can't afford it. Sometimes the idea is simply to take a good car to make an impression and they make arrangements with a dealer but when they get back they have to return it because they don't have the money to pay for it." Lola added that "it's every *méxicano*'s dream to go back with a new car or big truck in order to impress." It's like a visible, tangible symbol of one's success in *El Norte* (the United States).

I met Dora Méndez outside the Sacred Heart Parish in Turlock. She is an enterprising woman who makes extra money for her family by selling homemade potato chips and *chicharrones* (fried pork rinds) as people exit the mass. At the time of the interview, she had been in the United States for nearly 20 years. Like most people from Jalos, Dora was undocumented when she first came, but she now has her *papeles* (papers) and goes back to Jalos twice yearly. Dora sometimes goes in December or for Mother's Day, and she also goes for the August fiestas every year. She said that her children like going back for the fiestas and that they get along really well with the youth in Jalos.

Alma Castro, a 40-something catechism teacher, also migrated from Jalos nearly 20 years ago. She sees the August fiestas for the *Virgen* as *muy bonitas* (quite beautiful). Alma has four children and proudly shared that the oldest is 18, fully bilingual, and loves Jalos. Her daughter is intrigued by the fiestas and the old-fashioned dating customs like the *serenata*, which is a courtship ritual in which the boys and girls go around the plaza in opposite directions. A boy will stop and give a flower to a girl that he likes or who gets his attention or will ask her to take a turn around the plaza with him. Her daughter gets along well with the young people from Jalos, including her cousins, and she has even participated in the local *noviazgo* or dating rituals. She loves Jalos and, ironically, feels there is more freedom there because she can go out with her friends to the plaza or to eat with them.

Ofelia and Manuel Ruiz have been living in Turlock for eight years and would love to go back to Jalos for the fiestas but can't because they are without papers and have small children, ages 5 and 10. Manuel works as a *fiador* [feeder] at a dairy outside Turlock. When they lived in Jalos, they used to go to the fiestas every year, and Ofelia went back with the children for about five months when her mother was sick. The Ruizes would love to continue to go back to visit, but Manuel said, "*si podemos ir, ¿pero luego para regresar?*" [Yes, we can go. But what about the return?].

Despite making an exact replica of the Virgin and attempting to duplicate the Jalos celebration, the Turlock celebration differs in several important ways. First, and most importantly, it is primarily a religious event and family affair with little or no drinking or partying. Second, although organized by people from Jalos, the celebration extends out to the entire Mexicano community, not just to people from Jalos. Third, the event serves not only to intensify identification with Jalos but also to solidify the Mexican community of Turlock as a whole. The celebration is organized by people from Jalos but not limited to them and involves all of the parishioners at the Sacred Heart Parish. In fact, at one of the celebrations I attended, members of the Knights of Columbus marched at the front of the procession honoring the Virgin dressed in full uniform, sword, and formal regalia. Whether *los ausentes* are able to return to the fiestas in Jalos or not, it is evident that the celebrations remain a way for *los ausentes* to maintain a strong religious, social, and cultural bond to Mexico and their hometown. I was touched by the fact that in recreating the fiesta for the Virgin in the United States, the first order of business for the organization was to buy a replica of her statute, since the church provides no financial support for the event and it is all funded by private donations. This reinforces the religious importance and solemnity of the celebration. The inclusion of women and children also reinforces this as a family event, as does the banning of alcoholic drinks.

Father Mauricio, a Colombian priest from the parish of the Sacred Heart, commented during an interview on how impressed he was with how the first and second generations were so actively involved in the celebration to the Virgin in Turlock, and how important it was that parents were passing on their customs and traditions to their children. The Turlock fiesta also represents a form of reverse assimilation in bringing Mexican culture and Jalos rituals and traditions to the United States, as even non-Mexican Catholics have begun to partake in the celebration. Finally, unlike in Jalos, there is no alcohol, and sadly no fireworks to honor *La Virgen* or light up the beautiful Turlock sky.

Although thousands of people also attend *Carnaval*, *ausentes* are more critical of this event because it caters more to the youth and encourages the massive consumption of alcoholic beverages. A number of people I interviewed for the study expressed concern with the amount of drinking that goes on at the Jalos fiestas, especially *Carnaval*. Alma Castro, a catechism teacher, mentioned that she normally attends the August fiestas but not *Carnaval*. One of her brothers never misses *Carnaval* and brought a video that he took this past year. Alma was alarmed by the video—"What I saw in the film [video] more than anything, the one my brother brought, it looks like they drink too much. I asked myself, am I going to let my son go? [he was 16 at the time] Perhaps he will want to go. It's like, I wouldn't be able to let him go by himself. Go? No, because they sell [alcohol] in excess."

One of the local priests at the Sacred Hear Parish in Turlock, Padre Gustavo, believed that the fiestas had been a great success, but as he talked about people from Jalos, it was clear that he also saw alcohol and alcoholism as problems. He remarked that people from Jalisco seemed to drink a great deal, but it was more or less the custom. It was a problem both in Jalos and in Turlock. Alcohol was the cause of many "infractions and problems with the law," and for this reason he was opposed to selling beer at the Turlock fiesta. Father Gustavo believed that it helps that the age limit for drinking alcohol in California is 21, but in Mexico young children are able to drink.

Señora Lola added that at the Jalos fiestas, "*tomán como si se fuera a terminar*" [They drink as if it was going to run out]. A number of other people saw alcohol and drugs as major problems in Jalos. Two of the priests I interviewed in Jalos believed that drinking and drugs are a problem. Father Marcos, the priest at the San José Parish in one of the poorer sections in Jalos, noted that since the 1990s, when the United States started to cut down on the drug trade, drugs have become readily available to youth in the Los Altos region. Father Patricio, a priest at a more affluent parish, also saw drugs as a hidden problem in Jalos. Another problem is that children start drinking at a very early age and even drink openly during *Carnaval*, which lasts for approximately ten days, caters primarily to younger people, and can be characterized as a continuous party with non-stop music, dancing, and drinking.

Intergenerational Differences Within Families

Because of transnational migration, there are often intergenerational cultural differences in families not only between but also within generations. In the López family, for example, older siblings who were born in Jalos are often seen as belonging to a "Mexican" generation because they were born and spent their formative years in Mexico and were largely Spanish-language speakers. Those born in the United States or who came to the United States at a young age but lived part of their lives in Jalos and part in the United States, in turn, are seen as belonging to a Mexican American or Chicana generation—a hybrid generation that is fully bilingual and bicultural. Sociologists sometimes refer to this generation as the "one and one-half" generation because they are between the first or immigrant generation and the second generation, which was born and raised in the United States. Finally, people who were born and raised in the United States are often seen as the "American" generation because they speak Spanish but clearly prefer English, have never lived in Mexico, and define Jalos largely as a place they like to visit but where they would not want to live. It should be noted, however, that even members of the American generation closely identify with their culture and as being from Jalos.

One of the respondents, Sofia, an amazing and remarkably articulate ten-year-old born and raised in Turlock, is an example of the American generation. She frankly lamented her lack of fluency in Spanish, noting:

> My Spanish is not as good as it should be. I get corrected a lot in *Mexíco* so I don't speak that much. I love my culture. I love the way we get together at the Rancho[3] every Friday with the family, and how they talk and drink. And people saying this is one of your cousins.

Members of the Turlock focus group also strongly identified with being from Jalos. Although they had some friends who were not from Jalos, they were enthusiastic and took pride in being from Jalos.

The cross-generational component of families is especially evident in large families in which there is a substantial age gap among siblings. One of the key families in the Turlock sample, the López family, has 11 siblings and a 25-year age gap among them. There is a paradox within the family in that they themselves identify distinct generations within the family. The older generation tends to complain about the younger one, referring to them jokingly in the third person as "The Americans" or *los americanos*. They see the younger generations as a bit spoiled, lazy, and "less driven" than the immigrant generation. They also believe that this generation has not had to suffer or face the hardship or adversity faced by the first or the second generation. It is less driven. This generation "had it too easy, is less motivated, and they don't want to work."

---

[3]The *rancho* is a small piece of land in an unincorporated rural section of Turlock purchased by the children for their father, who is in his late 70s. His dream, like that of many migrants, was to go back to Jalos and buy a rancho. The purchase of the rancho was prompted by his wife's pronouncement that she would not move back to Jalos and face separation from her children and grandchildren.

Armando, a holistic doctor and one of the older siblings in the González family, remarked:

> The ones who were born in the United States had not lived "the Mexican experience." Those born in Mexico, on the other hand, knew what it's like to live in Mexico and understand the culture, whereas from Lupe down to the twins, that's the time we moved here. The twins were born here. Socorro was born here. My brother Manuel was born here also, so they never lived the Mexican experience.

He added, "The only way they learned was through us; stories we would tell them." Most importantly, living the Mexican experience included living in Mexico and understanding all of the struggles [you go] through in crossing the border illegally.

The fiestas serve not only to promote a Jalos identity, familism, solidarity, and collective pride but also to affirm the Mexicano version of the American Dream of success. There are a number of visible emblems of success during the fiestas, such as the large number of late-model shiny SUVs and trucks with license plates from various states in the United States, primarily California and Texas, but also many other states, including Maine, Iowa, Michigan, and Nebraska, as well as American clothes, toys, cell phones, iPads, electronic devices, and other consumer goods.

The Youth

Socorro, the youngest of the López children, also loves Jalos and identifies closely with her culture, even though she was born in Turlock, is a U.S. citizen, and has never lived in Jalos. Her brother Armando, the holistic doctor, said proudly:

> She likes mariachi music, she likes *banda*, she likes going to Mexican concerts. She is very proud you know. When she graduated from college she wore the Mexican flag. She speaks Spanish and she loves the Mexican culture, you know, the overall values, marriage, and family.

While people on both sides of the border "loved Jalos," distinctions and invidious comparisons were often made between *presentes* and *ausentes*. The youth in Jalos, for example, tended to view *ausentes* as keeping to themselves when they visited and being a bit standoffish. Lisa Blanco, a young woman from Jalos, said that *los pochitos*[4] tended to seek one another out and to "hang out together." Other respondents noted that while *ausentes* and *presentes* generally got along well, it was *los ausentes* who set the tone and created the mood for the fiestas. They were usually seen as having the dollars and as coming to Jalos to drink and to party. There was consensus among the youth in the Jalos focus groups that folks from the United States came to Jalos to drink and get crazy. The Jalos youth noted that several of the

[4] *Pocho* was historically a derisive term used by Mexicans to refer to US Mexicans who were assimilated and did not speak Spanish, or spoke it poorly. The diminutive *pochito* is less derisive and more of an endearment.

clubs in town were only open during the fiestas, and that the Americans (i.e., *ausentes)* patronized the clubs and kept them going.

Language was seen as a critical factor in retaining the culture. Although language is largely retained by youth in Turlock, it was generally believed that once you lost the language you lost the culture. Cheno González mentioned that once children lost the language or felt uncomfortable speaking Spanish, they would begin to feel embarrassed and to seek out people who only spoke English. Miguel González, Cheno's cousin, believed that language was the major divide between people from Jalos and those in the United States. He was very proud of the fact that his children and grandchildren spoke Spanish and admonished them not to speak English during the fiestas. He felt that the fact that they could speak Spanish made speaking English in Jalos that much more unacceptable. On the other hand, Miguel noted that sometimes when you go to the fiestas, "they [the Jalos side] act like they don't know you. There is a separation, a distance."

There were also significant differences in the way gender and sexuality play out on either side of the border. With regard to the perception of women, youth in the Turlock focus group felt there were clear differences between themselves and Jalos girls.

They typically described the Jalos girls as *fresas*, or literally strawberries, which meant that they were dainty and delicate and extremely feminine (Mirandé, 2014). Jalos girls are seen as *chiqueadas*, as spoiled or pampered. One of the boys in the Turlock focus group said that dating is very different, noting, "There it's for marriage. It's like you know how the parents get married young," One of twins added, "Yeah, it's not just to be with them, it's to marry them. They're focusing on marriage." Julia added:

It's like over there guys are really serious. Over here they don't much care. Girls are more respected and well treated than over here."

Güero, one of the boys in the focus group, commented that over here, girls will say, "'What?' and in Jalos it's like, 'Ay, no' ['oh, no,' mimicking a girl's voice] …. It's way different. The way they talk, it's sometimes like annoying."Another focus group member, Julia, stated, "Yes, it's like if you meet somebody and they say where are you from and you say Jalos, they would say, '*O, ya sabemos de las muchachas de Jalos*' [Oh yeah, we know about the girls from Jalos]." Everybody does say that the Jalos girls are *fresas,* which means they're delicate and spoiled. "They're stuck up and conceited." Güero added, "*Como dicen* [As they say], they are 'girlie.'"

Paradoxically, while the Mexican youth believed that there is much more freedom in the United States, the Turlock youth contended that there is much more freedom in Jalos. Güero said that even though he felt at home in Jalos, it's like a different world there. He noted, "yes, I feel like I'm at home, I don't know. It's just different. I mean it's a whole different world there. The rules are, man, like little kids can go to the store and buy cigarettes or alcohol."

Socorro and the Turlock focus group members believed there is more freedom in Jalos because you can drink or smoke without an ID, and you can go out late with no curfew. You can even drive a car without a license even if you are underage. Julia remarked, "Heck yeah! It's like driving, like there's people in back of trucks and stuff, so it's not really like over here." Jalos girls are different because they are very feminine, "girly girls," who wear make-up, have long fingernails, are more serious about marriage, and are looking for a husband. One of the boys pointed out that "they don't pay attention to you if you don't fit the profile. They are looking for someone that can take care of them and bring them to the United States."

Despite these differences, Turlock boys are also searching for a traditional wife who cooks, does housework, and takes care of the children; so they are attracted to the girls from Jalos. Girls from the United States, however, don't want to serve and be ordered by men. Young people from Turlock generally love visiting Jalos but would not want to live there. Girls like the gender freedom they have in the United States and would not want to be dependent on men. The youth also acknowledge there is nothing to do in Jalos after the fiestas.

Jalos youth, on the other hand, generally believe that there is more sexual freedom in the United States and that the girls on the American side of the border are more permissive. Interviews with the youth focus groups suggested that changes in gender roles and courtship practices were slow to take hold in Jalos. Most of the young people from Jalos said they carried on a traditional courtship. Couples were even slow to hold hands, and relations were characterized as *manos mojaditas*, or "sweaty palms." Couples did not date as such, but they might go to get an ice cream on Sunday or downtown for a walk in the square. It was the more affluent youth in the *Prepa* focus group who would go to the clubs drinking or on excursions to the beach in another city.

Social Capital and Social Support Networks

One of the major findings of this study is that Jalos residents who migrate to *El Norte*, especially the first generation, have been incredibly successful. Perhaps what is most impressive about this success is that it occurred at all, despite the fact that members of the first or pioneering generation were largely undocumented and had little or no education, and in many instances could not read or write. Like previous studies, the findings point to the importance of social networks and social capital in explaining their success. Social capital refers to the network of relationships that a person has, whereas cultural capital refers to values, norms, and ideology.

According to Coleman, "social capital is created when the relations among persons change in ways that facilitate action" (1990, 304). Social ties of friendship and kinship provide few advantages to people seeking to migrate in and of themselves, but "once someone in a person's network has migrated, however, the ties are transformed into a resource that can be drawn upon to gain access to foreign employment and all that it brings" (Massey et al., 2002, 19). Jalos migrants always had family or friends in the area who aided newcomers in finding not only employment but also housing, schools, and places to shop for food. In some cases, they even aided in helping them cross the border illegally by lending them false documents. An additional form of social capital that is critical to Jalos migrants and is examined in this study is religion and Church, and the social life surrounding it.

Recently migrated Jalos residents invariably had friends and/or family who helped in the transition and facilitated their integration into a new language, life, culture, and world. Father Gustavo León, a priest assigned to *El Sagrado Corazon* Parish in Turlock, noted that the impact when one culture clashes with another is considerable, especially when you have such a large, hegemonic culture like the American one. Father "Gus" felt that the transition is always difficult, but in Turlock, "persons [from Jalos] are better received, at least the ones I see," although "they have to be prepared for the reality of his country because many of them are undocumented."

Father Gus noted that there was a movement to make Jalos and Turlock sister cities, which was already in process. He felt it was a natural connection because there were so many people from Jalos here. The municipal president of Jalos had visited the previous year and wanted to enter into a formal accord between the two cities. There was a proposal to make Jalos and Turlock sister cities, adding that many people from Jalos who are living here participate in the celebration in honor of the *Virgen de la Asunción* in August.

Socorro López, the youngest of the López children, said that when people come from Jalos there isn't as much cultural shock as you would get if you went to another city like Los Angeles. The unique thing about Turlock, she added, is that when you come here, there are already a lot of people from Jalos and there is a strong support network.

El Que Quiere, Puede: The Mexican Version of the American Dream

Jalos migrants subscribe to an ideology that appears to greatly enhance their success in the United States. As previously noted, a common belief among migrants, especially members of the older generation, is that the United States is the land of opportunity and that in this country there are many opportunities to succeed. These beliefs are captured in the common refrain best articulated by Cheno González, a successful farmer, who said: "*El que quiere, puede*" [You can do it, if you try].

Cheno and his two brothers came to Turlock in the 1960s. The González brothers were among the first migrants to settle with their families in Turlock, and over the years they helped and supported many other Jalos migrants who subsequently settled there. Cheno not only believes that there are many opportunities for people here but has been extremely successful and is now the proud owner of his own farm—González Farms. He has literally lived out the Jalos version of the American Dream, and his remarks capture an ideology that is prevalent among Jalos migrants. The ideology, a form of cultural capital, basically holds that there are many opportunities to succeed in this country and that those who are motivated will succeed. Because the United States is the land of opportunity, moreover, if people don't succeed, it's probably because either they don't want to succeed or because they have imposed barriers on themselves that kept them from succeeding. A related theme is that subsequent generations, born in the United States with more opportunities, are somehow "less hungry" or motivated to succeed because they have not had to endure as much adversity as earlier generations. José Portillo, another pioneer, articulated a similar view and said there was plenty of work for folks from Jalos so that "the person who doesn't work, doesn't want to work."[5]

The pioneer generation also tended to minimize or deflect the importance of discrimination in impeding one's progress, often expressing the view that what people perceived as discrimination might be a way that they are limiting themselves or blaming others for their failures or limitations. Cheno said that he was helped and learned a lot from the Americans. Cheno's cousin, Miguel González, also minimized the importance of discrimination. When asked how Mexicans were treated in Turlock and whether he had witnessed discrimination, he said that he hadn't seen discrimination and that he had not experienced it, except, ironically, in the Church, where there was discrimination against Spanish speakers.[6] Although he had to struggle with English and experienced discrimination against Spanish speakers in the Church, he never saw a situation in which someone went somewhere and was not served because they were Mexican. José Portillo acknowledged discrimination, but felt that sometimes one brings it on oneself because of self-doubts and feeling that one is not capable. He pointed to the fact that he hasn't learned English, after so many years in this country. Sometimes you feel inadequate, "but it's not because they are discriminating against you."

But for Señor López there had always been discrimination, and he believes that the government always helped the Anglos more. Employers also gave Americans more latitude at work and protected them more. The foreman at a winery where he worked, for example, harassed him for many years because he supported the Union, but he learned his rights and was able to defend himself. Although there is always discrimination, he also believes that "sometimes it is one's own fault because of the way you behave." Señora López added:

> We know we are in a country that is not ours," adding that people complain about discrimination a great deal, she believed, but if you are a hard-working and responsible person that accepts your responsibilities, you shouldn't have any problems.

A final factor that promotes the success of Jalos residents in Turlock and other locales is the cultural capital that Jalos residents bring with them, including not only a strong Jalos identity, pride in their heritage, and identification with the land and agricultural work, but what I have termed the "Jalos work ethic." Jalos residents bring with them an ideology and a very strong work ethic, which extols the virtues of working and respecting the land. Jenny Rivera, a woman with a very successful

[5] This ideology generally parallels the Immigrant Paradox, which is the finding that first-generation immigrants in the United States generally tend to be healthier, are less likely to suffer from emotional and psychological problems, have lower divorce rates, and are less likely to be incarcerated than the second, third, and fourth generations.

[6] It was clear that when respondents were asked about discrimination, it elicited images of overt, formal, and legally sanctioned discrimination, rather than more subtle, indirect, less formal discrimination.

realty business, perhaps said it best when she described her father's tie to the land. It seemed ironic, she said, that given that her parents were migrant farm laborers, she and her siblings all owned land and had farms now. In addition to her business, she and her husband have a farm outside of Turlock and grow peaches and almonds. She added, "Yes, I think that one of the things that I am really thankful to my parents for is that despite the fact that we worked in the field and that we were migrant laborers, they taught us to love it, to love the field. That is, we never ever looked down on it."

When people asked her father what kind of work he did, he would always say, with a.

great deal of pride, "*que trabajaba en el fil*" [that he worked in the field]. And so her father made them feel like working in the field was *digno*, or dignified, and that it was honorable work. Another thing that she always remembers is that he taught them to be proud of where they were born. He would tell them to not tell people that they were born in Guadalajara and then moved to an anonymous rural area or a rancho. He would say to take pride in where they were born even if it was a small village. Since they were children, he taught them to be proud and to never be ashamed of who they were and where they came from or to be ashamed of their culture or what they did for a living. Since they were little, he showed them how to do the work, to struggle and strive.

Jenny related a story that she remembered vividly because being the oldest she always had to watch her siblings. One time her father came home from picking grapes. It was really hot in the valley, and she asked if he would please shut the window so that she could turn on the cooler. But as he lowered the window, the glass shattered and he cut his left hand badly, severing the tendons. They took him to the hospital, and the doctor asked what type of work he did and then told him that he would have to stay home from work because you need two hands in order to pick grapes. Jenny had mixed feelings. Although she was sorry her father was hurt, she was happy that he would be at home with her and her siblings. She was only nine and was really surprised when her father woke her up the next morning at five to go to work with him in order to help him. She recalls how he instructed her, "Put the pail on the ground. I am going to cut with the knife and you are going to prune and move the pail."

While this may have seemed insignificant at the time, Jenny uses this as an example with her children to show them that nothing is impossible and that you can succeed even if you face impediments. In other words, even if they cut our arms off, "with whatever we have left we are going to use it to move forward and to succeed!"

Conclusions and Policy Implications

This study found that transnational migrants from the town of Jalostotitlán in Jalisco, Mexico, were able to retain an almost primal transnational Jalos identity regardless of their place of birth, age, or immigration status. The findings have important policy implications, and challenge the prevailing assimilationist, Anglo conformity model which predicted the inevitable acculturation, incorporation, and integration of immigrant groups into the dominant society (Briggs, 1978; Gordon, 1964). According to this model, the first generation of immigrants are driven to assimilate and to quickly learn their new language, customs, and culture, and are readily incorporated and assimilated into the dominant society and culture.

While the study focused on a particular community with a long history of migration to the United States, the findings have broader policy implications consistent with emergent theories which are challenging the assimilationist paradigm and point to a growing number of trans-migrants who have migrated from one nation to another while successfully retaining familial, social, cultural, and political ties with their country of origin.

There has, for example, been a long simmering resentment and a sense among many Californians that "illegal aliens" are claiming rights in the United States to which they are not entitled. This resentment has been fueled by the presence of many Mexican flags during political protests, both during the anti-Prop. 187 marches in the 1990s (Gutiérrez, 1998) and more recent pro-immigrant protests. This feeling was exacerbated by the number of Latinos who have participated in these protests and by the large number of protesters chanting in Spanish and waving Mexican flags. Some were especially offended when reports surfaced that three horn players at the march had burst into an impromptu mariachi-style version of "The Star-Spangled Banner" (Gutiérrez, 1998). The U.S. soccer team has similarly found that playing in Los Angeles, or even Chicago, is not a home game for the United States national team. These findings challenge the assimilationist paradigm as well as popular expectations about how immigrants were supposed to adapt, adjust, and eventually "melt" into the mainstream of American life and point to the existence of a fictive third space between the host society and the sending nation that enables trans-migrants to remain connected across geographic boundaries.

References

Agius, V. (2012). *Barrios to burbs: The making of the Mexican American middle class*. Stanford University Press.

Berry, J. W. (2019). *Acculturation: A personal journey across cultures (elements in psychology and culture)*. Cambridge University Press.

Blauner, R. (1972). *Racial oppression in America*. Harper & Row.

Briggs, J. W. (1978). *An Italian passage: Immigrants to three American cities, 1890–1910*. Yale University Press.

Coleman, J. S. (1990). *Foundations of social theory*. Harvard University Press.

Glick Schiller, N. (1999). Transmigrants and nation-states: Something old and something new in the U.S. immigration experience. In C. Hirschman, P. Kasinitz, & J. De Wind (Eds.), *The handbook of international migration: The American experience* (pp. 94–119). Sage.

Glick Schiller, N. (2003). The centrality of ethnography in the study of transnational migration: Seeing the wetlands instead of the swamp. In N. Foner (Ed.), *American arrivals: Anthropology engages the new immigrant* (pp. 99–128). School of American Research Press.

Gordon, M. (1964). *Assimilation in American life: The role of race, religion, and national origins* (1st ed.). Oxford University Press.
Gowricharn, R. (Ed.). (2006). *Caribbean transnationalism: Migration, pluralization, and social cohesion*. Lexington Books.
Gutiérrez, D. G. (Ed.). (1996). *Between two worlds: Mexican immigrants in the United States*. Scholarly Resources.
Gutiérrez, D. G. (1998). Migration, emergent ethnicity, and the "third space": The shifting politics of nationalism in greater Mexico. *Journal of American History.*, *86*, 481–517.
Handlin, O. (1951). *The uprooted: The epic story of the great migration that made the American people* (2nd ed.). University of Pennsylvania Press.
Massey, D. S., Durand, J., & Malone, N. J. (2002). *Beyond smoke and mirrors: Mexican immigration in an era of economic integration.*
Massey, D. S., & Goldring, L. (1994). Continuities in transnational migration: An analysis of nineteen Mexican communities. *American Journal of Sociology, 99*(6), 1492–1533. https://doi.org/10.1086/230452
Mirandé, A. (2014). *Jalos USA: Transnational community and identity*. University of Notre Dame Press.
Newland, K. (2009). Circular migration and human development. Human Development Research Paper (HDRP) Series, Vol. 42. https://mpra.ub.uni-muenchen.de/19225/1/MPRA_paper_19225.pdf
Park, R. E., Burgess, E. W., & McKenzie, R. (1925). *The City*. University of Chicago Press.
Portes, A., & Bach, R. L. (1985). *Latin journey*. University of California Press.
Portes, A., & Rumbaut, R. G. (Eds.). (2006). *Immigrant America: A portrait* (3rd ed.). University of California Press. https://doi.org/10.1525/9780520940482
Rojas, E. (2008). Original Mexican settlers. In H. A. Hohenthal & J. Edwards Caswell (Eds.), *Streams in a thirsty land: A history of the Turlock region* (pp. 120–125). Turlock Historical Society.
Smith, R. C. (2006). *Mexican New York: Transnational lives of new immigrants*. University of California Press. https://doi.org/10.1525/j.ctt1ppp41
Thomas, W. I., & Znaniecki, F. (1996). *The Polish peasant in Europe and America*. Student Edition. University of Illinois Press.
Vecchio, D. (2007). Ties of affection: Family narratives in the history of Italian immigration. In E. R. Barkan (Ed.), *Immigration, incorporation and transnationalism* (pp. 105–119). Transaction Publishers.
Warner, W. L., & Srole., L. (1945). *The social systems of American ethnic groups*. Yale University Press.

Chapter 11
Adaptation and Identity Formation in the Israeli Diaspora

Steven J. Gold

Jewish migrants and their descendants from Eastern Europe have been a paradigmatic case of successful assimilators in many Western countries. They have adopted the nationality and culture in points of settlement, had little propensity to return to the country of origin, and overcame discrimination to attain levels of educational, economic, and social prominence matching those of the native-born elite (Pew, 2013). While identifying with the host society, at the same time, Jewish migrants have also retained political, familial, residential, religious, and cultural characteristics that distinguish them from other middle-class members of the host society (Glazer & Moynihan, 1963; Gold & Phillips, 1996; Hyman, 1998; Lipset, 1990; Waterman & Kosmin, 1986). This pattern is to a large extent a product of Jewish migrants' de facto or de jure status as refugees who had little connection to their country of origin, and hence, strong reasons to join the settlement country (Gold & Phillips, 1996).

This record of assimilation is most often associated with Russian Jewish migrants of the late nineteenth and early twentieth centuries. However, several more recently arrived Jewish migrant groups to the U.S. and other countries, including those who entered from Europe before and after World War II (WWII), and contemporary migrants from the former Soviet Union, South Africa, Iran, and Latin America appear to be maintaining this pattern as well (Breitman & Kraut, 1987; Friedman, 2007; Gold, 2016; Helmreich, 1992; Orleck, 1999; Simon, 1997). While Israeli emigrants—who are now among the largest Jewish migrant populations in many Western societies—share social, cultural, occupational, and residential characteristics with native-born Jews, they reveal dissimilar patterns of adaptation (Gold, 2002; Lev Ari, 2012; Rebhun & Lev-Ari, 2010; Shokeid, 1988). Most notable is their reluctance to identify with host societies. While they may remain in host societies for extended periods, achieve impressive records of social and economic mobility, raise their children in such settings, and acquire citizenship, they seldom describe themselves as host country nationals, socialize almost exclusively with other Israelis, frequently describe their intentions to return home, and often do so.

Migrants' hesitation to identify with host societies is not an unprecedented finding. However, a sizable literature suggests that migrants' lack of identification with

S. J. Gold (✉)
Michigan State University, East Lansing, MI, USA
e-mail: gold@msu.edu

D. J. Johnson et al. (eds.), *Re/Formation and Identity*, Advances in Immigrant Family Research, https://doi.org/10.1007/978-3-030-86426-2_11

their new home is a result of their encounters with discriminatory treatment (Abdi, 2015; Goldring, 1998; Portes & Rumbaut, 2001; Piore, 1979; Shryock et al., 2011; Waters, 1999). Israelis—who are generally White, well-educated and Western in outlook—describe few encounters with prejudice (Herman, 2000). Hence, their resistance to identification with the host society cannot be attributed to racism. The reason for Israeli emigrants' reluctance to identify with the country of settlement is that they have been socialized into Israeli culture and citizenship, a central tenet of which is that Jews and especially Israelis should live in, and identify with, Israel (Shokeid, 1988; Gold, 2002).

Despite their generally strong identification with the Jewish State and their reluctance to become diaspora Jews, many Israelis nevertheless bear aspirations, outlooks, familial relations, cultural and economic skills, and social ties that link them to major Western cities, including New York, Los Angeles, the San Francisco Bay Area, London, Berlin, Sydney, and Toronto (Gold, 2002). A fraction of Israelis act on these aspirations and, in so doing, confront strong and conflicting bases of identity associated with Western Jewish communities, on the one hand, and Israel, on the other. As they consider their identities in places of settlement, they cannot help but be aware that their emigration is condemned in the eyes of at least some members of those social groups—Israelis and diaspora Jews—with whom they bear significant social ties (Cohen, 1986). This chapter describes how Israeli emigrants and their children address conflicting identities associated with the country of origin and the host society.

Segmented and Transnational Identity Formation

Instructive in itself, the case of Israelis' confrontation with dilemmas of membership offers an opportunity for evaluating contemporary theories about migrants' national and ethnic identities. In recent years, scholars have contributed a large body of theorizing and research to the exploration of identity formation among migrant populations in Western societies. Noting the inadequacy of straight-line assimilation to explain the experience of contemporary migrants, two bodies of work have come to the fore (Gans, 1992; Park, 2013). The first is studies of segmented assimilation. The second body of work is associated with transnationalism and diasporas.

These two bodies of work share several similarities. Both are concerned with the adaptation of non-White, occasionally undocumented and sometimes non-Christian, migrants into racialized, economically polarized environments. In contrast to earlier models of immigrant adaptation which assumed ethnic identity to be primordial, both segmented and transnational approaches understand that ethnic identity is contextually determined, often reactive, a product of migrants' subjectivity, and associated with membership in networks (Cornell & Hartmann, 2007). Following from this, both models emphasize that migrant populations are themselves stratified by age, educational level, gender, ideology, phenotype, class, occupation, point of settlement, and legal status. As such, migrant subgroups encounter disparate experiences of adaptation according to their social characteristics and contexts of resettlement. Beyond these similarities, the two models of identity formation are quite distinct.

Segmented assimilation studies are primarily concerned with the children of recent immigrants to the United States and other locations (Alba & Foner, 2015; Portes & Zhou, 1993). Studies representing this approach generally describe how migrants create identities and deploy co-ethnic resources in ways that either facilitate advancement (through educational achievement or the creation of ethnic economies) or preclude upward mobility (via oppositional acculturation, rejecting school, involvement in gangs). Many involve proletarian populations and refugees—groups from Mexico, the Caribbean, Vietnam, the Middle East, Africa, Turkey, and South Asia—but more recent studies also concern more affluent groups from India, China, and Korea (Kasinitz et al., 2008; Louie, 2000; Rumbaut & Portes, 2001; Rumbaut, 2002). These countries of origin offer expatriates few structural incentives for the maintenance of continuing contact. Moreover, families included in this work generally lack the funds and legal status for frequent trips home or for the extensive cultural socialization of their children (Light & Gold, 2000). In his study of transnational ties among immigrant children in the United States, Rumbaut (2002) found that three out of four respondents had never visited the homeland or had done so only once or twice in their lives.

Such research has only begun to focus on the role of religion—which is among the most central avenues of transnational identification and connection—in forging migrant identities (Ebaugh & Chafetz, 2000; Kibria, 2011; Min, 2010). Consequently, segmented assimilation scholarship generally downplays the enduring influence of country of origin factors in shaping long-term identities among the majority of migrant populations. In their comprehensive study of recent immigrant children, Rumbaut and Portes (2001) assert that the national identities of today's children of immigrants are products of their experience in the United States as opposed to being cultural forms retained from the country of origin.

In contrast, research that emphasizes transnationalism often discounts the centrality of assimilation within a nation state. Instead, it sees multiple, often deterritorialized forms of belonging and social solidarity as foundations of identity (Appadurai, 1996; Levitt & Jaworsky, 2007; Rebhun, 2014).

Transnational migrants engage in frequent travel, and even those who do not themselves cross borders with regularity are described as residing within transnational social spaces where their outlooks and values are influenced by information, norms, and social resources brought in from other continents or deterritorialized collectives (Faist, 2000; Hirsch, 1999; Pessar, 1997; Portes et al., 1999; Waldinger, 2015). While the transnational view acknowledges that migrants might confront various forms of oppression, many versions of transnational theorizing see the economic, political, and geographical mobility associated with border crossing as capable of granting migrants a high degree of agency and flexibility to avoid forms of oppression associated with specific localities. Sarah Mahler asserts: "Through transnational processes everyday people can generate Creole identities and agencies that challenge multiple levels of structural control" (Mahler, 1998, p. 68). Galchinsky

(1998) refers to a specifically Jewish legacy that is relevant to Israelis as he cites the work of social theorists who understand Jews' historic involvement with diasporas as a means of eluding identity politics.

Finally, transnationalism poses a temporal frame distinct from that of segmented assimilation studies. Defining migration as an ongoing process in a globalized world, it is concerned with migrants themselves rather than their offspring and tends to see the border-crossing consciousness as enduring rather than generational (for an alternate view, see Lev Ari & Cohen, 2018). In contrast, most segmented assimilation research retains a focus that considers migrant generations, especially the "1.5 generation" (persons born abroad who arrived in the host society prior to age 12) and the second generation (born in the host society), who eventually identify with the host society.

These two approaches offer much to our understanding of the adaptation experience and identity formation of Israeli emigrants in Western societies. However, neither model in itself presents a comprehensive interpretation of the complex feelings of connection, loyalty, and the multiple opportunities and obligations that many contemporary migrants encounter. While the segmented assimilation model does a good job of showing how Israelis, like other Jewish migrants, join and adapt to host societies, it fails to capture the extent to which Israeli emigrants' identities and behaviors continue to be shaped by their country of origin even as they reside abroad (Gold, 2002, 2015, 2018). And the transnational perspective, in my view, pays insufficient attention to the practical and existential difficulties Israeli emigrants confront as they seek to raise a family, earn a living, and maintain a social life while negotiating between often-conflicting sets of obligations, identities, relationships, networks, and opportunities associated with the country of origin, on the one hand, and the country of settlement, on the other (Anish, 2003; Ray, 2013; Rebhun, 2014).

To better understand the patterns of adaptation and identification of Israeli migrants, I draw upon segmented assimilation and transnational theories and review the impact of various national, religious, and ethnic forces on Israeli migrants and host communities. This analysis suggests that with time abroad, Israeli emigrants and especially their children increasingly downplay national identities and, instead, emphasize Jewish peoplehood as a means of reconciling their Israeli identity with their presence in the diaspora.

Method

Data for this chapter were collected in the form of in-depth interviews and participant observation fieldwork with Israeli immigrants in several settings intermittently between 1991 and 2016 (Gold, 2002; Gold & Hart, 2013). A major source were 194 in-depth interviews (conducted in both Hebrew and English) with Israeli immigrants and others knowledgeable about their community in the Los Angeles area (87 women and 107 men, including both the wife and the husband of 20 couples).

We also interviewed 50 Israeli emigrants in London, 14 in Paris, and 30 returned emigrants in Israel. Most interviews were audio-recorded and transcribed. To facilitate rapport and openness, we did not use closed-end questionnaires. To ensure that we would collect data on pertinent topics, we selected questions from prepared lists involving family adaptation, self-employment, and community activism among Israeli migrant populations (Gold, 1994). Interview quotes are presented directly from transcripts (with most grammatical errors intact) in order to reflect the speech patterns of respondents. Pseudonyms are used throughout to protect interviewees' identities.

To supplement in-depth interviews, we conducted participant observation research at a variety of religious and secular community activities and other Israeli settings such as parties, restaurants, Israeli-oriented synagogues, schools, youth programs, shops, and the pool of a San Fernando Valley apartment complex with many Israeli residents. We visited numerous meetings of communal organizations in Los Angeles and London as well as Yom Hatzmaut (Israeli Independence Day) festivals in Los Angeles and Metropolitan Detroit. Field observations were recorded in written field notes and sometimes via photography.

Referrals to subjects were obtained from a variety of sources, including Jewish communal agencies, Israeli Consulates in several cities, representatives of various Israeli associations, published lists of Israeli organizations and ethnic businesses, research assistants' personal contacts, and snowball sampling (whereby the people we interviewed referred us to additional respondents from their social networks). Unless otherwise specified, the quotations used throughout were selected because they represented what appeared to be a widely held position among the respondent population.

We sought to contact respondents who would represent various subgroups and categories. Because visible activists, professionals, and self-employed businesspeople are among the established elite of the Israeli emigrant community, we also interviewed the recently arrived and those of humble means to gain an inclusive picture of the population. Our sample intentionally incorporated Israelis from a variety of backgrounds. These included including Ashkenazi (Northern European), Mizrahi (Yemenite, Iraqi, Turkish, Moroccan, Syrian, Egyptian, and Persian), Kibbutzniks (Israelis who had lived in rural collectives), persons from multi-national families, and those maintaining both secular and religious orientations and right- and left-wing ideological outlooks.

Long-term and in-depth contacts with Los Angeles' Israeli population, together with the results of data collected by the U.S. Census, Jewish Population Surveys, and feedback from Jewish and Israeli community leaders, give us a degree of certainty regarding our understanding of the size and nature of this population (Gold, 2016). Nevertheless, there may be reason to question the inclusiveness of our sample. Population estimates are imprecise, and Israelis are especially difficult to identify because, beyond fluency in Hebrew, they lack distinctive names or other defining social features, and a sizable proportion were not born in the Jewish State.

For these reasons, extensive reliance upon random sample surveys would be impractical. Accordingly, involvement in Israeli émigré networks became a major means of identifying and contacting Israelis. We made concerted efforts to contact diverse subgroups, but the extent to which the sample is inclusive or representative is unclear.

Results and Discussion

Five Israeli women, one Hebrew-speaking American woman, and one English-speaking man (the author) contributed to the interviewing and fieldwork in Los Angeles, London, Paris, and Israel. In all cases but one, these researchers were themselves engaged in graduate training in social science fields and were involved in their own scholarly research about the experience of Israeli emigrants.

Fluent in Hebrew, well versed in the study of migration and ethnicity, and themselves part of Israeli emigrant communities, they had excellent rapport with the communities they were studying and offered many insights into their characteristics and dynamics. As women, they had access to and comprehension of aspects of the Israeli emigrant community that I, an American man, did not. At the same time, because I am not Israeli, respondents may have felt free to express certain viewpoints that they might have been unwilling to share with co-nationals. By relying on multiple methods and a team of field researchers characterized by diversity, we sought to generate a well-rounded view of Israeli emigration and return.

Israel and Emigration

Israel was envisioned as a homeland for the world's Jews in the 1890s, in order to end their long saga of oppression in other people's lands, and was brought into being in 1948, following the Holocaust (Goldstein, 1995). Zionists—Jews who support settlement in Israel—depict the diaspora as a place of isolation, degradation, insecurity, and displacement (Zerubavel, 1995). Accordingly, certain forms of Israeli ideology hold that Jews who live in the diaspora should move to Israel and condemn emigration from there.

During the country's early years and to some degree until the present, those who leave were regarded as traitors and losers, and media reports and academic research often described the unsuccessful, even corrupt lives of those who went abroad (Sabar, 2000). During the 1970s, Israeli politicians such as Prime Minister Yitzhak Rabin were especially vitriolic on this issue, describing Israeli emigrants weaklings and insects (Ritterband, 1986).

While many actions intended to discourage Israeli emigration involve Israel-based peer pressure in the form of unflattering articles, condemnation by political leaders, and derisive media portrayals, the Israeli government has also used its leverage with host Jewish communities and their communal organizations to withhold aid from Israeli emigrants. In order to discourage further emigration and impede the settlement of those abroad, from the early 1970s until the late 1980s, the Israeli Consulate in New York "repeatedly urged the Federation [the leading American Jewish community organization] to provide no special services to Israelis" (Tugend, 1989, p. 2).

Accordingly, diaspora Jewish communities offered little of the financial and social support and resettlement assistance that they have customarily extended to Jewish immigrants and refugees (Gold & Phillips, 1996; Howe, 1976). In her report "The Israeli Corner of the American Jewish Community," Sherry Rosen (1993) asserted that the host group's response has been to deny Israeli émigrés' position as Jewish immigrants seeking to settle in the United States.

In spite of their mutual support of Israel, diaspora Jews and Israelis have differing national allegiances. While early Zionists depicted the diaspora as a place of degradation and exile, they failed to convince a sizeable fraction of the world's Jews. This is especially the case for those in affluent and democratic nations of the West. Instead, diaspora Jews chose to stay put, contributing political and financial backing from afar (Rosenthal, 2001). Moreover, since about 2000, some observers have argued that the religious and ideological orientation of Israel, on the one hand, and significant parts of diaspora Jewish communities, on the other, have diverged (Pewforum, 2017).

The former has become increasingly conservative, largely abandoning the goal of making peace with the Palestinians, and has granted increasing power to the ultra-Orthodox in directing the country's culture and religion. In contrast, significant fractions of diaspora Jewry, especially the younger generation of non-Orthodox Jews, maintain liberal social and political values. Several studies contend that this results in a lessening degree of identification with or support for Israel (Beinart, 2010; Gold, 2013).

Differences Between Israeli and Diaspora Jewish Identities

In addition to their ideological differences, the divergence between Israeli and diaspora Jewish notions of membership occurs because the basic group identities associated with being Israeli, on the one hand, and diaspora Jewish, on the other, are rooted in particular cultural/national contexts (Gold, 2002; Pewforum, 2017; Rebhun & Lev Ari, 2010). Israelis and diaspora Jews speak distinct languages, maintain different cultural norms and practices, eat different kinds of food, have contrasting political outlooks, and like different kinds of sports, music, and entertainment (Lev-Ari, 2012). Israelis are said to identify as citizens of their own country, while diaspora Jews, especially those in affluent Western democracies, regard themselves as members of the host society and, at the same time, as participants of a religious/ethnic subpopulation.

In the following interview excerpt, an Israeli woman living in California describes her feeling of distance from American Jews:

> There is a big inability to relate to American Jews … If I meet an East Coast kind of typical Jew, I don't know what I should do. I feel that there is a minority mentality there that I can't decipher. It is very embarrassing for me. He is trying to communicate in a way that is fathomable to another American Jew and I can't figure it out.

As the quote suggests, Israeli and diaspora Jewish groups often express their common religious identification in disparate ways (Liebman & Cohen, 1990). For many Israelis, ethnic identity is secular and nationalistic. While they celebrate Jewish holidays and speak Hebrew, they often connect these behaviors to "Israeliness" rather than to Jewishness. The secular majority is less accustomed to participating in organized religious activities than is the case for many diaspora Jews. In fact, secular Israelis are sometimes hostile to religion, as they object to the religious and political agendas of Orthodox parties in the Jewish State (Gold, 2002).

Israelis are unfamiliar with the sizable financial burden diaspora Jews must accept in order to maintain a Jewish life and raise Jewish children in Christian societies. Instead, Israelis rely upon the larger society and public institutions to provide a Jewish education for their children (Rosenthal & Auerbach, 1992). As a consequence, there exists a difference between the secular/nationalist identity shared by many Israelis and the synagogue-affiliated religious character of U.S. Jews (Mittleburg & Waters, 1992).

Early Reactions of Israeli Emigrants to Diaspora Identities

During fieldwork, we found ample evidence of the widely asserted claim regarding Israeli migrants' ambivalence about being outside of the Jewish State and their pervasive, if unfulfilled, desire to return home. In the words of an Israeli woman who, with husband and three children, spent eight years in Palo Alto:

> Israelis have a lot of problems about staying here. They say, "we will stay here for the time being." I went to school here [in the United States] growing up and now my child is going to school here too. I have a lot of good feelings about it [U.S. schools] compared to the Israeli school system. But we want to go back now. I have come to the conclusion that I can't bear this permanent sojourn anymore. Hopefully, we will be back in less than a year.

Many studies of Israeli emigration report that migrants—including those who have spent a decade or more in the host society—continue to "sit on their suitcases" and refer to their imminent return (Shokeid, 1988). According to this body of research, migrants make such displays in order to emphasize their identity as Israelis, while denying their status as stigmatized emigrants (Rebhun & Lev-Ari, 2010; Shokeid, 1988). Several respondents who resided in Canada, France, England, and Australia as well as the United States asserted that while they were not mistreated in these points of settlement, at the same time they didn't enjoy being social outsiders. Consequently, they either planned to return home or actually did so. In the words of a woman who had returned to Israel from the United States:

> I felt that the United States is a society that opened its arms for us. I think we had wonderful opportunities there and we took advantage of them. We had a very good life, but I was still an outsider. I would have remained an outsider and that's why we decided to come back. We didn't want to be outsiders for the rest of our lives.

Israeli parents are often concerned about their children being deprived of a childhood similar to their own. Further, because Israeli culture emphasizes a series of shared experiences as key to the development of national identity (including service in the Israeli Defense Forces for men and women alike), a child's growing up abroad means that she or he will be excluded from forms of engagement that are essential for both social membership and occupational success.

The Changing Context of Israeli Emigration

While spending extended periods of time in host societies would, in itself, be likely to increase Israelis' identification with the local community, since the late 1980s, both Israel and diaspora Jewish agencies have taken a more conciliatory approach toward Israeli expatriates, removing some of the stigma associated with emigration. and opening a path for incorporation into Jewish communities in host societies. In 1991, Prime Minister Yitzak Rabin recanted his famous condemnation of Israeli émigrés in an interview in the Israeli American newspaper *Hadashot LA* (Rosen, 1993).

Further, because of the significance it attaches to the return of Israeli emigrants to Israel, in 1992 the Israeli government took responsibility for "re-aliyah" and offered a package of benefits, including cash assistance, low-cost airfare, suspension of import duties, education, assistance in finding jobs and housing, financial aid for school tuition, and a reduction in military duty for Israelis and their family members who return (Cohen, 2009).

Upon seeing co-religionists in need, realizing the human and financial resources available within the émigré community, and heeding Israel's call to assist emigrants, segments of host country Jewish populations began to acknowledge both the existence of the Israeli immigrant community and the importance of outreach to them. Accordingly, since the early 1990s, major Jewish communities (Paris, London, Toronto, New York, Los Angeles, Chicago, Miami, and the San Francisco Bay Area) have created various endeavors to aid and incorporate Israelis. These include social activities, secular, Israeli-style education programs, and Israeli divisions of community social/philanthropic organizations (Cohen, 2010; Shokeid, 1988; Uriely, 1994). Such activities are provided by Israeli government agencies, members of the Israeli emigrant community, and a range of local Jewish organizations, including Jewish community centers, YM/YWHAs, synagogues, and Jewish federations (Gold, 2002, 2016). While those provided by members of the Israeli community and the Israeli government are regarded as more authentic, Israel emigrants rely on all of these sources to expose their children to Israeli and Jewish environments (Gold, 2018).

Reflecting the institutional acceptance of Israeli immigrants, Jewish population studies conducted during the past 20 years in New York (2009), Los Angeles (1997), Toronto (2005), London (1998) Greater Miami (2015), and the San Francisco Bay Area (2018) have enumerated Israelis, their presence being one of the few positive tendencies in a general trend of shrinking Jewish demographics (Cohen & Veinstein, 2009; Herman, 1998; Horowitz, 1993; Schmool & Cohen, 1998; Shahar & Rosenbaum, 2005). Finally, research notes that successful and confident Israeli migrants have created a variety of their own organizations, social activities, businesses, synagogues, media outlets, and other institutions in points of settlement (Gold, 2016).

While there are Israeli emigrants in a number of nations, by far the largest fraction reside in the United States. The 2011–2013 American Community Survey of the U.S. Census Bureau estimated there were 139,980 Israel-born persons in the United States. Drawing on U.S. and Israeli Census data, Rebhun and Lev Ari (2010) assert that the total population of Israelis in the United States, including those born in Israel, those born in other countries, and their U.S. born children and American spouses, is 250,000. Accordingly, U.S. programming has the greatest impact on Israeli emigrants.

Israel's modified attitude toward its citizens abroad was not a random event. Rather, it developed in a context of unprecedented demographic and economic growth. In 1989—the time of the last major spike of anti-emigrant editorializing in Israeli newspapers—Israel was suffering economic stagnation, had a rate of inflation near 20 percent, an ongoing fear of war, and an inability to retain many of its best and brightest (Moaz & Temkin, 1989), and the country's major export was citrus.

Only a few years later, conditions had changed dramatically. Due to the massive migration from the former Soviet Union that occurred during the 1990s, the Jewish State's population had increased by close to 20 percent or almost 1,000,000 persons, many of whom were highly educated. Its inflation rate was below 3 percent, and it had the greatest number of engineers per capita of any nation [almost double that of the second-ranking United States (Richtel, 1998)]. Israel had become a center of high technology, earning the nickname "Start Up Nation" (Senor & Singer, 2009). Thus, it could offer its more affluent citizens a standard of material life equal to that of the industrialized West. These political, economic, technological, and demographic developments transformed Israeli society, making it better able to tolerate population losses to emigration. Further, economic advancement made Israel ever more in need of a globalized workforce—including the overseas involvement of Israelis in centers of technology and venture capital like Boston's Route 128, New York City, and Silicon Valley—to facilitate the continued growth of its economy (Gold, 2018).

Finally, Israeli policymakers realized that emigration is not always permanent. In fact, former Israelis often travel to the Jewish State for business, pleasure, and family reasons, and are much more likely to settle there than other segments of diaspora Jewry (Gold, 2002). This fact was not missed by Israeli officials, who altered policy toward emigrants to encourage return rather than punish them for going abroad (Cohen, 2009; Gold, 2013).

The Impact of Israeli Emigrants' New Status on Mobility and Assimilation

New levels of tolerance extended by both the country of origin and host societies have increased Israeli emigrants' prospects for success. Given that emigration is no longer stigmatized, a growing number now feel freer to explore opportunities abroad. Many are in possession of high levels of education and technical skill, an entrepreneurial orientation, and English-language ability These characteristics are associated with quick and successful adaptation to affluent Western societies for both migrants and their children (Bean & Stevens, 2003; Cohen & Haberfeld, 2003; Gold, 2016). Greater levels of toleration for Israeli emigrants also allow members of this population to develop friendly relations with native-born Jews, whose connections, resources, and institutions can advance their adaptation to the host society in social, educational, and cultural dimensions.

Cohen and Haberfeld (2003) conclude that the educational and occupational resources possessed by Israeli immigrants, along with the welcoming environments in which they settle, mark them as among the most successful of recent migrant groups to the United States. In many cases, Israeli immigrants' children's (1.5 or second generation) achievements outstrip those of their parents. "Like other successful immigrant groups, the economic success of Israeli immigrants and their offspring in the United States is due not only to their high level of education, but also to unmeasured traits that help them earn more than demographically comparable natives" (Cohen & Haberfeld, 2003, p.156). That being said, Israeli emigrants' ability to find good jobs and comfortable living conditions in multiple settings can make their lives more complex. Their cognizance of the costs and benefits associated with life at home and abroad, the relatively easy process of migration and return, and their enduring immersion in international networks cause the beckoning prospect of greener pastures to be especially hard to resist. Moreover, when trouble arises in the form of war, political crisis, family dissension, or recession, relocation offers a means to get away until it is safe to return home or travel elsewhere. Consequently, more than a few Israeli emigrant families follow a cycle of continued travel.

Reconsidering Identity

As a consequence of their experiences abroad, many of those who have lived outside of Israel for extended periods of time described how they began to re-evaluate their identities as Jews and Israelis in the diaspora. While continuing to emphasize the primacy of their Israeli origins, they also began to acknowledge the benefits of a less-bounded existence in cultural, financial, religious, and other realms.

Several respondents referred to changes in Israeli society that occurred while they were abroad as altering their identification with the Jewish State. A leftist former kibbutznick (member of a rural collective) residing in London described his anger at the increasing Americanization and materialism of Israeli culture: "When I go back, I find that the advertising on the radio and the whole scene—the McDonald's and the Toys-R-Us and the whole thing—it's coming in from America, and I don't like it." A woman who returned to Israel following her divorce felt that the country she left no longer existed:

> I am and always was very Israeli in all my being. And although during these years (abroad) I felt less and less of that sense of belonging, I still felt I am an Israeli and felt very strong about my identity. However, when I returned, I felt I did not belong here anymore because things have happened here that I was not a part of. I belonged to that country that I left behind years ago. This country does not exist anymore apart from its existence in my memory.

Child Rearing and National Identity

While the first generation generally retains a strong identity based in Israel, they confront considerable difficulty in transmitting this outlook to their kids, who are growing up in Toronto or LA (Lev-Ari & Cohen, 2018). Hence, the experience of having children abroad caused many Israeli immigrant parents to reflect on the nature and meaning of their national identity. Most concluded that if they do nothing, their children would acquire the Christianity-tinged outlook prevalent in the host society. Yet, if they involve their children in local Jewish activities as a means of retaining some connection to Judaism, then their offspring are likely to adopt another unfamiliar identity—that of diaspora Jews. In the following excerpt, Ron describes his family's reaction to U.S. Judaism:

> Their Judaism is quite impressive, but it is treated to their needs and so I think they have a very nice sense of community, but for us, it didn't fit us. It is very foreign for us. It is weird, it is bizarre for us, you know, to see the rabbi that was a woman and she was a lesbian and all. But, I have lots of respect for their community and the way they live and the sense of community that they had. (Gold 2002, p. 209)

Israeli emigrant parents seek to resolve these identificational dilemmas through participation in Israel-oriented activities of their own creation as well as those provided by Israeli organizations and the local Jewish community. They include sing-alongs, parties, Israeli dancing, after-school Hebrew classes, Tzofim (Israeli scouts), and trips back to Israel.

Another problem cited by émigré parents is that migration hinders their ability to share a common language, culture, and life-shaping experiences with their children. A woman living in London described this:

> By the time they went to school, they started talking in a language that is alien to me, singing songs I do not know, experiencing things which are unfamiliar to me. That was difficult for me. I know it sounds selfish, but I want my kids. I want to be part of their lives. I wanted them to have something in common with me. I wanted them to have the same identity as mine. (Gold, 2002, p. 115)

Many of the refugee and proletarian populations associated with segmented assimilation research lack the personal and communal resources required to immerse themselves and their children in the country of origin's culture and language (Rumbaut, 2002). In contrast, the relatively affluent Israeli population visits Israel frequently, partakes of Hebrew language social and cultural activities, and—with the assistance of local Jews—provides various Jewish- and Israeli-style child-rearing activities for their children. In this way, Israeli emigrants, together with local Jews, consciously foster transnational identities among their kids to a greater extent than is the case among less affluent groups (Lev Ari, 2012). In this, Israeli emigrants' high levels of education, legal status, relative affluence, and ability to settle in largely White suburbs distinguish them from migrant populations and settings most commonly associated with the patterns of segmented assimilation. Instead, certain aspects of Israeli emigrants' patterns of settlement in the West can be likened to those depicted in studies of Asian ethnoburbs (Portes & Rumbaut, 2014; Zhou, 2013).

Activities intended to help Israeli emigrant children maintain their national identity abroad were valued. Nevertheless, they were considered as poor substitutes for Israeli life itself. In the words of an Israeli-born psychologist who writes a column in a Los Angeles Hebrew newspaper, "You can't raise an Israeli child in California, for goodness sake." Several respondents described their immersion in local Jewish life was simply a way of retaining a Jewish identity, socializing children, and gaining access to a co-ethnic community. Accordingly, émigré parents often described their religious association with local Jews as a means to an end. If and when they returned to Israel, they planned to resume their secular lifestyles. In the words of a woman who recently returned to Israel from Australia: "We went to the synagogue there almost once a month, just for the social aspect of it. We understood that this is what the synagogue is for to the local Jews – a social club around the religious issues."

Jewish Ethnicity/Peoplehood

Despite national, ideological, and religious differences between Israeli and diaspora Jews, Israelis do feel connections to other Jews and see themselves, Jewish migrants from other countries, and native-born Jews in settlement countries as members of the same ethnic and religious group. According to Israeli anthropologist Natan Uriely (1994), they refer to "a common past, the observance of the same holidays, rituals and ceremonies, the retelling of the same myths and similar responses to common symbols" (Uriely, 1994, p. 30; Liebman & Cohen, 1990). Accordingly, Jewishness offers a flexible form of identification that allows Israelis to make connections with other Jews and to consume communal resources without repudiating their Israeli identities.

Few members of the first generation renounced their connection to Israel to become flag-waving Canadian, British, or American Jews. A considerable number did, however, begin to describe themselves as members of the global community of the Jewish people. In this way, they were able to reconcile their connections to two or more nationalities without appearing disloyal to either. Moreover, by referring to their connection to Jewish peoplehood, Israeli emigrants can present themselves as experts in Hebrew and Israeli culture (a positive attribute) rather than exiles from Israeli society (a negative one).

In this way, many Israeli emigrants view the Jewish religion and Jewish ethnicity as sources of linkage to potential friends and allies. An Israeli communal activist who works to bring his co-nationals into American Jewish organizations refers to this sense of peoplehood as a motive for his community work:

> Take for example my own family. I don't see that just because somebody's grandmother left the same village in Poland that my grandmother lived in 80 years ago and came to New York, and my relatives came to Israel, that I'm that different from that person. We are the same people.

Engaging in diaspora-based forms of Jewish identification and religious practice in order to provide their kids with a kernel of Jewish identity, certain Israelis became entranced. They planned to continue this outlook either in the host society or upon returning to the Jewish State. Such was the case for Yossi, whose family returned to Israel after several years in North America:

> My kids went there to a Jewish religious school. They learned a lot about Judaism and practiced much more than we did at home (in Israel). They did the prayers before and after the food and in the morning and in the evening, and many other things that my kids did not get to see in our home. Even if I wanted to give them that knowledge, I could not, because I did not have it myself. My wife and I are both from secular homes, after all. So, we actually learned from our kids all about religious practices, and we tried to do some at home so that they would not feel alienated from what the school is teaching them. That is how we ended up going to the synagogue.

Second-Generation Identity

One of the most pointed critiques of transnational identity is that it is short-lived (Levitt & Jaworsky, 2007). Even those who question the concept admit that the members of foreign-born migrant generations are inclined to identify with and keep connected to the country or origin. Later generations, they submit, are much more likely to lose the cultural and linguistic knowledge associated with the country of origin and, instead, identify with the host society. For example, even expressions of an extra-national resistance identity by the children of immigrants are viewed as the consequence of their exposure to American youth culture, not the retention of the origin country's ethos (Rumbaut, 2002; Rumbaut & Portes, 2001).

A growing body of evidence about the national identification of second-generation Israelis is mixed. One body of published research, as well as the comments of certain respondents, suggests that émigré youngsters, particularly those who arrived at a very early age or were born abroad, are likely to connect with

Jewish members of the host society (Lev-Ari, 2012; Lev-Ari & Cohen, 2018; Rebhun, 2014). Summarizing her fieldwork in Toronto, Rina Cohen argued,

> Interviews with students who are the sons and daughters of Israeli migrants indicate that the younger generation is losing both its fluency in Hebrew and its national affiliation with Israel … the second generation is converging into the established Jewish community. (Cohen, 1999, pp. 121, 133)

Confirming this finding, Israeli sociologist Lev-Ari (2012) and colleagues found that the second generation felt most comfortable in the host society. While they sought to interact with Israelis and absorb Israeli culture, they exhibited strong tendencies toward assimilation to the host society (Lev-Ari & Cohen, 2018).

In contrast to these findings, another body of data suggests that some children of Israeli emigrants retain a positive view of Israel, if not a fully realized Israeli identity, in diaspora settings. Parents encouraged this by speaking Hebrew, socializing with other Israeli emigrants, enrolling the children in Israeli-style youth programs, and sending them on regular trips to Israel. Levi-Ari and Cohen (2018, p. 357) assert, "The vast majority of our interviewees expressed strong preferences for the cultural traditions of their parents' homeland."

Uriely (1994) found that second-generation Israelis in Chicago continue to call themselves "Israelis" rather than Americans by a wide margin, while Shokeid (1988) encountered a similar pattern among Israeli emigrants' children in New York. Lev-Ari (2012) found that members of the 1.5 generation (those born in Israel who moved to the abroad prior to their 12th birthday) were more likely to have a diasporic outlook focused on their country of origin, were more likely to feel estranged from local Jewry, and tended to socialize within the Israeli emigre community.

Finally, a third pattern of adaptation maintained among the children of Israelis is synthetic and transnational. Young people who often travel back and forth between Israel and points of settlement may develop a multi-national outlook. Israeli sociologist Zvi Sobel, in his pre-travel survey of 117 Israeli emigrants (most of whom planned to enter the United States), found clear evidence of a transnational outlook. About one-half of respondents repudiated the notion "that leaving Israel and moving to the U.S. was an act of emigration." Instead, they defined the journey as "temporary" or "commuting." Moreover, "almost all interviewees denied that their leaving meant a cessation of contributing to the development of Israel … Almost all saw their departure as … to Israel's good" (Sobel 1986, p.196). Sol, a garment entrepreneur who returned from Los Angeles to Israel, illustrates this position as he described his family's experience in and access to two nationalities as providing his children with the best of both worlds:

> My son Bar was born in LA and lived here for almost 16 years. Then we moved back. He is graduating high school this year and I told him that he has a choice either to come to go to college here or to go to the Army and he decided to go to the Army in Israel. Even though Bar and his brothers grew up in the United States, they love Israel and follow what's happening. I think for the kids, it was a very good move.
>
> Bar was just in LA and he was able to keep in touch with his friends over here, so when he comes here, he has a lot of friends and he can see a lot of people. So, in a way, I feel that my

> kids have a taste of both worlds and, I hope that they'll learn to live in this big village we call the world. Yeah, they have a taste of both and I hope it will be for the beneficial.

> All of them are bilingual and they speak pretty good English and Hebrew and, when they'll grow, they'll decide what they want to do. But I felt it was my obligation to show them a little bit of Israel. I'm an Israeli and I thought this was the most important thing. (Gold, 2002, p. 140)

The transnational approach is further exemplified in the following excerpt from an interview with a British-born child of Israelis in London, "I feel, I'm a person who wants to fit in everywhere. That's what I want. I don't know how much this can be true, but I want to travel back to see both sides." In fact, émigré parents frequently described how their children went through periods of changing national allegiance in which they identified with various environments and practices. For example, a woman who had lived in London for almost 30 years elaborated about how each of her three children—at their own insistence—had been educated in both Jewish and secular schools in England, attended youth programs in Israel, and served in the Israeli armed forces. Now grown, one lives in Israel, while the others are attending graduate school in the United States and England. Rather than developing discrete, nationally bounded identities, they feel connections to multiple locations and groups.

Conclusions

Jewish migrants to Western countries have been depicted as rapid and enthusiastic assimilators, who identify strongly with their host societies. Israelis, for their part, are also patriotic, but express their allegiance to the Jewish State and often learn to berate the diaspora as a location unfit for Jewish life. Hence, Israeli emigrants in Western countries find themselves at the juncture between mutually exclusive narratives of identity and national membership. This was especially the case prior to the 1990s, when the emigration of Jewish Israelis was harshly condemned by both Israel and settlement communities (Cohen, 1986; Shokeid, 1988).

More recently, both Israel and host communities have treated Israeli emigrants in a more benign manner (Cohen, 2009). As a consequence, emigrants have developed their own co-ethnic communities and increasingly participate in local Jewish activities in points of settlement, while simultaneously retaining various links to the Jewish State. In order to negotiate between conflicting claims of the nations with which they are affiliated, they often emphasize their membership in a deterritorialized Jewish community that spans both Israel and the diaspora.

In this manner, we see the actions of Israeli emigrants as partly compatible with elements of both segmented assimilation and transnational views of migrant identity, but fully consistent with neither view. Rather, their identities are simultaneously shaped by both the context of settlement (as predicted by the former) and their links with the country of origin (as suggested by the latter). The most accurate understanding of Israeli emigrants' outlook is captured by a synthesis of the two approaches.

It is important to realize that Israeli migrants' patterns of adaptation and identification are not simply a product of their own sentiments and ideologies. Instead, actions such as Israel's condemnation of or outreach to its emigrants, host Jewish communities' reactions to the Israelis in their midst, as well as political, social, and economic developments—economic booms and busts, wars and political conflicts in points of origin and settlement—also have considerable impacts on migrants' outlooks and life chances.

Prior to the latter half of the twentieth century, Jews were to a large extent a marginalized population wherever they lived. Accordingly, significant fractions of Jewry could be characterized as peripatetic cosmopolitans who crossed national boundaries in order to flee persecution and to join societies that offered a safe and comfortable existence.

Despite Jews' ability to now regard themselves and to be regarded as full citizens of the countries in which they live, their long-held traditions of extra-national identity have not been fully abandoned. Owing to both Israel's and host societies' recently increased tolerance of transnational lifestyles, pressures to live and identify with a single nation state have eased. The national, ideological, and religious differences between Israeli and diaspora Jews notwithstanding, both groups feel connections to each other and see themselves, Jewish migrants from other countries, and native-born Jews in settlement countries as belonging to the same ethnic and religious group.

It would appear that the whole notion of being an Israeli versus a diaspora Jew is not nearly as clear-cut a distinction as the literature on international migration often suggests. Instead, such factors as flexible notions of ethnic and national identity, access to and participation in border crossing, social and occupational networks, and the ability to sustain cultural competence and legal status in more than a single society allow Israelis to maintain meaningful forms of involvement in multiple national settings at one time.

As this study of Israeli emigrants demonstrates, it is precisely because of their enduring connections to the hospitable climate of Israel, on the one hand, and of Western states of settlement, on the other, that this group continues to find the extra-national identity of Jewish peoplehood to be particularly functional. By emphasizing their ethnic and religious connections to Jews in both Israel and the diaspora, Israelis can maintain connections with co-religionists without violating mutually exclusive ideologies of national membership.

Finally, as noted by many scholars, the formation of migrant identity is the result of an array of disparate contextual and life course concerns. Accordingly, the topic merits further investigation (Lev-Ari, 2012; Lev-Ari & Cohen, 2018; Portes & Rumbaut, 2014; Rebhun, 2014). While a synthetic model of migrant identity is especially appropriate for understanding Jewish Israelis—who are characterized by strong links to communities of settlement as well as extensive commitments to the country of origin—such an approach also provides insights into the experience of other contemporary migrant groups who, like them, maintain multiple forms of national, religious, and ethnic affiliation.

References

Abdi, C. (2015). *Elusive Jannah: The Somali diaspora and a borderless Muslim identity*. University of Minnesota Press.

Alba, R., & Foner, N. (2015). *Strangers no more: Immigration and the challenges of integration in North America and Western Europe*. Princeton University Press.

Aneesh, A. (2003). Between fantasy and despair: The transnational condition and high-tech immigration. In D. Gabaccia & C. W. Leach (Eds.), *Immigrant life in the US: Multi-disciplinary perspectives* (pp. 51–64). Routledge.

Appadurai, A. (1996). *Modernity at large: Cultural dimensions of globalization*. University of Minnesota Press.

Bean, F. D., & Stevens, G. (2003). *America's newcomers and the dynamics of diversity*. Sage.

Beinart, P. (2010). The failure of the American Jewish establishment. *The New York Review of Books*. http://www.nybooks.com/articles/archives/2010/jun/10/failure-american-jewish-establishment/?pagination=false

Breitman, R., & Kraut, A. (1987). *American Refugee policy and European Jewry 1933–1945*. Indiana University Press.

Cohen, N. (2009). Come home, be professional: Ethno-nationalism and economic rationalism in Israel's return migration strategy. *Immigrants & Minorities, 27*, 1–28. https://doi-org.proxy1.cl.msu.edu/10.1080/02619280902895843

Cohen, N. (2010). From legalism to symbolism: Anti-mobility and national identity in Israel, 1948–1958. *Journal of Historical Geography, 36*, 19–28.

Cohen, R. (1999). From ethnonational enclave to diasporic community: The mainstreaming of Israeli Jewish migrants in Toronto. *Diaspora, 8*(2), 121–136. https://doi.org/10.1353/dsp.1999.0018

Cohen, S. M. (1986). Israeli Émigrés and the New York federation: A case study in ambivalent policymaking for Jewish communal deviants. *Contemporary Jewry, 7*, 155–165. https://doi-org.proxy1.cl.msu.edu/10.1007/BF02967952

Cohen, S. M., & Veinstein, J. (2009). *Israeli Jews in Greater New York: Their numbers, characteristics and patterns of Jewish engagement*. UJA Federation of New York Report.

Cohen, Y., & Haberfeld, Y. (2003). Economic integration among children of Israeli immigrants in the United States. *International Migration, 41*, 141–159. https://doi-org.proxy1.cl.msu.edu/10.1111/1468-2435.00255

Cornell, S., & Hartmann, D. (2007). *Ethnicity and race: Making identities in a changing world* (2nd ed.). Pine Forge Press.

Ebaugh, H. R., & Chafetz, J. S. (2000). *Religion and the new immigrants: Continuities and adaptations in immigrant congregations*. Altamira.

Faist, T. (2000). *The volume and dynamics of international migration and transnational social spaces*. Oxford University Press.

Friedman, K. (2007). "On Halloween we dressed up like KGB agents": Re-imagining Soviet Jewish refugee identities in America. In J. Gerson & D. Wolf (Eds.), *Sociology confronts the Holocaust: Memories, identities, and diasporas* (pp. 236–259). Duke University Press.

Galchinsky, M. (1998). Scattered seeds: A dialogue of diasporas. In D. Biale, M. Galchinsky, & S. Heschel (Eds.), *Insider/outsider: American Jews and multiculturalism* (pp. 185–211). University of California Press.

Gans, H. (1992). Second generation decline: Scenarios for the economic and ethnic futures of the Post-1965 American immigrants. *Ethnic and Racial Studies, 15*(2), 173–192.

Glazer, N., & Moynihan, D. P. (1963). *Beyond the melting pot*. The MIT Press.

Gold, S. (1994). Patterns of economic cooperation among Israeli immigrants in Los Angeles. *International Migration Review, 28*(105), 114–135. https://doi-org.proxy1.cl.msu.edu/10.2307/2547028

Gold, S. (2002). *The Israeli diaspora*. Routledge/U Washington Press.

Gold, S. (2013). Enhanced agency for recent Jewish migrants to the US. *Contemporary Jewry, 33*(1–2), 145–167. https://doi-org.proxy1.cl.msu.edu/10.1007/s12397-013-9100-z

Gold, S. (2015). Adaptation and return among Israeli enclave and infotech entrepreneurs. *Research in the Sociology of Work: Special Issue on Immigration and Work, 27*, 203–229. https://doi.org/10.1108/S0277-283320150000027022

Gold, S. (2016). Patterns of adaptation among contemporary Jewish immigrants to the US. In A. Dashefsky & I. M. Sheskin (Eds.), *American Jewish yearbook* (Vol. 115, pp. 3–44). Springer. https://doi.org/10.1007/978-3-319-24505-8

Gold, S. (2018). Israeli infotech migrants in silicon valley. *RSF: The Russell Sage Foundation Journal of the Social Sciences, 4*(1), 130–148. https://doi.org/10.7758/rsf.2018.4.1.08

Gold, S., & Hart, R. (2013). Transnational ties during a time of crisis: Israeli emigration, 2000 to 2004. *International Migration, 51*(3), 194–215. https://doi-org.proxy1.cl.msu.edu/10.1111/j.1468-2435.2009.00574.x

Gold, S. J., & Phillips, B. A. (1996). Mobility and continuity among Eastern European Jews. In S. Pedraza & R. G. Rumbaut (Eds.), *Origins and destinies: Immigration, race and ethnicity in America* (pp. 182–194). Wadsworth.

Goldring, L. (1998). The power of status in transnational social fields. In M. P. Smith & L. E. Guarnizo (Eds.), *Transnationalism from below* (pp. 165–195). Transaction.

Goldstein, J. (1995). *Jewish history in modern times*. Sussex Academic Press.

Helmreich, W. B. (1992). *Against all odds: Holocaust survivors and the successful lives they made in America*. Simon and Schuster.

Herman, P. (1998). *Los Angeles Jewish population survey '97*. Jewish Federation of Los Angeles.

Herman, P. (2000, December). *The Jews of the Jews: Characteristics of Los Angeles households of Israelis by birth and Israelis not by birth*. Paper presented at the Thirteenth World Congress of Jewish Studies.

Hirsch, J. S. (1999). En el norte la Mujer Manda: Gender, generation and geography in a Mexican transnational community. *American Behavioral Scientist, 42*(9), 1332–1349.

Horowitz, B. (1993). *The 1991 New York Jewish Population Study*. United Jewish Appeal—Federation of Jewish Philanthropies of New York. https://doi-org.proxy1.cl.msu.edu/10.1007/BF02986639

Howe, I. (1976). *World of Our fathers*. Harcourt.

Hyman, P. E. (1998). *The Jews of modern France*. University of California Press.

Kasinitz, P., John, H., Mollenkopf, J. H., Waters, M. C., & Holdaway, J. (2008). *Inheriting the city: The children of immigrants come of age*. Sage.

Kibria, N. (2011). *Muslims in motion: Islam and national identity in the Bangladeshi diaspora*. Rutgers University Press.

Lev-Ari, L. (2012). North Americans, Israelis, or Jews?: The ethnic identity of immigrants' offspring. *Contemporary Jewry, 32*, 285–308. https://doi.org/10.1007/s12397-012-9091-1

Lev-Ari, L., & Cohen, N. (2018). Acculturation strategies and ethnic identity among second-generation Israeli migrants in the United States. *Contemporary Jewry, 32*(3), 345–364. https://doi.org/10.1007/s12397-018-9258-5

Levitt, P., & Jaworsky, N. B. (2007). Transnational migration studies: Past developments and future trends. *Annual Review of Sociology, 33*(1), 129–156. https://doi-org.proxy1.cl.msu.edu/10.1146/annurev.soc.33.040406.131816

Liebman, C. S., & Cohen, S. M. (1990). *Two Worlds of Judaism: The Israeli and American experiences*. Yale University Press.

Light, I., & Gold, S. J. (2000). *Ethnic economies*. Academic Press.

Lipset, S. M. (1990). A unique people in an exceptional country. In S. M. Lipset (Ed.), *American pluralism and the Jewish community* (pp. 3–29). Transaction.

Louie, A. (2000). Re-territorializing transnationalism: Chinese Americans and the Chinese motherland. *American Ethnologist, 27*(3), 645–669. https://doi.org/10.1525/ae.2000.27.3.645

Mahler, S. J. (1998). Theoretical and empirical contribution towards a research agenda for transnationalism. In M. P. Smith & L. E. Guarnizo (Eds.), *Transnationalism from below* (pp. 64–100). Transaction.

Mittelberg, D., & Waters, M. C. (1992). The process of ethnogenesis among Haitian and Israeli immigrants in the United States. *Ethnic and Racial Studies, 15*(3), 412–435.
Min, P. G. (2010). *Preserving ethnicity through religion in America.* NYU Press.
Moaz, S. & Temkin, (1989). *Olim and Yordim.* Jerusalem Post. May 9.
Orleck, A. (1999). *The Soviet Jewish Americans.* Greenwood Publishing Group.
Park, J. (2013). The socioeconomic prospect of immigrant children in the twenty-first century. In S. J. Gold & S. J. Nawyn (Eds.), *The Routledge international handbook of migration studies* (pp. 307–318). Routledge.
Pessar, P. R. (1997). *Caribbean circuits: New directions in the Sdudy of Caribbean migration.* Center for Migration Studies.
Pew Research Center. (2013). *A portrait of Jewish Americans findings from a Pew Research Center Survey of U.S. Jews.* http://assets.pewresearch.org/wp-content/uploads/sites/11/2013/10/jewish-american-full-report-for-web.pdf
Pewforum. (2017). *American and Israeli Jews twin portraits from Pew Research Center.* http://www.pewforum.org/essay/american-and-israeli-jews-twin-portraits-from-pew-research-center-surveys/
Piore, M. J. (1979). *Birds of passage.* Cambridge University Press.
Portes, A., & Rumbaut, R. G., (2014). Immigrant America: A portrait. University of California Press.
Portes, A., Guarnizo, L. E., & Landolt, P. (1999). Introduction: Pitfalls and promise of an emergent research field. *Ethnic and Racial Studies, 22*(2), 217–327.
Portes, A., & Rumbaut, R. G. (2001). *Legacies: The story of the immigrant second generation.* University of California Press.
Portes, A., & Zhou, M. (1993). The new second generation: Segmented assimilation and its variants. *The Annals of the American Academy of Political and Social Science, 530*, 74–96.
Ray, M. (2013). The global circulation of skill and capital – Pathways of return migration of Indian entrepreneurs form the United States to India. In T. T. Yong & M. Rahman (Eds.), *Diaspora engagement and development in South Asia* (pp. 75–102). Palgrave Macmillan.
Rebhun, U. (2014). Immigrant acculturation and transnationalism: Israelis in the United States and Europe compared. *Journal for the Scientific Study of Religion, 53*(3), 613–635. http://www.jstor.org/stable/24644230
Rebhun, U., & Lev Ari, L. (2010). *American Israelis: Migration, transnationalism, and diasporic identity.* Brill.
Richtel, M. (1998). New Israelis get computers to aid assimilation. *New York Times* (January 21).
Ritterband, P. (1986). Israelis in New York. *Contemporary Jewry, 7*, 113–126.
Rosen, S. (1993). *The Israeli corner of the American Jewish community.* Issue Series #3. Institute on American Jewish-Israeli Relations. The American Jewish Committee.
Rosenthal, M., & Auerbach, C. (1992). Cultural and social assimilation of Israeli immigrants in the United States. *International Migration Review, 99*(26), 982–991. https://doi.org/10.2307/2546973
Rosenthal, S. T. (2001). *Irreconcilable differences? The waning of the American Jewish love affair with Israel.* Brandeis University Press.
Rumbaut, R. G. (2002). Severed or sustained attachments? Language, identity, and imagined communities in the post-immigrant generation. In M. C. Waters & P. Levitt (Eds.), *The changing face of home: The transnational lives of the second generation* (pp. 43–95). Sage.
Rumbaut, R. G., & Portes, A. (2001). Conclusions—The forging of a New America: Lessons for theory and policy. In R. G. Rumbaut & A. Portes (Eds.), *Ethnicities: Children of immigrants in America* (pp. 301–317). University of California Press.
Sabar, N. (2000). *Kibbutznicks in the diaspora.* SUNY Press.
Schmool, M., & Cohen, F. (1998). *A Profile of British Jewry: Patterns and trends at the turn of a century.* Board of Deputies of British Jews.
Senor, D., & Singer, S. (2009). *Start-up nation: The story of Israeli's economic miracle.* Twelve Books.

Shahar, C. & Rosenbaum, T. (2005). 2005 Toronto (Greater)- Jewish life in Toronto survey. UIA Federations Canada and UJA Federation of Greater Toronto.
Shokeid, M. (1988). *Children of circumstances: Israeli immigrants in New York*. Cornell University Press.
Shryock, A., Abraham, N., & Howell, S. (2011). The terror decade in Arab Detroit: An introduction. In N. Abraham, S. Howell, & A. Shryock (Eds.), *Arab Detroit 9/11 Life in the terror Ddcade* (pp. 1–25). Wayne State University Press.
Simon, R. J. (1997). *In the golden land: A century of Russian and Soviet Jewish immigration in America*. Praeger.
Sobel, Z. (1986). *Migrants from the promised land*. Transaction.
Tugend, T. (1989). Peretz: Integrate Yordim into Jewish community. *Jerusalem Post, 29*, 2. http://ezproxy.msu.edu/login?url=https://www-proquest-com.proxy1.cl.msu.edu/newspapers/peretz-integrate-yordim-into-jewishcommunity/docview/320973121/se-2?accountid=12598
Uriely, N. (1994). Rhetorical ethnicity of permanent sojourners: The case of Israeli immigrants in the Chicago area. *International Sociology, 9*(4), 431–445. https://doi-org.proxy1.cl.msu.edu/10.1177/026858094009004003
Waldinger, R. (2015). *The Cross-border connection: Immigrants, emigrants, and their homelands*. Harvard University Press.
Waterman, S., & Kosmin, B. (1986). The Jews of London. *Geographical Magazine, 58*, 21–27.
Waters, M. C. (1999). *Black identities: West Indian immigrant dreams and American realities*. Harvard University Press.
Zerubavel, Y. (1995). *Recovered roots: Collective memory and the making of Israeli national tradition*. University of Chicago Press.
Zhou, M. (2013). Understanding ethnicity from a community perspective. In S. Gold & S. Nawyn (Eds.), *The Routledge international handbook of migration Ssudies* (pp. 250–259). Routledge.

Part IV
Identity as a Protective Factor: Coping with Depression, Distress, and Adjustment Stress

Chapter 12
First- and Second-Generation Black Immigrant Experiences with Ethnic Misidentification, Acculturation, and Psychological Distress

Annette V. Clarke, Celeste H. Poe, Constance Rose, and Sita G. Patel

> *When I am misidentified, I feel as though I don't belong to any ethnic group. It makes me feel as though my ancestors who have survived slavery are not within me or not recognized for surviving within me, and that makes me feel profoundly sad—as though by assimilating through immigration and through being mixed, I am not treating that legacy of survival and resilience properly.* (Second-generation female participant from Puerto Rico)

This personal account describes the sadness and confusion associated with the devaluing of one's ethnic identity. As a Black Puerto Rican female, other people misidentifying her ethnicity creates a feeling of belonging nowhere and the implication that she does not have a right to claim the adversity and triumphs of her people.

Black immigrant (BI) experiences with microaggressions (brief daily indignities to people of color) or with acculturation (negotiation between majority and traditional cultures) are not well known. In particular, the miscategorization of Black immigrants as African American is a particular type of microaggression, termed ethnic misidentification in this study, which indicates a misalignment between self-reported ethnicity and identification by others. Among other ethnic groups, this type of experience has been shown to prompt negative emotional reactions and is related to higher rates of suicidality, substance use, discrimination, physical health, and mental health disparities (Ahluwalia & Pellettiere, 2010; Campbell & Troyer, 2007; Millender, 2012). Research on misidentification among Black immigrants is very limited, given the frequent assumption of a universal Black American experience. To address this gap in the research, the present study gathered quantitative and qualitative data from Black immigrant adults who were either foreign-born (first generation) or U.S.-born to immigrant parents (second generation) and examined their experiences of ethnic misidentification and the relationships between ethnic misidentification, acculturation strategy, acculturative stress, and psychological

A. V. Clarke (✉) · C. H. Poe · C. Rose · S. G. Patel
Palo Alto University, Palo Alto, CA, USA
e-mail: aclarke@paloaltou.edu; cpoe@paloaltou.edu; crose@paloaltou.edu; spatel@paloaltou.edu

D. J. Johnson et al. (eds.), *Re/Formation and Identity*, Advances in Immigrant Family Research, https://doi.org/10.1007/978-3-030-86426-2_12

distress. As acculturation strategy (ways in which individuals negotiate their identity in a new environment) evolves from one generation to the next, examination of first- and second-generation Black immigrant experiences of, and responses to, ethnic misidentification can increase knowledge about these unique experiences and their implications for overall well-being.

Identity Complexity for Black Immigrants in America

There are questions as to the ownership of identity, whether it is an inherent phenomenon or determined through context and setting, and who has the authority to make such determinations. Clarke (2017) asked, "Is identity who *you* say you are or who *others* say you are?" (p. 1). Perhaps, just as important as understanding one's own identity is having that identity recognized, understood, and validated by others. However, the idea that identity is self-determined is often challenged by differing views among people from various races and cultures (Kaplan, 2015). This poses a challenge for immigrants encountering a host society and is particularly complex for Black immigrants, who face complex adversity——both as immigrants and as Black people (Clarke, 2017).

Black immigrants (BI) represent a diverse set of people from around the world. In the United States, anyone having origins in any of the Black racial groups in Africa are classified as Black or African American (U.S. Census Bureau, 2011). This may include anyone who identifies as Black or African American (also Afro-American, Nigerian, Haitian, etc.) singly or in combination with any other racial group (i.e., Black and White, Black and Asian; Grieco, 2009; U.S. Census Bureau, 2011). BI number around 4.2 million in the United States, making up about 9% of the foreign-born population and 10% of the Black population, which represents a 3% increase since the 1980 census (Anderson & López, 2018). These numbers have been steadily increasing over the past 100 years and have been observed between 1970 and 2016 (Anderson & López, 2018). This growth in the BI population has important implications for BI interactions with people of different races and the Black American community, in particular (Tormala & Deaux, 2002).

While BI may come from any country, there are specific regions that the majority of immigrants call home. Approximately 49% hail from Caribbean countries with the top three countries of origin being Jamaica, Haiti, and Trinidad and Tobago. Countries in Africa represent 39% of immigrant nationalities with Nigeria, Ethiopia, and Ghana being the top countries of origin. People from countries in Europe, Asia, and North America make up 12% of the Black foreign-born population combined (Anderson, 2015). Given that the majority of BI are coming from predominantly Black countries with different histories and relationships to race, it is important to understand how the concepts of race and ethnicity impact the ways in which BI are integrated into the American system of racial and ethnic categorization.

The "One Drop Rule" and Ethnoracial Classification in the United States

Racial discourse in the United States has a long and pervasive history of being conceptually dichotomous—for example, being Black or White (Tormala & Deaux, 2002). Newly arrived immigrants are often viewed within this racial dichotomy and so are vulnerable to categorization by individuals and institutions that represent the majority (Fisher & Model, 2012; Tormala & Deaux, 2002). Grouping people based on appearance is a natural tendency in a phenotypically focused society such as the United States, where there is a system of racial categorization that permeates every level of society, institutions, and government (Cutler, 2008; Haozous et al., 2014). Consistent with the variable understanding of the term "race," racial boundaries, and what it means to be a particular race in America, are also constantly evolving. In the case of African Americans, this process has evolved over a 400-year history in this country as a function of their economic, political, social, and psychological conditions (Tormala & Deaux, 2002).

While post-migration identities may build upon pre-migration identities, they are unique. Often, immigrants adopt narrow identities connecting them to a region or language, which may grow into larger collectives to enhance the power a particular group may yield—explaining how Neapolitans became Italians, Litvaks became Jews, and Trinidadians became West Indians. According to the practice of hypodescent or the "one drop rule," immigrants with any physical trace of African heritage are automatically labeled Black and grouped with African Americans (Fisher & Model, 2012). While the term "African American" labels descendants of North American Blacks, it can also be a pan-ethnic term, where anyone having African ancestry may be labeled African American. This pan-ethnic understanding is bolstered by the shared experience of discrimination among all Black people, the reluctance of White Americans to acknowledge ethnic differences among Blacks, and the pressure from U.S. descendants of enslaved people for Black immigrants to de-emphasize their distinctiveness (Waters, 1999). However, BI, some of whom are not considered Black in their homelands, do not always agree with this label. This denial of ascribed identity may have an unknown impact on Black immigrants.

From Openly Racist to Microaggressive

Racism can be demonstrated on an individual level as well as a systemic level. The negative impact of racism and discrimination on an individual's physical and mental health outcomes has been well documented. Racism has pervasively permeated American society such that almost everyone born and raised in the United States inherits its systemic biases and those who migrate here find themselves subject to it as well (Sue et al., 2007). The relationship between racism-related constructs, such as perceived discrimination and poor mental health, has been supported by meta-analysis of as many as 110 studies (Pascoe & Smart Richman, 2009). Relationships have been drawn between racial discrimination and high blood pressure, sleeping problems, substance use, eating disorders, psychological stress, and depression (Nadal, 2011). The connection between perceived discrimination and psychological distress, including depression, has been supported in African-American and Latino populations (Torres et al., 2012).

Often considered demonstrations of racist ideals, microaggressions are brief everyday exchanges that intentionally or unintentionally communicate subtle denigrating messages to people of color but are presumed to be innocent and innocuous (Sue et al., 2007; Nadal, 2011). Recognition of microaggressions emerged after the Civil Rights Movement challenged the racist status quo and called for equality, justice, change, and human rights provisions (American Psychological Association, 2012; Sue et al., 2007). People are now more aware of racism, more politically correct, and tend to avoid overt racist acts (Nadal, 2008). Nevertheless, microaggressions are pervasive and automatic in daily conversations and interactions (via individual perpetrators) or present within a particular setting or environment where they can communicate invalidation of a person's racial identity by the exclusion of representative artifacts or images. Similar to "microinequities," which refer to a pattern of being overlooked, under-respected, and devalued because of one's race or gender, microaggressions are also dismissive in nature and can include subtle snubs or dismissive looks, gestures, and tones (Sue et al., 2007).

Microaggressions are detrimental because their invisibility has harmful implications for the well-being, self-esteem, and standard of living of people of color and may have more influence on racial anger, frustration, and self-esteem than traditional overt forms of racism (Sol6rzano et al., 2000; U.S. Department of Health and Human Services, 2001). Microaggressions can impair performance in a multitude of settings by sapping the psychic and spiritual energy of recipients and by creating inequities (Franklin, 2004; Sue, 2007). While there is quantitative research on this topic, much of the prior research has used qualitative methods, which has been beneficial in exploring these emergent phenomena (Nadal, 2011). Research has been conducted with Asian Americans (Sue, 2010), Latin Americans (Rivera et al., 2010), indigenous people (Hill et al., 2010), multiracial individuals (Johnston & Nadal, 2010; Nadal, 2011), religious minorities, students of color (Sue et al., 2009), and African Americans (Sue et al., 2007, 2009).

For example, a study examining racial microaggressions and coping strategies among a group of ten Black American corporate leaders (Holder et al., 2015) found that women reported several themes of microaggressions including environmental manifestations (e.g., exclusion from representation, second-class status of ethnic markets, and tokenism), stereotypes of Black women (e.g., aggressive Black woman, assumption of criminal status, mammy-caretaker), exclusion (e.g., exclusions from work and social meetings, and career opportunities), and invisibility (e.g., being ignored or dismissed through the use of body language). They also reported assumed universality of the Black experience (Holder et al., 2015). Their White counterparts assumed that all Blacks shared the same limited experiences, opinions, and interests, and that a single person could act as the spokesperson for all Blacks because of their shared racial background.

Ethnic Misidentification and Misclassification in Minorities

Ethnic misidentification is a type of microinvalidation that is akin to "Assumption of Similarity/Exoticization," the assumed similarity to or difference from other members of one's perceived group. While ethnic misidentification may not be seen as outwardly negative, it communicates a disregard for an individual's heritage and culture, negating their unique experiences. As Sue et al. (2007) noted, microinvalidations are not necessarily aggressions, but taken in context, hearing these statements frequently can make the recipient experience them as aggressions and, perhaps, later begin to exhibit adverse physical and mental health outcomes as evidenced with overt racism. Ethnic misidentification has been documented with several groups with similarities to Black immigrants (i.e., small concentrated minority and immigrant populations) outside the microaggression literature and demonstrates the deleterious effects of being misidentified.

One study examined misclassification of Mayan immigrants (Millender, 2012). Misclassification as Latino, Hispanic, or Spanish places Mayan immigrants, who actually descend from the native people of Central America and self-identify as indigenous native Mayans, was found to increase risk for mental illness, suicide, and substance use. Another immigrant group that has faced adversity because of their misclassification is Sikhs. In a post 9/11 society, Sikhs are sometimes mistakenly identified as Muslim or Arab, viewed as strange or dangerous, and associated with terrorist activity. These misclassifications have led to increases in discrimination, hate crimes, and racial and religious profiling (Ahluwalia & Pellettiere, 2010). Misidentification is most often documented among Native American populations. Making up approximately 1% of the American population, Native Americans are easily misidentified because they are a relatively small community, and ignorance of their cultures is prevalent. There are many reports of the health and mental health problems experienced by Native Americans such as social isolation and suicide, with suicide rates being 50% higher than the general population. These mental health disparities may be a function of Native American's straddling of multiple cultures and lack of clarity about their identities as well as being routinely misclassified (Campbell & Troyer, 2007). Having a strong racial-ethnic identity is important for self-esteem among ethnic-racial minority groups. Moreover, others' validation of these important identities is critical in social interactions, as it can become difficult when one's identity is constantly in question.

Misidentification and Black Immigrants

Incidences of misidentification of Black people, in general, and Black immigrants, in particular, are not well recorded in the microaggression literature. To our knowledge, only one study has been conducted specifically with Black immigrants. A related microaggression, assumption of universality of the Black American experience, was recorded wherein Black people were thought to be interchangeable and individuals were often viewed as Black representatives with the ability to speak for all Black people and their experiences (Sue et al., 2007). In one example, a Black man was "consulted" by his White coworker on whether or not to refer to another coworker as Black or African American. That participant expressed frustration at the question and wondered how he was supposed to know what someone else would prefer. This highlights differences in the way individuals within the same racial group may self-identify, and it is erroneous to assume universality of experience among all Black people, whether immigrant or non-immigrant. When misclassified, young Black adults have been shown to have higher rates of suicidal ideation, suicide attempts, and used more counseling services (Campbell & Troyer, 2007).

What is the psychological impact when misclassification is based on ethnicity or culture instead of race? Black immigrants are grouped with African Americans though they may have very different cultural practices. While Native Americans or Latinos may suffer from ambiguity, Black immigrants may experience problems with similarity. Unlike Native Americans, Black immigrants may be less likely to identify as multiracial, preferring to identify as Black, in keeping with a system of hypodescent that is prevalent in U.S. society (Haozous et al., 2014). Depending on their ethnic make-up, Black immigrants may have different options for how they identify. For example, Ferguson & Bornstein (2014) expands the bicultural perspective of acculturation to propose a tridimensional perspective, which takes into account influences from mainstream European-American culture, African-American culture, and one's native heritage. Latino and Hispanic are ethnic terms for a heterogeneous people including White, Indian, Black, Asian, and mixed-race individuals. However, they are increasingly treated as a racial group, which enables Afro-Latino people to avoid the stigma associated with being identified as African American, a historically stigmatized group stretching back to slavery (Fisher & Model, 2012). For immigrants who do not have Latino roots but are also not necessarily considered Black in their home countries, this option does not exist. Such is the case for Cape Verdean immigrants, who come from a highly mixed-race nation.

Misclassification may overlook specific mental health risk factors that could be brought about by acculturative stress (Millender, 2012). There may also be some utility in the grouping: similarity in appearance may make it easier to blend in with the general populace for some, such as White immigrants (de Anda, 1984). Although Black immigrants cannot blend in with the White community, they may be able to blend in with the Black American community, which could be either beneficial or harmful. Regardless, blending in with the community is not always an option for Black immigrants, who make a variety of choices when it comes to alignment. In turn, these choices can lead to a variety of mental health and success outcomes (Ferguson & Bornstein, 2014). Deaux (2004) describes social identity as those aspects of a person that are defined in terms of group memberships. There is an overall American identity, but there are also subgroups of Asians, Africans, and Latinos. National identity is flexible and subjective, so a person can choose to be one, many, or none. However, diversity in the Black community is not always understood. Exploring immigration processes, such as acculturation, can provide a richer story.

Immigrant Acculturation Strategies and Psychological Outcomes

Acculturation is a gradual process of adaptation whereby group- (cultural-) and individual- (psychological-) level changes occur during continuous intercultural contact (Berry, 2003). Berry (1992) describes individual-level acculturation as a multidimensional process of learning about a host country's cultural norms and a negotiation between adopting those norms and maintaining one's group cultural customs when differing cultural groups come into continuous contact. Individuals can retain and/or relinquish ethnic society customs while simultaneously adopting and/or rejecting majority society values, reflecting the presence of varying choices or preferences in the process. However, research examining acculturation among people of African descent is limited. For this population, acculturation has been studied as it relates to suicide, substance abuse, health-promoting behaviors, management of cultural diversity, and racial identity (Obasi & Leong, 2009). Research on individual groups revealed generational differences in acculturation, whereby certain customs of the majority society are acquired, and certain ethnic customs are relinquished with successive generations (Stephenson, 2000). As a result, generational differences in acculturation styles may affect experiences of microaggressions, acculturation-related stress, and subsequent psychological distress.

Stress can help explain the relationship between perceived discrimination and mental health (Pascoe & Smart Richman, 2009). Among African Americans, it was found that experiences of discrimination exacerbate other stressors such as negative life events (Torres et al., 2012). Even when considering the spectrum of findings across outcomes for immigrants (Berry, 2006; Rudmin, 2006), for many immigrant groups, other stressful life events may include acculturative stressors such as cultural adaptation, pressures in learning a new language, and the complex task of negotiating differing cultural values (Berry, 2003; Ferguson & Bornstein, 2014; Torres et al., 2012). For misidentified Black immigrants, acculturative stress may include managing problems with maintaining their own identity, while being categorized differently and assuming new hybrid identities. Acculturative stress has been linked to psychological consequences including depression and suicide among Asian, Latino, and African Americans. There are few studies examining acculturative stress among Black people in general, and immigrant Blacks in particular, as much of the research focuses on Asians and Latinos (Walker et al., 2008). Black immigrants are a unique group because they experience the stress of immigration and acculturation as well as racism and microaggressions.

Ethnic Misidentification and Acculturation Among Black Immigrants

While Black immigrant populations are growing in the United States, they often remain hidden among the African-American population. Significant empirical gaps remain, as current research tends to focus on either African-American, or Asian and Latino populations. Essentially, their immigrant status is not recognized, and their lived experiences and mental health are not widely studied. It is likely that much of the research on African Americans includes misidentified Black immigrants. However, this categorization may not match their personal identification, which leads to misidentification from others, a type of microaggression that has been shown to have negative emotional and psychological outcomes such as depression, suicidality, and substance use among other ethnic groups. Black immigrants may experience discrimination based on both race and their first-generation status. The acculturation process they experience likely leads to acculturative stress as a result, which may also negatively impact psychological outcomes. The relationship between overt racism and acculturation has been studied. However, the relationship between microaggressions and acculturation remains an area for further study. Additionally, microaggression studies among immigrants in general, and Black immigrants in particular, are limited. Since acculturation and microaggressions have a demonstrated impact on physical and mental health outcomes, it is important to examine Black immigrant samples for these phenomena (Clarke, 2017).

To address these gaps, the present study used a mixed-methods approach to explore Black immigrant experiences with ethnic misidentification, examine its relationship to acculturation, and investigate how ethnic misidentification and acculturative stress may interact to predict psychological distress. For the quantitative part of the study, we hypothesized that (1) ethnic misidentification would be a significant predictor of psychological distress among both first- and second-generation Black immigrants, (2) ethnic misidentification would be a significant predictor of acculturative stress for first-generation Black immigrants, (3) acculturative stress would be a significant predictor of psychological distress for first-generation Black immigrants, (4) the relationship between ethnic misidentification and psychological distress would be mediated by acculturative stress, (5) the relationship between ethnic misidentification and acculturative stress would be moderated by acculturation strategy, and (6) there would be an indirect effect of ethnic misidentification on psychological distress through acculturative stress that is moderated by acculturation style (Clarke, 2017). Qualitative data were used to illustrate and deepen quantitative results.

Method: Quantitative and Qualitative Data

Participants

In order to reach the most diverse sample of Black immigrants, the first author employed several recruitment techniques including personal contact over the Internet and in person using business cards across six states (California, New York, Florida, Georgia, Nevada, and Texas). Recruitment materials were emailed to personal contacts, immigrant, student, social, and professional groups, including the Latino Psychology Association, and a university community. Social media sites, such as Facebook, Twitter, and Craigslist, were used to distribute the online survey. Snowball sampling was used as a technique to help reach hidden populations, and participants were encouraged to share the survey link with other Black immigrants in their networks (Griffith et al., 2016). Participants accessed a Qualtrics survey via personalized Qualtrics email links, anonymous link, and anonymous Quick Response code (printed on business cards), and were offered a $25 raffle incentive from Amazon.com upon completion of the survey (Clarke, 2017).

Ultimately, 106 English-speaking, Black-identified (African, African American, Afro-Caribbean, Afro-Latino, Multiracial, etc.) adult immigrants living in the United States participated in the survey. The sample was multiethnic, with the most frequent ethnicities being African American, Black, or Haitian (9.4% each), followed by Afro-Caribbean, Caribbean, and West Indian (6.6% each). Many identified regionally (i.e., African, Caribbean) or nationally (i.e., Jamaican, Trinidadian) or in combination with race (i.e., Afro-Latino). Participants were either first generation ($N = 42$) or U.S.-born to immigrant parents (second-generation; $N = 64$) and from 13 different states. Participants were mostly women ($N = 85$, 80.2%), heterosexual ($N = 97$, 91.5%), and highly educated, with two-year college degrees or higher accounting for 88.7% ($N = 63$) of the sample. Most were unmarried (i.e., single/never married, separated/divorced, widowed, or in a relationship; 64.2%; $N = 68$) and had no children (54.7%; $N = 58$).

After accessing the survey via computer or mobile phone, participants confirmed their age and acknowledged informed consent. They were then presented with the demographics questionnaire, followed by measures of ethnic misidentification, general microaggressions, psychological distress, acculturation strategy, and acculturative stress. At the completion of the survey, referral information to the National Suicide Prevention Lifeline and a hyperlink to the $25 Amazon.com gift card raffle were presented.

Culturally Appropriate Instrumentation

Demographics

U.S. Census questions were supplemented by researcher-generated questions (e.g., *What is your zip code?*; *What is your country of origin?*) and collected basic demographic information such as birth year, sex, gender, sexual orientation, marital status, and level of education. Using categorical responses rather than open-ended questions forces people to fit predetermined categories and may inadvertently be microaggressive (Johnson & Nadal, 2010; Nadal, 2011). Thus, the present study used a combination of questions to allow participants to freely self-identify (e.g., *Black* and *Other* (with text entry) were options in the race question; *"What is your ethnicity"* was open ended). The immigrant generational status demographic question was drawn from the Measurement of Acculturation Strategies for People of African Descent (MASPAD; Obasi, 2004) measure, which included this question along with the original measure (e.g., *I was born outside of the United States* and *My parents and I were born in the U.S.; All grandparents were born outside the U.S.*).

Ethnic Misidentification

Two strategies were used to measure ethnic misidentification. First, the Racial and Ethnic Microaggressions Scale (REMS-45; Nadal 2011) measures the number of times microaggressions were experienced in participants' daily lives over the past six months. Its six subscales can be used independently and reflects the original taxonomy by Sue et al. (2007) and microaggressions recorded in qualitative studies (Nadal, 2011). The Exoticization/Assumptions of Similarity subscale (REMS4) was closest to the conceptualization of misidentification and was used to measure ethnic misidentification. It addresses stereotyping and objectification according to race or ethnicity (e.g., "Someone objectified one of my physical features because of my race," and "Someone told me that all people in my racial group are all the same"; Forest-Bank & Jenson, 2015; Nadal, 2011). Previous researchers demonstrated the full measure's (a = 0.92) and Exoticization/Assumption of Similarity (a = 0.91) subscale's high internal consistency for a Black sample, which was maintained in the present study with the overall scale (a = 0.95) and the Exoticization/Assumption of Similarity subscale (a = 0.80).

The second strategy for measuring ethnic misidentification was to use a series of 14 researcher-generated items. A five-point Likert response scale (1 = Never, 2 = Rarely, 3 = Sometimes, 4 = Often, 5 = Always) collected answers on the first 12 statements, which were about misidentification in different contexts, feelings about being misidentified, and the importance placed on being correctly identified (e.g., "My ethnic identity [nationality, cultural background, etc.] is correctly identified by others.") The last two items were open-ended questions that solicited examples of misidentification experiences and related feelings (e.g., "Describe an experience you have had, in which your nationality or cultural background was misidentified. How did this experience make you feel?"). These questions provided the qualitative exploration of ethnic misidentification experiences and supplemented the quantitative data.

Confirmatory factor analysis determined that the set of 12 questions was not a good model fit for use as a single-scale measure of ethnic misidentification. Thus, an exploratory factor analysis was conducted to determine the possibility of using individual questions as measures for ethnic misidentification. Individual items were not a good model fit, but highly correlated items were used as measures of ethnic misidentification in the main analyses. Item 1 was used independently, and three separate composites were formed from items 2–3, 4–5, and 7–8, which were highly correlated. These items are referred to as the ethnic misidentification (EMID; Clarke, 2017) questions.

Psychological Distress

Psychological distress was assessed with the 18-item Brief Symptom Inventory-18 (BSI-18; Derogatis, 2000). It measures the three most prevalent psychiatric domains: somatization, depression, and anxiety. Commonly used in diverse clinical and community settings, the BSI-18 yields consistently high internal consistencies (Cronbach's alphas of 0.91, 0.92, and 0.97 for Black samples), which was supported in the current study with Cronbach's alphas of 0.92 (total scale), 0.87 (Depression subscale), 0.82 (Anxiety), and 0.79 (Somatization subscale).

Acculturation Strategy

The Measurement of Acculturation Strategies for People of African Descent (MASPAD; Obasi, 2004) is a 45-item, bidimensional measure based on Berry's acculturation theory and identifies traditionalist, integrationist, assimilationist, and marginalist acculturation strategies (Berry, 1992; Obasi & Leong, 2009). It distinguishes between acculturation beliefs and behaviors using two orthogonal dimensions that vary independently (i.e., preference for maintaining one's own heritage ethnocultural group [Traditionalist subscale; D1] and preference for participating in the society of a different ethnocultural group [Assimilationist subscale; D2]) such that medium-to-high scores on D1 and low scores on D2 indicate a traditionalist style, medium-to-high scores on both dimensions indicate an integrationist style, low scores on D1 and medium-to-high scores on D2 indicate an assimilationist style, and low scores on both dimensions indicate a marginalist style (Obasi & Leong, 2009). The MASPAD has been normed specifically on multi-generational Black immigrants and purports adequate internal consistency of both scales (D1: $a = 0.87$; D2: $a = 0.75$; Obasi & Leong, 2009). The present study produced

Cronbach's alphas of 0.84 (Traditionalist subscale; D1) and 0.72 (Assimilationist; D2), respectively.

Acculturative Stress

The commonly used 30-item Social, Attitudinal, Familial, and Environmental Acculturative Stress Scale-Revised (SAFE-R; Hovey, 2011) assesses acculturative stress across social, attitudinal, familial, and environmental contexts including perceived discrimination. This questionnaire was appropriate only for the first-generation sample, as determined by the generational status question. The present study had comparable reliability ($a = 0.90$) to other studies involving samples of African-American and Black college students, and Haitians and Haitian Americans, with reliability coefficients ranging from 0.87 to 0.89 (Joiner & Walker, 2002).

Data Collection and Analyses

To reach the most participants in the fastest way possible, quantitative and qualitative data were collected concurrently through the online survey. The mechanisms that influence the relationships between ethnic misidentification, acculturation strategy, acculturative stress, and psychological distress were examined in the quantitative data and analyzed using SPSS 24.0 and Mplus 7.0, a structural equation modeling software package (Muthén & Muthén, 1998). The qualitative portion of the study elicited narratives of ethnic misidentification and emotional responses to these experiences. These data were reviewed and coded for the presence of ethnic misidentification and specific emotional reactions. Qualitative reports of ethnic misidentification were used to supplement quantitative data and as illustrations to provide context.

Collecting and Interpreting Interview Data

The first author collected qualitative data consisting of two open-ended items that gathered narratives of ethnic misidentification and emotional reactions. These data were collected concurrently with quantitative data via the online survey. The second and third authors used a directed content analysis approach (Hsieh & Shannon, 2005) to identify common patterns within the data. Coders met together three times: once, to become familiar with participants' responses; again, to agree on coding themes and create detailed descriptions for each; and once after individually manually coding all interviews to agree on codes.

Quantitative Results

Relationships in the Quantitative Data

Multiple regression analyses were used to test the direct effect that ethnic misidentification has on psychological distress among Black immigrants (hypothesis 1). The model was run five times, using the five measures of ethnic misidentification. For both first- and second-generation Black immigrants, ethnic misidentification was a significant predictor (with REMS4 as a measure of ethnic misidentification) of psychological distress, [$F(1,102) = 16.009$, $p < 0.000$] with an R^2 of 0.132. The relationship did not reach significance when ethnic misidentification (EMID) questions were used.

Acculturative stress could be measured only in the first-generation sample so relationships between that and ethnic misidentification or psychological distress included only the first-generation sample. Multiple regression analyses were used to test the direct effect that ethnic misidentification has on acculturative stress (hypothesis 2). When EMID questions were used as the predictor variable, the relationship did not reach significance. However, when REMS4 was the predictor variable measure, the direct relationship reached significance ($b = 23.30$, SE(b) = 6.42, z = 3.63, $p = 0.00$), demonstrating that ethnic misidentification is a significant predictor of acculturative stress.

Multiple regression was used to test the direct effect that acculturative stress has on psychological distress for first-generation Black immigrants (hypothesis 3). With each predictor measure, acculturative stress was a consistent significant predictor of psychological distress for first-generation Black immigrants. Table 12.1 shows the significant relationships for hypothesis 3.

Mediation, moderation, and moderated mediation tests again ran the model five times, using five measures of ethnic misidentification in Mplus 7.0 (Muthén & Muthén, 1998). Recommended by Preacher, Rucker, and Hayes (2007) as a resampling procedure for hypothesis testing of indirect and conditional indirect effects, which does not assume a normal distribution, bootstrapping procedures using 2000 bootstrapped samples with replacement were used to construct the confidence intervals and accounted for skewed and kurtotic distributions in the present study (Clarke, 2017; Efron & Tibshirani, 1993; Field, 2009). Acculturative stress emerged

Table 12.1 Results of b-path direct effect of acculturative stress on psychological distress for Black immigrants ($N = 37$)

Model	Beta	Z	*P*-value
EMID1	0.254	4.883	0.000
EMID2/3	0.240	3.926	0.000
EMID4/5	0.241	4.341	0.000
EMID7/8	0.249	4.624	0.000
REMS4	0.231	3.555	0.000

as a significant mediator in the relationship between ethnic misidentification and psychological distress (hypothesis 4), but only when REMS4 was used to measure the predictor variable. Since the relationship between ethnic misidentification and acculturative stress was significant (path a) and the relationship between acculturative stress and psychological distress was consistently significant (path b) and ethnic misidentification predicted psychological distress (c-path), partial mediation was found. The 95% confidence interval (46.13–64.20) did not contain 0, indicating a significant indirect effect ($b = 5.39$, $SE(b) = 2.13$, $z = 2.53$, $p = 0.012$). Acculturative stress partially explains the relationship between ethnic misidentification and psychological distress.

Hypotheses 5 and 5a–d expected that acculturation style would have a conditional effect on the relationship between ethnic misidentification and acculturative stress for Black immigrants and that specific acculturation styles would dictate the strength of the relationship. However, conditional effects approached but did not achieve significance and provided no support for hypotheses 5 and 5a–d. Finally, the overall model (hypothesis 6) expected that the indirect effect of ethnic misidentification on psychological distress through acculturative stress would be moderated by acculturation style for Black immigrants. This relationship failed to reach significance and did not provide support for hypothesis 6 (Clarke, 2017).

Qualitative Results

What Interviewees Told Us About Ethnic Misidentification

Qualitative findings revealed eight major themes depicting the context of ethnic misidentification experiences, as well as personal reactions to these experiences (Clarke, 2017). Four major themes depicted the various contexts in which Black immigrant individuals may experience ethnic misidentification. These themes included misidentification based on stereotype, the source of the misidentification (the person who misidentified the participant), misidentification based on lack of knowledge, and misidentification coupled with denial or dismissal of the individual's self-ascribed ethnicity. Personal reactions or approaches to experiences of ethnic misidentification also revealed four themes following misidentification including negative reactions, neutral reactions, positive reactions, and the feeling of being unseen or devalued.

Contextual Experiences of Ethnic Misidentification

Stereotypes. Participants described experiences of ethnic misidentification based on stereotypical phenotypes and stereotypical customs/norms. For example, when misidentified based on phenotype, participants noted attributes such as their skin tone and hair texture as being the cause of the misidentification. A second-generation Jamaican and Costa Rican–identified woman shared her experience of being misidentified and stereotyped due to her appearance/phenotype. She stated, "People in America always give me double takes when I mention my family is from Costa Rica because I don't look like a stereotypical fair-skinned, straight haired 'Latina.' " Another second-generation female from Trinidad and Tobago shared a similar experience, highlighting a misconception about the language that she spoke along with her being misidentified due to her appearance/phenotype. She shared, "Due to my hair and skin tone, I get mistaken for Dominican, Panamanian, or 'Black Puerto Rican' often. People come up to me speaking Spanish and are surprised that I'm 'just black.' " When misidentification occurred based on stereotypical customs or practices, participants cited attributes such as their language, the way they dressed, and the origins of their names as the cause of misidentification. For example, one second-generation man from Haiti expressed, "Because of my name, people ask me if I'm French or from Louisiana (Cajun)." Another assumption or encounter with misidentification can occur when one may not readily display cultural markers of their specific ethnic background. For example, not having an accent that may determine where one is from. One first-generation woman from Jamaica shared, "People assume I was born in the U.S. because I don't have an accent." As these examples show, participants with various ethnic backgrounds have faced misidentification rooted in stereotypes due to their outward appearances and accents, or lack thereof.

Sources of Misidentification. Participants also reported on their experiences of being ethnically misidentified by people who hold membership within their ingroup or an outgroup. A second-generation Puerto Rican–identified woman shared her experience of misidentification with people belonging to both her ingroup and outgroup. She stated, "I am light skinned/olive complexion, and I have had people suggest that I am white, or 'white bread' in the past—this has usually come from darker African Americans or Latinos." Similarly, a first-generation woman from Haiti expressed, "[I felt] disappointed that although we are from the same geographical island, he could be so rude and ignorant about the ethnic makeup and diversity." Other participants reported experiences of being misidentified, specifically by outgroup members. For example, a second-generation Nigerian-identified woman expressed frustration for being ethnically misidentified and fetishized by White men during online dating and expressed a need to separate herself from stereotypical media images of African Americans:

> I've done a lot of online dating sites. I've had Caucasian men reach out to me and say things such as "I've never been with a black girl before but I want to experience you." I've had more than 20 Caucasian guys say this statement to me. They assume that I'm like the black girls depicted on BET and shows like Love and Hip Hop.

Lastly, a first-generation Jamaican-identified woman also expressed experiencing misidentification from her White counterparts: "In settings where dealing with those of the Caucasian race, it is not uncommon for them to label all black people as African-American. I usually make it very clear that I am Jamaican to clear up any confusion." As demonstrated, misidentification can occur from those whom we directly connect and share commonalities with (ingroup) as well as those with whom we do not (outgroup). These experiences can feel equally dismissive and invalidating.

Lack of Awareness. Participants highlighted the lack of knowledge or ignorance of others in understanding geography as well as the diversity within the Black Diaspora. For example, a second-generation Haitian-identified woman stated, "I've had people think Haiti was in Africa (mainland)." Another first-generation Antiguan-identified woman reported the tendency for others to group individuals from the Caribbean together although each is uniquely different: "As an Antiguan, I generally get lumped into black American or if I say I'm Caribbean, I'm generally considered to be Jamaican or from another larger well known Caribbean island." A Cuban-identified first-generation woman highlighted the lack of knowledge regarding ethnic diversity:

> When people find out I speak Spanish they are confused, I tell them I am Cuban (and it never fails) afterwards I hear "oh you're Cuban?! I thought you were black!" As if it had to be one or the other, as if I couldn't be a black Cuban.

These reports illustrate a dire need for greater awareness of ethnic and cultural diversity in order to approach others with curiosity instead of assumptions. While it is normative to lack knowledge about individuals and places with which we are less familiar, the assumption of universality and denial of diversity among BI appears to increase the likelihood for misidentification.

Denial of Identity. Participants also described experiences of ethnic misidentification that were marked by others questioning, denying, or dismissing their identity. For example, a first-generation Jamaican-identified woman expressed:

> In a class, the professor referred to me as African American to which I responded I wasn't. He was insistent on calling me AA though I said I prefer the term Black American if I were to be referred to as American at all.

Another Jamaican-identified, yet second-generation, woman described the tendency for others to assume that she is ashamed of her race when she corrects their ethnic misidentification:

> People just assume that I am African American because of my appearance. The fact that my cultural identity is not tied to the American South never occurs to anyone. When I raise it, people assume that I am somehow trying to not be black.

Similarly, a first-generation Grenadian-identified woman stated:

> As a child, I was asked where I am from in an effort to understand my French sounding last name. The person thought I was Haitian and when I explained that I was not from Haiti, they assumed I was lying or ashamed to say I was Haitian.

These accounts show that self-ascribed identity may be blatantly dismissed even when a misidentification is corrected. It begs the question of whose ascription is most important. Moreover, it robs BI of their voice and autonomy in naming and proclaiming their own identity.

Emotional Experiences of Ethnic Misidentification

Negative Reactions. The following participants' responses reflected negative experiences of ethnic misidentification. For example, a second-generation Haitian-identified woman reported:

> I was annoyed that he assumed that I had a particular upbringing. I understood how that could be misunderstood but at the same time knowing that this was someone who was educated (MD and PhD), and Indian American - I felt like he should know better.

Another second-generation Dominican- and Honduran-identified woman noted the pain associated with these similar experiences:

> When I was younger it hurt me a lot. Especially that I experienced this from other Latinos. It was so severe that I eventually stopped speaking Spanish and tried to hide my nationality. It wasn't until college that I became more comfortable with my identity.

A first-generation Kenyan-identified woman spoke about the anger often associated with these experiences: "Infuriated....why do I have to fit into a box of misconceptions for others to be comfortable?" These negative reactions to ethnic misidentification appeared to range from benign feelings such as confusion and disappointment, to more intense emotions such as feeling insulted or enraged.

Neutral Reactions. Some participants expressed reactions depicting more neutral or indifferent feelings following experiences of ethnic misidentification. Many of these participants explained they had become used to these experiences over time and likely no longer had strong negative or positive reactions to them. For example, a second-generation Belizean-identified woman expressed feeling "no different than any other day, I think it's just something I have gotten used to, which is probably just my own coping mechanism." Another participant, a second-generation Haitian-identified woman appeared to have a similar reaction: "It has happened so frequently in my life that it doesn't impact me emotionally. It has become part of my experience. I guess I would say indifferent." One second-generation Eritrean man expressed his understanding of the misidentification: "Indifferent, as it is an honest mistake. No malicious intent." As seen in these examples, some participants appeared to internalize misidentification experiences and took them as neither positive nor negative. As one participant mentions, this may indicate a necessary adaptation or way of coping in response to pervasive experiences of misidentification.

Positive Reactions. Some participants spoke to the positive experiences of being ethnically misidentified, expressing feelings of pride and flattery. For example, a second-generation Jamaican-identified woman expressed, "I was actually flattered." A first-generation Jamaican-identified woman expressed her pride in being able to express her ethnic identity and what she has overcome: "I am always proud to say I am Jamaican. Even more proud to say I am an immigrant. In my mind, I've defied many odds and my very presence in some settings is evidence of my resilience." A second-generation Sierra Leonean–identified woman expressed a similar sentiment: "I've learned to take pride in being African so if my ethnicity ever comes up in conversation, I tell people I'm African and I feel good." While experiences of ethnic misidentification can often have a negative impact on individuals, these participants demonstrate the importance of not assuming that all individuals find these experiences distressing. The examples of reframing misidentification into a positive light further support the argument of diversity among BI.

Unseen/Devalued. A group of participants expressed feeling as if their ethnic identity, or parts of it, were being ignored or undervalued. Some reported feeling unseen and invisible. A second-generation Dominican and Puerto Rican–identified woman stated, "It made me feel confused and erased." Another second-generation Puerto Rican–identified female explained how her family's history feels devalued during these experiences:

> When I am misidentified I feel as though I don't belong to any ethnic group. It makes me feel as though my ancestors who have survived slavery are not within me or not recognized for surviving within me, and that makes me feel profoundly sad - as though by assimilating through immigration and through being mixed, I am not treating that legacy of survival and resilience properly.

Other individuals noted feeling as if they did not belong anywhere, sometimes feeling torn between two identities. For example, a second-generation Ghanaian man expressed, "I always felt like I was in this gray area; not fully accepted by Blacks AND Whites. Floating between groups, never fully a part of either." As seen above, experiences of misidentification can at times be more complex than a simple positive or negative experience. For some, these experiences call into question one's sense of belonging and can make others feel as if a part of them has been ignored or erased. These feelings of invisibility and lack of validation may lead to more devastating and lasting consequences for these individuals.

Discussion

The use of interview narratives to supplement survey information gathered from Black immigrants in this study provided a nuanced understanding of the impact of ethnic misidentification on psychological distress. Quantitative survey data indicated that ethnic misidentification was a significant predictor of acculturative stress, acculturative stress was a significant predictor of psychological distress, and ethnic misidentification was a significant predictor of psychological distress for Black immigrants. These findings demonstrate that acculturative stress may explain the relationship between ethnic misidentification and psychological distress whereby ethnic misidentification may lead to psychological distress through the process of acculturative stress. However, acculturation strategy did not significantly impact the relationship between ethnic misidentification and acculturative stress (Clarke, 2017).

The specific microaggression of misidentification is understudied among Black people and, as such, it was important to collect narratives of ethnic misidentification. Supplementing quantitative results, contextual and emotional themes of participants' experiences with ethnic misidentification emerged in the qualitative data. The overwhelming majority of participants reported incidences of ethnic misidentification, and most reported negative emotional reactions associated with these experiences (Clarke, 2017). These narratives highlight subjective emotionally charged experiences of ethnic misidentification, which could potentially rise to the level of psychological distress (Clarke, 2017).

Ethnic Misidentification and Psychological Distress

A direct quantitative link was found between ethnic misidentification and psychological distress, comparable to findings from previous studies. While some previous research has linked ethnic misidentification directly to other emotional and mental health outcomes, few have included Black immigrants. It has been found that ethnic misidentification puts Mayans at risk for mental illness, suicide, and substance use; Sikhs at risk for discrimination, hate crimes, and racial and religious profiling; and Native Americans at risk for suicidality and poor health outcomes (Ahluwalia & Pellettiere, 2010, Campbell & Troyer, 2007, Millender, 2012). Misidentified Blacks were found at risk for higher rates of suicidal ideation and suicide attempts, and more frequent use of counseling services in one quantitative study that included them (Campbell & Troyer, 2007). It is important to consider that since ethnic misidentification is understudied in Black immigrants, the relationship between ethnic misidentification and psychological outcomes may be different for them (Clarke, 2017).

The narratives of ethnic misidentification in the present study are similar to those found in previous qualitative research, which documented instances of ethnic misidentification and negative emotional reactions under terms such as invalidation of interethnic differences, assumed universality, denial of personal differences, automatic connection among Black people, and invisibility (Holder et al., 2015; Sue et al., 2007; Watkins et al., 2010). Participants in previous studies reported negative emotional reactions to general microaggressions such as anger, frustration, belittlement, alienation, invisibility, and invalidation (Araújo Dawson, 2009; Sue et al., 2007, 2009). The present study uncovered similar emotional reactions to ethnic misidentification. These narratives revealed subclinical negative outcomes such as disappointment and frustration. Contrary to previous research, positive, neutral, and negative emotional reactions were uncovered, indicating that ethnic misidentification does not always have to lead to negative reactions. In particular, misidentification may be viewed positively when it results in an opportunity for one to demonstrate pride, or "belongingness," which is inclusive and welcoming, rather than "othering," which is exclusive and marginalizing. Interestingly, the qualitative data revealed that people who make a misidentification can be both similarly identified and dissimilarly identified from the targets of the misidentification, which may take place in different contexts (i.e., work, school, social settings), and can be based on a number of things ranging from appearance, to language, to education. Reports of misidentification were also linked to location or foreign countries where the population of Blacks was either small or more homogeneous, a finding that is similar to studies of Native Americans being misidentified in geographic locations with smaller Native American populations (Bertolli et al., 2007; Campbell & Troyer, 2007; Clarke, 2017; Jim et al., 2014).

Acculturation and Psychological Distress

Consistent with previous research with people of African descent, the most frequent acculturation strategies reported in the present study were integrationist and traditionalist (Obasi et al., 2009; Obasi & Leong, 2009, 2010). However, the traditionalist strategy was secondary when compared to previous research. An integrationist strategy may buffer stress and relates to better psychological outcomes, which may account for the low levels of acculturative stress and psychological distress among the present study participants (Berry, 2003; Liang et al., 2004; Obasi & Leong, 2009). This contradicts some findings from the study of Ferguson et al. (2014), where among behavioral outcomes, the integrationist strategy was disadvantageous. Acculturation strategy was a significant moderator in the relationships between confidence in mental health practitioners and psychological distress and the relationship between everyday discrimination and general stress, but acculturation strategy was not a significant moderator in the relationship between ethnic misidentification and acculturative stress (Araújo Dawson, 2009; Obasi & Leong, 2009). This finding failed to support Berry's (2006) proposal that acculturation strategy would moderate the relationship between perceived discrimination, a related variable, and acculturative stress (Clarke, 2017).

In the present study, acculturative stress was related to psychological distress, which is consistent with the literature. This study was one of two that found support for this relationship with Black samples. There was a direct positive relationship between acculturative stress and psychological distress in the present study, wherein increased acculturative stress predicted increased psychological distress. Being misidentified is a logical impediment to adjusting to a new culture, and the resulting stress plausibly relates to psychological distress. These results suggest that ethnic misidentification and acculturative stress are real problems for Black immigrants and that these relationships should be closely examined with them in mind (Clarke, 2017).

Summary and Key Points

The present study examined a very specific microaggression for Black people, one that has many different labels, and attempted to bring all of these labels under one umbrella. Ethnic misidentification or, simply, misidentification is more descriptive than the disparate terms and has the potential for widespread use and tracking. The present study also amplified Black immigrant voices, which have been missing from the literature, highlighting the implications of acculturative stress in the potential development of psychological issues following ethnic misidentification. The narratives of these first- and second-generation Black immigrants also depicted experiences of ethnic misidentification from both ingroup and outgroup members, based on stereotypes and lack of knowledge, and demonstrated a wide range of emotional reactions to these misidentifications including flattery and invisibility. For Black immigrants, ethnic identity may complicate the determination of who is in one's ingroup versus outgroup, and increase microaggressions, which are inherently "othering."

Findings showed that ethnic misidentification, acculturative stress, and psychological distress are real problems for Black immigrants with the potential for long-term clinical implications, which may become more of a concern as immigration continues to diversify the United States and challenge the status quo of how people are identified. It is important to consider how the current political and cultural climate of heightened racial tensions and anti-immigrant sentiment may be more stressful for Black immigrants and lead to more frequent reports of ethnic misidentification, acculturative stress, and psychological distress. Conversely, ethnic misidentification may be viewed as less negative because there may be some utility in appearing African American, rather than as a Black immigrant, as tensions rise. Further, the study allowed for Blacks to self-identify, and the responses indicated that many preferred ethnic designations when allowed to self-identify. Allowing people to self-identify may complicate research, but it may also allow for a deeper understanding of the populations studied and a richer understanding of psychological phenomena (Clarke, 2017).

While there may be challenges generalizing study results to other minority communities, later generations, or Blacks in other countries, the implication of acculturative stress as a critical factor in the development of psychological distress may create areas for intervention for other immigrant communities. Narratives demonstrating the wide diversity of experiences of ethnic misidentification and subsequent feelings may be relevant to other minority communities as well. Future researchers may examine differences in later generations, other minority groups within the United States, and Black immigrants in other countries to examine if ethnic misidentification has similar relationships in other populations (Clarke, 2017).

References

Ahluwalia, M. K., & Pellettiere, L. (2010). Sikh men post 9/11: Misidentification, discrimination, and coping. *Asian American Journal of Psychology, 1*(4), 303–314. https://doi.org/10.1037/a0022156

American Psychological Association, Presidential Task Force on Preventing Discrimination and Promoting Diversity. (2012). *Dual pathways to a better America: Preventing discrimination and promoting diversity. American Psychological Association.* http://www.apa.org/pubs/info/reports/promotingdiversity.aspx

Anderson, M. (2015). A rising share of the U.S. black population is foreign born. *Pew Social Trends*. https://www.pewsocialtrends.org/2015/04/09/a-rising-share-of-the-u-sblackpopulation-foreign-born/

Anderson, M., & López, G. (2018). Key facts about black immigrants in the U.S. *Pew Social Trends*. https://www.pewresearch.org/fact-tank/2018/01/24/key-facts-about-blackimmigrants-in-the-u-s/

Araújo Dawson, B. (2009). Discrimination, stress, and acculturation among Dominican immigrant women. *Hispanic Journal of Behavioral Sciences, 31*(1), 96–111. https://doi.org/10.1177/0739986308327502

Berry, J. W. (1992). Acculturation and adaptation in a new society. *International Migration, 30*, 69–85. https://doi.org/10.1111/j.1468-2435.1992.tb00776.x

Berry, J. W. (2003). Conceptual approaches to acculturation. In K. M. Chun, P. B. Organista, & G. Marin (Eds.), *Acculturation: Advances in theory, measurement, and applied research*. American Psychological Association.

Berry, J. W. (2006). Contexts of acculturation. In D. L. Sam & J. W. Berry (Eds.), *The Cambridge handbook of acculturation psychology* (pp. 27–42). Cambridge University Press. https://doi.org/10.1017/CBO9780511489891.006

Bertolli, J., Lee, L. M., & Sullivan, P. S. (2007). Racial misclassification of American Indians/Alaska Natives in the HIV/AIDS reporting systems of five states and one urban health jurisdiction, U. S., 1984–2002. *Public Health Reports, 122*, 382–392. https://doi.org/10.1177/003335490712200312

Campbell, M. E., & Troyer, L. (2007). The implications of racial misclassification by observers. *American Sociological Review, 72*, 750–765. https://doi.org/10.1177/000312240707200505

Clarke, A. V. (2017). *Black and immigrant in America: Ethnic misidentification, acculturation, and psychological distress* (Publication No. 10634389) [Doctoral dissertation, Palo Alto University]. ProQuest Dissertations Publishing.

Cutler, C. (2008). Brooklyn style: Hip-hop markers and racial affiliation among European immigrants in New York City. *International Journal of Bilingualism, 12*(1–2), 7–24. https://doi.org/10.1177/13670069080120010201

de Anda, D. (1984). Bicultural socialization: Factors affecting the minority experience. *Social Work, 29*(2), 101–107. https://doi.org/10.1093/sw/29.2.101

Deaux, K. (2004). Immigration and the Color Line. In G. Philogène (Ed.), *Racial identity in context: The legacy of Kenneth B. Clark* (pp. 197–209). American Psychological Association. https://doi.org/10.1037/10812-011

Derogatis, L.R. (2000). *The Brief Symptom Inventory–18 (BSI-18): Administration, scoring and procedures manual*. National Computer Systems.

Efron, B., & Tibshirani, R. (1993). *An introduction to the bootstrap*. Chapman and Hall.

Ferguson G., & Bornstein M. (2014). Tridimensional (3D) acculturation: Culture and adaptation of black Caribbean immigrants in the USA. In Dimitrova R., Bender M., van de Vijver F. (Eds.), *Global perspectives on well-being in immigrant families. Advances in Immigrant Family Research* (Vol 1, pp 31–51). Springer. https://doi.org/10.1007/978-1-4614-9129-3_3

Ferguson, G. M., Bornstein, M. H., & Pottinger, A. M. (2014). Tridimensional acculturation and adaptation among Jamaican adolescent-mother dyads in the United States. *Child Development, 83*, 1486–1493. https://doi.org/10.1111/j.1467-8624.2012.01787.x

Forest-Bank, S., & Jenson, J. M. (2015). Differences in experiences of racial and ethnic microaggression among Asian, Latino/Hispanic, Black, and White young adults. *Journal of Sociology, 42*(1).

Fisher, G. A., & Model, S. (2012). Cape Verdean identity in a land of Black and White. *Ethnicities, 12*(3), 354–379. https://doi.org/10.1177/1468796811419599

Field, A. P. (2009). *Discovering statistics using SPSS* (3rd ed.). Sage Publications Ltd.

Franklin, A. J. (2004). *From Brotherhood to manhood: How black men rescue their relationships and dreams from the invisibility syndrome*. Wiley.

Grieco, E. M. (2009). *Race and Hispanic origin of the foreign-born population in the United States: 2007*. United States Census Bureau. https://www.census.gov/library/publications/2010/acs/acs-11.html#:~:text=By%202007%2C%2080%20percent%20of%or%20Asia%20

(Figure%201).&text=The%20proportion%20of%20the%20total,population%20was%20 White%2C%20not%20Hispanic

Griffith, D. A., Morris, E. S., & Thakar, V. (2016). Spatial autocorrelation and qualitative sampling: The case of Snowball type sampling designs. *Annals of the American Association of Geographers, 106*(4), 773–787. https://doi.org/10.1080/24694452.2016.1164580

Haozous, E. A., Strickland, C. J., Palacios, J. F., & Arambula Solomon, T. G. (2014). Blood politics, ethnic identity, and racial misclassification among American Indians and Alaska Natives. *Journal of Environmental and Public Health, 2014*, 1–9. https://doi.org/10.1155/2014/321604

Hovey, J. D., (2011). *SAFE-Revised Acculturation Stress Scale (SAFE-R)*.(Version 3). www.psimh.com/researchclinicalinstruments. The University of Toledo, Ohio. United States.

Holder, A. M. B., Jackson, M. A., & Ponterotto, J. G. (2015). Racial microaggression experiences and coping strategies of black women in corporate leadership. *Qualitative Psychology, 2*(2), 164–180. https://doi.org/10.1037/qup0000024

Hsieh, H.-F., & Shannon, S. E. (2005). Three approaches to qualitative content analysis. *Qualitative Health Research, 15*(9), 1277–1288. https://doi.org/10.1177/1049732305276687

Hill, J. S., Kim, S., & Williams, C. D. (2010). The context of racial microaggressions against indigenous peoples: Same old racism or some- thing new? In D. W. Sue (Ed.), *Microaggressions and marginality: Manifestation, dynamics, and impact* (pp. 105–122). Wiley & Sons.

Jim, M. A., Arias, E., Seneca, D. S., Hoopes, M. J., Jim, C. C., Johnson, N. J., & Wiggins, C. L. (2014). Racial misclassification of American Indians and Alaska Natives by Indian health service contract health service delivery area. *American Journal of Public Health, 104*(S3), S295–S302. https://doi.org/10.2105/ajph.2014.301933

Johnston, M. P., & Nadal, K. L. (2010). Multiracial microaggressions: Exposing monoracism in everyday life and clinical practice. In D. W. Sue (Ed.), *Microaggressions and marginality: Manifestation, dynamics, and impact* (pp. 123–144). Wiley & Sons.

Joiner, T. E., & Walker, R. L. (2002). Construct validity of a measure of acculturative stress in African Americans. *Psychological Assessment, 14*(4), 462–466. https://doi.org/10.1037/1040-3590.14.4.462

Kaplan, C. (2015). Rachel Dolezal and racial illogic. *The Chronicle of Higher Education, 61*(40), B4–B5. https://www.chronicle.com/article/what-the-1920s-tell-us-about-dolezal-and-racial-illogic/

Liang, C. T. H., Li, L. C., & Kim, B. S. K. (2004). The Asian American Racism-Related Stress Inventory: Development, factor analysis, reliability, and validity. *Journal of Counseling Psychology, 51*, 103–114. https://doi.org/10.1037/0022-0167.51.1.103

Millender, E. (2012). Acculturative stress among Maya in the United States. *Journal of Cultural Diversity, 19*(2), 58–64. https://pubmed.ncbi.nlm.nih.gov/22924204/

Muthén L. K., & Muthén B. O. (1998). *Mplus: The Comprehensive Modeling Program for Applied Researchers: User's Guide*. Muthén & Muthén.

Nadal, K. L. (2008). Preventing racial, ethnic, gender, sexual minority, disability, and religious microaggressions: Recommendations for promoting positive mental health. *Prevention in Counseling Psychology: Theory, Research, Practice and Training, 2*, 22–27. https://www.researchgate.net/publication/303696592_Preventing_racial_ethnic_gender_sexual_minority_disability_and_religious_microaggressions_Recommendations_for_promoting_positive_mental_health

Nadal, K. L. (2011). The Racial and Ethnic Microaggressions Scale (REMS): Construction, reliability, and validity. *Journal of Counseling Psychology, 58*(4), 470–480. https://doi.org/10.1037/a0025193

Obasi, E.M. (2004). *Measurement of acculturation strategies for people of African descent (MASPAD) manual.* http://www.obasi.org/Documents/MASPAD_Manual.pdf Google Scholar

Obasi, E. M., Flores, L. Y., & Myers, L. J. (2009). Construction and validation of the worldview analysis scale (WAS). *Journal of Black Studies, 39*, 937–961. https://doi.org/10.1177/0021934707305411

Obasi, E. M., & Leong, F. T. (2009). Psychological distress, acculturation, and mental health seeking attitudes among people of African descent in the United States: A preliminary investigation. *Journal of Counseling Psychology, 56*(2), 227–238. https://doi.org/10.1037/a0014865

Obasi, E. M., & Leong, F. T. (2010). Construction and validation of the measurement of acculturation strategies for people of African descent (MASPAD). *Cultural Diversity and Ethnic Minority Psychology, 16*(4), 526–539. https://doi.org/10.1037/a0021374

Pascoe, E. A., & Smart Richman, L. (2009). Perceived discrimination and health: A meta analytic review. *Psychological Bulletin, 135*, 531–554. https://doi.org/10.1037/a0016059

Preacher, K. J., Rucker, D. D., & Hayes, A. F. (2007). Addressing moderated mediation hypotheses: Theory, methods, and prescriptions. *Multivariate Behavioral Research, 42*(1), 185–227. https://doi.org/10.1080/00273170701341316

Rivera, D. P., Forquer, E. E., & Rangel, R. (2010). Microaggressions and the life experience of Latina/o Americans. In D. W. Sue (Ed.), *Microaggressions and marginality: Manifestation, dynamics, and impact* (pp. 59–83). John Wiley & Sons Inc.

Rudmin, F. W. (2006). Debate in science: The case of acculturation. *AnthroGlobe Journal.* http://www.anthroglobe.info/docs/rudminf_acculturation_061204.pdf

Solórzano, D., Ceja, M., & Yosso, T. (2000). Critical race theory, racial microaggressions, and campus racial climate: The experiences of African American college students. *Journal of Negro Education, 69*, 60–73. https://www.jstor.org/stable/2696265

Stephenson, M. (2000). Development and validation of the Stephenson Multigroup Acculturation Scale (SMAS). *Psychological Assessment, 12*(1), 77–88. https://doi.org/10.1037/1040-3590.12.1.77

Sue, D. W. (2010). *Microaggressions in everyday life: Race, gender, and sexual orientation.* Wiley.

Sue, D. W., Bucceri, J., Lin, A. I., Nadal, K. L., & Torino, G. C. (2009). Racial microaggressions and the Asian American experience. *Asian American Journal of Psychology, S*(1), 88–101. https://doi.org/10.1037/1948-1985.s.1.88

Sue, D. W., Capodilupo, C. M., Torino, G. C., Bucceri, J. M., Holder, A. M. B., Nadal, K. L., & Esquilin, M. (2007). Racial microaggressions in everyday life: Implications for clinical practice. *American Psychologist, 62*, 271–286. https://doi.org/10.1037/0003-066x.62.4.271

Tormala, T., & Deaux, K. (2002). Black immigrants to the United States: Confronting and constructing ethnicity and race. In R. Mahalingam (Ed.), *Cultural psychology of immigrants* (pp. 131–150). Erlbaum.

Torres, L., Driscoll, M. W., & Voell, M. (2012). Discrimination, acculturation, acculturative stress, and, Latino psychological distress: A moderated meditational model. *Cultural Diversity and Ethnic Minority Psychology, 18*(1), 17–25. https://doi.org/10.1037/a0026710

United States Census Bureau. (2011). *Who is Black in America?* http://www.census.gov/newsroom/cspan/black/2012.07.06_cspan_blacks.pdf

United States Department of Health and Human Services. (2001). *Mental health: Culture, race, and ethnicity-A supplement to mental health: A report of the surgeon general.* https://www.ncbi.nlm.nih.gov/books/NBK44243/

Walker, R. L., Wingate, L. R., Obasi, E. M., & Joiner, T. E. (2008). An empirical investigation of acculturative stress and ethnic identity as moderators for depression and suicidal ideation in college students. *Cultural Diversity and Ethnic Minority Psychology, 14*(1), 75–82. https://doi.org/10.1037/1099-9809.14.1.75

Waters, M. C. (1999). *Black identities: West Indian immigrant dreams and American realities.* Harvard University Press.

Watkins, N. L., LaBarrie, T. L., & Appio, L. M. (2010). Black undergraduates experience with perceived racial microaggressions in predominantly White colleges and universities. In D. W. Sue (Ed.), *Microaggressions and marginality: Manifestation, dynamics, and impact* (pp. 25–58). Wiley.

Chapter 13
Longitudinal Linkages Between Ethnic-Racial Identity and Discrimination-Related Distress Among Mexican-Origin Girls

Mayra Y. Bámaca-Colbert, Griselda Martinez, Peter S. Y. Kim, Lorena Aceves, Jasmin E. Castillo, Dawn P. Witherspoon, and Melinda Gonzales-Backen

Latino youth are at higher risk of experiencing depressive symptoms compared to youth from other ethnic-racial backgrounds (Anderson & Mayes, 2010; Centers for Disease Control and Prevention, 2018), and their rate of depression has been rising in recent years (Substance Abuse and Mental Health Services Administration [SAMHSA], 2018). Among Latino youth, girls consistently report higher levels of depressive symptoms than boys (Romero-Acosta et al., 2014; McLaughlin et al., 2007), a pattern that is also present among Mexican-origin youth, who are the focus of the current study (Arora & Wheeler, 2018; Orth et al., 2014). Past work with Latinos also reports an association between nativity and depressive symptoms, such that U.S.-born individuals report more mental health problems compared to their foreign-born counterparts (Alegría et al., 2008). However, the overall evidence on this association is inconsistent. For instance, among a Mexican-origin adolescent sample, Arora and Wheeler (2018) reported that being foreign-born was correlated

M. Y. Bámaca-Colbert (✉)
University of California, Merced, CA, USA
e-mail: mbamaca@ucmerced.edu

G. Martinez · P. S. Y. Kim · L. Aceves · J. E. Castillo
The Pennsylvania State University, Pennsylvania, PA, USA
e-mail: gxm212@psu.edu; peterkim@psu.edu; lqa5151@psu.edu; Jec542@psu.edu

D. P. Witherspoon
The Pennsylvania State University, Pennsylvania, PA, USA
e-mail: dpw14@psu.edu

M. Gonzales-Backen
Florida State University, Tallahassee, FL, USA
e-mail: mgonzalesbacken@fsu.edu

D. J. Johnson et al. (eds.), *Re/Formation and Identity*, Advances in Immigrant Family Research, https://doi.org/10.1007/978-3-030-86426-2_13

with higher levels of depressive symptoms compared to being U.S.-born. However, Polo and Lopez (2009) did not find any differences in depressive symptoms by nativity status with a different Mexican-origin youth sample. Altogether, the above patterns point to Mexican-origin girls' vulnerability for long-term depressive-related problems and suggest that there may be differences by nativity status. Therefore, longitudinal research that examines risk and protective factors associated with depressive symptomatology among Mexican-origin girls is warranted.

Exposure to cultural-minority stressors such as discrimination has been considered detrimental to mental health (Brondolo et al., 2009), whereas culturally informed developmental processes such as ethnic-racial identity (ERI) have been found to promote better mental health outcomes among minoritized youth (Rivas-Drake et al., 2014). Within a minoritized context like the United States, the development and adaptation of Latina girls occur within a sociopolitical context that exposes them to multiple forms of discrimination. The discriminatory behaviors (e.g., being ignored) and messages (e.g., "speak English!") to which these youth are exposed take place at a time when they are forming and reforming a sense of self as part of an ethnic-racial minoritized group (Cheon & Yip, 2019; Meca et al., 2020) whose heritage, race, and culture are persistently devalued (García Coll et al., 1996). Further, there is evidence that during adolescence gender differences emerge in the developmental progression of some aspects of ERI (i.e., exploration and resolution), with girls' ERI formation showing more changes than boys' (Umaña-Taylor et al., 2009). Therefore, Latina girls' long-term depressive symptomatology may be jointly informed by exposure to discrimination and the development of an ethnic-racial identity; yet, the mechanism by which these two factors may jointly contribute to depressive symptoms among Latina girls is not well understood. In this study, we used data from a longitudinal project focused on Mexican-origin girls to test a mediational model where three dimensions of ERI (i.e., exploration, resolution, and affirmation) were examined as mediators of the longitudinal association between discrimination and depressive symptoms. Differences in the proposed mechanisms by nativity status were also explored.

Ethnic-Racial Discrimination and Depressive Symptoms

Ethnic-racial discrimination experiences are unique stressors implicated in poor health among ethnic-racial minoritized adolescents (Benner et al., 2018; Causadias & Korous, 2019). These discrimination experiences are based on larger societal views associated with differential treatment that ethnic-racial minoritized groups encounter and are prevalent among Latinos. A recent report by the Pew Research Center (2018), for example, showed that discrimination experiences (e.g., being told to go back to their home country or being called offensive names) were prevalent among foreign- and U.S.-born Latino adults. There is limited knowledge on the actual prevalence of discrimination among Latino youth, but a recent study that incorporated data from three different samples from the Southwest of the United

States with over 1,000 Latino youth of ages 12 and older indicated that experiences of discrimination were prevalent among girls and boys (although higher from boys) and similar across generational statuses (Zeiders et al., 2018). Specific to girls, Zeiders et al. (2018) found that experiences of discrimination ranged from 47% to 70% for those 12 to 18 years of age. These figures illustrate a widespread experience of discrimination among Latina girls.

The extant literature on discrimination has consistently shown deleterious longitudinal effects on the mental health of Mexican-origin youth, including depressive symptoms (e.g., Arora & Wheeler, 2018; Stein et al., 2019; Zeiders et al., 2015). A recent meta-analysis, for example, reported small- to moderate-effect sizes on the discrimination-depressive symptoms link among adolescents (Benner et al., 2018). More specific to the current study, past research with Mexican-origin teen mothers reported that ethnic-racial discrimination experiences were associated with teen moms' depressive symptoms one year later (Zeiders et al., 2015). Another longitudinal study with Mexican-origin adolescents found that discrimination experiences were associated with changes in depressive symptoms (Arora & Wheeler, 2018). The aforementioned studies, however, did not test for differences between U.S.-born and Mexico-born youth. In fact, in a recent meta-analysis, Benner et al. (2018) concluded that studies rarely examined whether the discrimination-well-being link differs across individual characteristics such as nativity (U.S.-born vs. foreign-born). In a recent study, Sirin et al. (2015) found that discrimination-related stressors predicted more symptoms of depression among Latino youth with U.S.-born parents compared to youth with foreign-born parents. This may suggest that U.S.-born Latino youth may be more at risk for discrimination and its deleterious effects on adjustment. In the current study, we examined the longitudinal link between discrimination and depressive symptoms and potential differences between U.S.-born and Mexico-born adolescent girls. Due to the sample size, however, we were not able to examine potential differences by nativity of parents given that the majority of participants' parents were foreign-born.

Altogether, the discrimination experiences that Latino/a youth encounter have serious implications for the mental health of these youth, and past work points to significant short-term and long-term associations. In search of factors that buffer (i.e., protect) the effects of discrimination, researchers have increasingly turned to identifying cultural protective factors such as ERI. However, fewer studies have examined culturally bound mediating mechanisms that may be informed by discrimination experiences and ultimately more directly reduce or enhance the risk imposed by discrimination (Delgado et al., 2019). Relevant to the current study focus, a large body of research has examined the protective effect of ethnic-racial identity—a subjective sense of belonging to an ethnic and/or racial group (Umaña-Taylor et al., 2002), but limited work has examined the potential mediating role that ERI may serve in the association between discrimination and mental health outcomes (Meca et al., 2020). This is surprising, given that negative group evaluations such as discrimination not only influence how one feels about herself, but how one identifies with her ingroup (Branscombe et al., 1999).

A Potential Mechanism Between Discrimination and Depressive Symptoms: Ethnic-Racial Identity (ERI)

ERI is a multidimensional construct that is guided by both the Eriksonian theory of identity formation (Erikson, 1968) and the conceptualization of Umaña-Taylor et al. (2014) that encompasses both the process and affect toward one's ethnic group membership. According to Erikson's (1968) theory of ego identity development, identity formation progresses through a period of exploration to achievement in which the individual has made identity commitments. Based on Erikson's theoretical notions, Umaña-Taylor et al. (2014) contend that individuals similarly attain their *ethnic-racial group* membership through a process of *exploration* and *resolution* (i.e., an understanding of the meaning of one's ethnicity for their identity). Additionally, they argue that *affect* about ethnic-racial group membership, operationalized as affirmation (Umaña-Taylor et al., 2014), is distinguishable from the process of identity development (i.e., exploration and resolution), as it describes positive or negative feelings toward one's group. Because ERI is best conceptualized as a multidimensional construct, studies tend to incorporate various components concurrently in the same study. This can be especially important for studies that examine the association between perceptions of discrimination and mental health problems in adolescence, a time when ERI shows developmental changes which could result in ERI components to be distinctly shaped by discrimination experiences (Cheon & Yip, 2019; Meca et al., 2020) or differentially related to mental health (Brittian et al., 2015; Zeiders et al., 2019).

Discrimination and Ethnic-Racial Identity

Social identity theory posits that in stratified societies, the process of forming an identification with one's own group (when perceived to be inferior) occurs against the backdrop of a conflict. Specifically, one's own identity development works in opposition to the social hierarchy of groups, thereby encouraging identity exploration and self-reflection (Tajfel, 1981). Tajfel also stressed the lasting importance of social context in the development of ERI when social influences work in opposition to developing an identification with one's own ethnic-racial group. Elaborating on the importance of context, the rejection-identification model (Branscombe et al., 1999) suggests that awareness of discrimination against one's ingroup can lead to increased identification with that group in order to protect one's sense of belonging and acceptance.

Empirical research, although limited, provides support for the above premises. In a study investigating trajectories of ethnic-racial identity among Black and Latino adolescents, Pahl and Way (2006) found that perceptions of discrimination were associated with increased exploration of ERI. In another study with a predominantly Latino sample of undergraduate students, researchers found that students who perceived more negative societal views and unfair treatment of their ethnic-racial group engaged in more ethnic-racial exploration than those who did not (Brittian et al., 2015). Not all studies have found this pattern, however. For example, in a study with a sample of Mexican-origin youth, Romero and Roberts (2003) found results that contradicted the rejection-identification model, specifically that discrimination was actually associated with lower ethnic-racial exploration and affirmation. Also, a study that examined the concurrent and longitudinal associations between discrimination and ERI dimensions (Zeiders et al., 2019) among Mexican-origin teen mothers found that increases in an individual's perceived discrimination predicted decreases in both future resolution and affirmation and were unrelated to exploration. Other work suggests that the link between discrimination and ERI can be bidirectional and distinct for different ERI components (Meca, et al. 2020).

There is evidence that nativity matters for both discrimination and ERI (Meca et al., 2020), but studies dedicated to understanding the discrimination-ERI relation among Latina girls generally control for nativity rather than examine potential differences by nativity (Toomey et al., 2013; Zeiders et al., 2019). A recent study with both genders examined potential differences by nativity, and results indicated no significant differences in associations between discrimination and ERI components by nativity groups (Meca et al., 2020). Romero and colleagues (2014) conducted more purposive analyses that considered potential differences by nativity status and gender and found neither had an effect. Because there is minimal evaluation about whether the discrimination-ERI association differs by nativity status and there is a need for systematically evaluating this possibility (Cheon & Yip, 2019; Meca et al., 2020), we explored potential differences by nativity status.

Ethnic-Racial Identity and Depressive Symptoms

The framework on culture development and psychopathology underscores the need to consider individual- and social-level cultural processes that conceptualize culture through a developmental perspective to better understand mental health outcomes (Causadias, 2013). ERI can be considered an individual- and social-level cultural process given that ERI may be informed by personal (e.g., gender) and contextual (e.g., discrimination) experiences. Specifically, there is some evidence that during adolescence gender exacerbates growth across ERI dimensions for girls (exploration, resolution, and affirmation [affect]), whereas for boys only growth in affirmation is evident (Umaña-Taylor et al., 2009). A different study, not specific to Latino youth, found that among boys but not girls, being in the moratorium stage (low on affirmation, high on exploration) was associated with more substance use and drug use problems (Jones & Galliher, 2007). These findings postulate the possible differences by gender that may be observed in the association between ERI and adjustment outcomes. Further, a recent meta-analysis supports the premise that experiences with discrimination can stimulate ERI (Benner et al., 2018). Finally, there is also robust support that links the multiple dimensions of ERI to overall positive adjustment, including lower levels of depressive symptoms (see Rivas-Drake et al., 2014 for a review). For instance, in a study with a sample of ethnic-racial minoritized immigrant-origin youth, ERI was found to be associated with fewer depressive symptoms (Tummala-Narra, 2015). In a longitudinal study on Latino adolescents, findings indicated that ERI (centrality, private regard, public regard) was associated with lower levels of depressive symptoms concurrently and across time (Cross et al., 2018). A study using the same sample from the current study found that ERI (exploration, resolution, affirmation) increased across adolescence and that different dimensions of ERI were differentially associated with levels of depressive symptoms (Gonzales-Backen et al., 2016). Altogether, there is substantive empirical evidence that points to the integration of discrimination and ERI in developmental research aimed to better understand the processes by which cultural-related minoritized experiences (i.e., exposure to discrimination and forming an ethnic-racial identity) can promote or reduce depressive symptoms among girls.

Ethnic-Racial Identity: A Mediating Process?

A majority of literature that looks at ERI and discrimination in relation to adjustment does so by examining the protective (moderating) effect of ERI in the link between discrimination and mental health outcomes (Romero et al., 2014). For instance, positive ethnic-racial affect reduced the positive association between increases in discrimination and depressive symptoms across time among a diverse sample of adolescents (Stein et al., 2016), suggesting that the effects of discrimination on adjustment, in the presence of ERI, can be diminished.

We know less about ERI's potential mediating role in the discrimination-mental health link during adolescence, but identifying mediating factors is particularly important to elucidate mechanisms that are set in motion when individuals experience stressful experiences (Delgado et al., 2019). Considering that discrimination is implicated in the development of ERI and increased depressive symptomatology, ERI can function as a mediating factor in the association between discrimination and mental health during adolescence. At a time when girls experience significant developmental progression in forming and reforming different aspects of their ethnic-racial identity (Gonzales-Backen et al., 2016; Umaña-Taylor et al., 2009), which may be the consequence of discriminatory encounters or messages (Cross & Cross, 2008), this mediating process can have important implications for girls' mental health.

Conceptually, the rejection-identification model (Branscombe et al., 1999) suggests that discrimination experiences may affect depressive symptoms indirectly by stimulating individuals' identification with their ethnic-racial group (i.e., mediator), and this aspect of one's overall identity formation can contribute more directly to one's overall well-being. Although there is a dearth of empirical research examining the mediating role ERI could play in the link between discrimination and mental health, the research considering this role generally conceptualizes ERI as a risk reducer. That is, based on the rejection-identification hypothesis, one could expect that experiences with discrimination would prompt positive identification with one's ingroup, promoting a person's exploration, resolution, and affirmation with their ethnic-racial group, which would ultimately diminish the negative direct contribution of discrimination on adaptation. Contrary to expectations however, a study with a diverse sample of college students showed that ethnic discrimination was associated with lower ERI affirmation (Brittian et al., 2015). Further, Brittian et al. (2015) found that the link between ethnic-racial discrimination and depressive symptoms was positively mediated by ERI affirmation for Latino young adults, but not for Black young adults. In addition, higher levels of ERI resolution were indirectly related to fewer depressive symptoms via ERI affirmation, and ethnic-racial discrimination was positively associated with ERI exploration (Brittian et al., 2015). These findings highlight the potential mediating role of ERI in the link between discrimination and depressive symptoms among Latino youth and young adults, but also potential differences in patterns of association. Altogether, the current body of literature calls for additional studies that examine the compensatory or mediating role that ERI may serve in the link between discrimination and depressive symptoms.

Present Study

Using data from a longitudinal project with three time points, the overall aim of this study was to examine the mediating role of three dimensions of ERI (i.e., exploration, resolution, affirmation) at Time 2 (T2) in the association between perceptions of discrimination at Time 1 (T1) and depressive symptoms at Time 3 (T3), among a sample of Mexican-origin adolescent girls. Further, we examined differences between immigrants and U.S.-born adolescents in this mediating process. We controlled for depressive symptoms and ERI dimensions at T1. The following hypotheses and research questions were examined:

RQ_1: Do ethnic-racial identity components mediate the link between discrimination at T1 and depressive symptoms at T3?

H_1: T1 perceived discrimination would predict more T3 reports of depressive symptoms, after controlling for T1 ERI and perceived discrimination.

H_2: T1 perceived discrimination would predict more T2 ERI dimensions after controlling for T1 ERI.

H_3: Components of T2 ERI would mediate the association between T1 discrimination and T3 depressive symptoms.

RQ_2: Does the mediating process differ between immigrant and U.S.-born adolescents?

H4: The mediating process may be more salient for U.S.-born adolescents.

Method

Participants and Procedures

Data came from a longitudinal research project focused on cultural and developmental factors related to the well-being of Mexican-origin adolescent girls. At baseline (T1), participants (*N* = 338) were in seventh (*n* = 170) and tenth (*n* = 168) grades and were recruited from predominantly Latino public schools in a large Southwestern U.S. metropolitan area. Adolescents participated 2.5 and 3.5 years later for two additional assessments. To be eligible to participate in the study at baseline, both parents had to be of Mexican descent, and the mother figure had to be willing to participate at T1. Mothers were not invited for the follow-up study (for a detailed description of recruitment procedure, see Bámaca-Colbert et al., 2014).

One hundred seventy-five (51.8%) adolescents participated in at least two data collection time points. The retention rate at Time 2 (T2) was 45.3% (seventh grade *n* = 78, tenth grade *n* = 75) and 43.2% (seventh grade *n* = 75, tenth grade *n* = 70) at T3. Most attrition was due to loss of contact with participants (e.g., change of residence with no forwarding address, disconnected or incorrect phone numbers).

At T1, the seventh-grade sample ranged in age from 12 to 14 years (*M* = 12.27, *SD* = 0.48) and 63.6% (*N* = 82) reported being U.S.-born, and the tenth-grade sample ranged in age from 14 to 17 years (*M* = 15.21 years, *SD* = 0.46) and 60.6% (*N* = 86) were U.S.-born. At T1, 62.8% of adolescents lived in households with both their biological mother and father (64.3% and 61.3% for seventh and tenth graders, respectively), but other family constellations were reported (e.g., 23% of seventh graders and 18% of tenth graders lived with only their biological mother). Participants were compensated for their time, and all measures that were not originally available in Spanish were translated by the first author and back-translated by a researcher of Mexican origin.

Measures

Perceived Discrimination

A revised version of the Perceived Discrimination Scale (Whitbeck et al., 2001) was used at T1 to assess the degree to which adolescents perceived to have experienced discrimination. The scale was composed of ten items that examine global discrimination, authority discrimination, and school discrimination. Sample items include "How often has someone yelled racial slurs or racial insults at you?" and "How often have you encountered teachers who didn't expect you to do well because you are Hispanic/Latino?" Items were rated on a four-point Likert scale, with end points ranging from almost never (1) to very often (4). An average score across all items was created with higher scores indicating greater frequency of perceived discrimination. The alpha obtained for the current study was 0.85.

Ethnic-Racial Identity

The Ethnic Identity Scale (Umaña-Taylor et al., 2004) was used to assess adolescents' ERI. This scale is composed of 17 items that examine three different dimensions of ethnic-racial identity formation: exploration (seven items; e.g., "I have attended events that have helped me learn more about my ethnicity"), resolution (four items; e.g., "I have a clear sense of what my ethnicity means to me"), and affirmation (six items; e.g., "I wish I were of another ethnicity," reverse coded). Participants responded to items on a four-point Likert scale ranging from (1) does not describe me at all to (4) describes me very well. For the current study, the alphas at T1 were 0.78, 0.83, and 0.86 for exploration, resolution, and affirmation, respectively, and at T2 were 0.81, 0.88, and 0.78 for exploration, resolution, and affirmation, respectively.

Depressive Symptoms

The Center for Epidemiologic Studies Depression Scale (CES-D; Radloff, 1977) was used to assess depressive symptoms. This 20-item scale asked adolescents to think about the past week. Items were rated on a four-point Likert scale ranging from (0) rarely or none of the time to (3) mostly or almost all the time. Higher values signify a greater frequency of depressive symptoms. Sample items include: "I could not get going" and "I had crying spells." For the current study, the alpha values were 0.90 at T1 and 0.92 at T3.

Nativity

At Time 1, adolescents were asked about their birth place. Given that the sample focused on Mexican-origin individuals, response options included being born in either the United States (1) or Mexico (2).

Plan of Analysis

Bivariate correlations between study variables were examined. Next, we used structural equation modeling (SEM) in Mplus 7.4 (Muthén & Muthén, 1998–2012) to test for meditation by ethnic-racial identity exploration, resolution, and affirmation. We used direct and indirect effects modeling and bias-corrected bootstrapped confidence intervals in Mplus (MacKinnon et al., 2004). In the final model, we added only covariate paths that were significant at the bivariate level.

Multi-group SEM analyses were conducted to determine whether the full model differed due to adolescent nativity. The first step was to run a fully unconstrained model in which all structure paths, variances, and covariances were freely estimated for both groups. Second, a fully constrained model in which all covariances and paths were set to be equal across groups was estimated. Finally, the fully unconstrained model was compared to the fully constrained model to determine if the models were different due to adolescent nativity.

To account for missing data, full information maximum likelihood (FIML) estimation was used to fit the models to the raw data. The data analytical approach was deemed appropriate given that the current data is missing completely at random (MCAR test) ($\chi^2(20) = 22.12$, $p = .33$). When compared to other methods of handling missing data such as list wise or pairwise deletion, FIML produces less biased and more reliable results (Schafer & Graham, 2002). To determine good model fit, we used the chi-square test (χ^2; non-significant), the confirmatory fit index (CFI; $\geq$ 0.95; Bentler, 1990), and root mean square error of approximation (RMSEA; $\leq$ 0.06; Browne & Cudeck, 1992). For studies with large sample sizes like this study, slight changes from a perfect model can result in a significant χ^2. Thus, we focused primarily on CFI values of 0.95 or greater (Hu & Bentler, 1999) and RMSEA for assessing model fit.

Results

Preliminary Results

Bivariate correlations, means, and standard deviations of study variables are presented in Table 13.1. Correlation results demonstrated that T1 discrimination was negatively associated with T2 ERI affirmation such that adolescents who reported higher levels of discrimination at T1 reported lower ERI affirmation at T2. T1 discrimination was also significantly and positively associated with T3 depressive symptoms. T2 ERI affirmation was negatively associated with T3 depressive symptoms such that more ERI affirmation at T2 was associated with fewer depressive symptoms at T3.

Full Model Results

The full model controlled for adolescent age, depressive symptoms at T1, and ethnic-racial identity dimensions at T1. Fit indices (CFI = 1.00, TLI = 1.00; RMSEA = 0.00; $\chi^2(12) = 10.73$, $p = 0.55$) indicated that the full model was a good fit to the data (see Table 13.2). The model explained 11% of the variance in T2 ERI exploration, 5% of the variance in T2 ERI resolution, 16% of the variance in ERI affirmation, and 21% of the variance in T3 depressive symptoms. For hypothesis one, results revealed that the association between T1 discrimination and T3 depressive symptoms was not significant. For hypothesis two, the results revealed that T1

Table 13.1 Correlations of study variables

	1	2	3	4	5	6	7	8	9	10	11
1. Nativity	–										
2. Age T1	0.23**	–									
3. ERI exploration T1	0.07	0.14*									
4. ERI resolution T1	0.03	0.08	0.60**	–							
5. ERI affirmation T1	−0.01	0.10	−0.06	−0.04	–						
6. Depressive symptoms T1	−0.10	0.07	−0.07	−0.09	−0.14*	–					
7. Discrimination T1	−0.09	0.04	0.07	−0.04	−0.13*	0.38**	–				
8. ERI exploration T2	0.03	0.11	0.34**	0.19*	0.05	0.08	0.00	–			
9. ERI resolution T2	0.02	0.16*	0.15	0.13	0.16	0.02	0.09	0.45**	–		
10. ERI affirmation T2	−0.09	−0.17*	−0.03	0.17*	0.22**	−0.13	−0.16*	0.07	0.25**	–	
11. Depressive symptoms T3	−0.07	−0.03	−0.17*	−0.15	−0.07	0.34**	0.19*	0.06	−0.12	−0.32**	–
M	11.74	13.73	2.68	3.00	3.53	0.92	1.33	2.84	3.26	3.35	0.86
SD	2.92	1.54	0.67	0.80	0.70	0.52	0.44	0.63	0.71	0.29	0.58

Note. *ERI* Ethnic-racial identity
$p < 0.05$*, $p < 0.01$**

Table 13.2 Parameter estimates for the full mediation model

Parameter estimates	β (SE)
Discrimination T1→ERI exploration T2	−0.01(0.14)
Discrimination T1→ERI resolution T2	0.18(0.16)
Discrimination T1→ERI affirmation T2	**−0.12(0.06)^**
Discrimination T1→depressive symptoms T3	0.05(0.12)
ERI exploration T2→depressive symptoms T3	**0.16(0.09)^**
ERI resolution T2→depressive symptoms T3	−0.11(0.08)
ERI affirmation T2→depressive symptoms T3	**−0.42(0.18)***
Control variables	
ERI exploration T1→ERI exploration T2	**0.31(0.07)*****
ERI resolution T1→ERI resolution T2	**0.11(0.07)^**
Age T1→ERI resolution T2	**0.06(0.03)^**
ERI affirmation T1→ERI affirmation T2	**0.09(0.03)****
ERI resolution T1→ERI affirmation T2	**0.07(0.03)****
ERI exploration T1→depressive symptoms T3	**−0.16(0.07)***
Depressive symptoms T1→depressive symptoms T3	**0.29(0.09)*****

Note. ERI Ethnic-racial identity. CFI = 1.00, TLI = 1.00; RMSEA = 0.00; $\chi^2(12) = 10.73$, $p = 0.55$
$p < 0.10$^, $p < 0.05$*, $p < 0.01$**, $p < 0.001$***

discrimination was only associated with T2 ERI affirmation at a trend level ($\beta = -0.12$; *S.E.* = 0.06; $p = 0.054$).

To test hypothesis three, the full model tested meditation using (1) T2 ERI exploration, (2) T2 ERI resolution, and (3) T2 ERI affirmation. Findings indicated no significant mediation in the indirect ($\beta = 0.03$; *S.E.* = 0.05; $p = 0.57$) and direct effects ($\beta = 0.08$; *S.E.* =0.12; $p = 0.49$). Although the bootstrap meditation test did not meet significance, the individual paths did reveal that T2 ERI affirmation was significantly associated with T3 depressive symptoms ($\beta = -0.42$; *S.E.* = 0.18; $p = 0.02$). Altogether, findings suggested that ERI did not function as a mediator in the association between T1 discrimination and T3 depressive symptoms, but individual paths suggested otherwise.

Multi-group Analyses by Adolescent Nativity

The χ^2 difference test revealed that the fully unconstrained model ($\chi^2(42) = 53.11$, $p = 0.12$) did not have a better fit to the data than the fully constrained model ($\chi^2(55) = 67.99$, $p = 0.11$, $\Delta\chi^2(13) = 14.88$, $p = 0.32$). These results suggest that the associations tested in the full model did not differ by adolescent nativity. Although

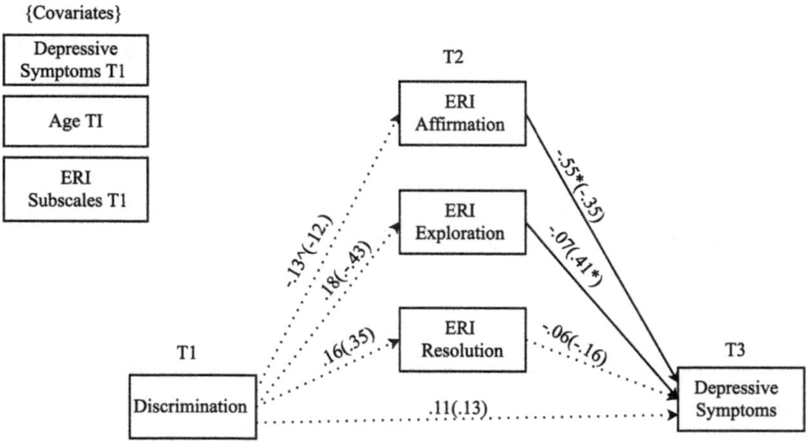

Fig. 13.1 Statistical model tested by adolescent nativity. *ERI* Ethnic-radical identity. Model fit: CFI: 0.91, TLI: 0.87, RMSEA: 0.04. $x^2(42) = 53.11, p = 0.12$. Presentation of values: b-weight for U.S.-born (b-weight for born in Mexico). Solid lines represent significant paths for one or both groups. Dotted lines represent no significance. $p \leq 0.10$^, $p \leq 0.05$*, $p \leq 0.01$**

the chi-square difference test did not reveal differences between the fully constrained and fully unconstrained group models, some paths were significant in the freely estimated model for one group only (see Fig. 13.1 for unconstrained nativity model indicating paths that were significant for one group only), and in fact, the potential indirect mediating mechanism appears to be present only among U.S.-born girls.

Discussion

The cumulative evidence that perceived discrimination is concurrently and prospectively associated with depressive symptoms and other health-related outcomes has motivated numerous research efforts that aim to uncover potential protective factors. Past work has found considerable support for the generally salubrious protective (moderating) role of ERI against the effects of discrimination on depressive symptoms. Additionally, from an applied perspective, unveiling mediating mechanisms by which risk factors may impact child development is of utmost importance (Roosa, 2000), and theoretical work points to an alternate mechanism by which ERI can inform the discrimination-depressive symptoms link. Specifically, discrimination can contribute to depressive symptomatology via its direct influence on ERI (Branscombe et al., 1999), which can more directly contribute to an individual's depressive symptomatology. In general, however, there is limited knowledge on whether ERI serves a mediating role in the discrimination-depressive symptoms association, but given that Latina girls show increasingly more changes in their ERI formation during adolescence than boys, examining the role of ERI as it is forming and reforming during adolescence is critical for this population. Further, it is plausible to assume that specific encounters that disparage one's ethnic-racial group, implicitly or explicitly through action or messaging, set in motion ethnic-racial identity processes that are nonlinear and cyclical (Cross & Cross, 2012), which in turn differentially impact well-being. In the current study, using a sample of Mexican-origin female adolescents, we aimed to test ERI as a potential mediating mechanism using longitudinal data across three time points. Past work has also found nativity differences such that individuals who are U.S.-born are likely to report more mental health problems compared to foreign-born individuals (Alegría et al., 2008). We also examined potential differences in associations between U.S.-born and Mexico-born adolescent girls, as differences in discrimination experiences have been reported by U.S.-born and Mexico-born individuals (Nair et al., 2013), and ERI may function differently for U.S.- and Mexico-born youth (Benner et al., 2018).

In discussion of the overall model, the results indicated that the hypothesized model was a good fit to the data, based on fit statistics, but that the significance test of mediation failed to reach statistical significance. Despite the lack of statistically significant mediation, we interpret trend-level and significant direct and indirect associations, as we believe these emerging associations within a longitudinal design of girls have important implications for future work. Yet, we caution readers with our interpretation of findings.

First, despite the non-significant mediation test within the full model, extrapolating from individual path coefficients suggests one potential indirect mechanism from discrimination at T1 to depressive symptoms at T3 via ERI affirmation at T2, controlling for age, depressive symptoms, and ERI at T1. Specifically, higher levels of perceived discrimination at T1 contributed to lower levels of ERI affirmation at T2 (marginal significance), which, in turn, significantly predicted higher levels of depressive symptoms at T3. This finding points to an alternate risk enhancer mechanism of influence (Delgado et al., 2019) whereby discrimination experiences increase the risk for later depressive symptoms by contributing to lower levels of ERI affirmation, which more directly predicted subsequent depressive symptoms. These findings are in line with past studies that have found that ethnic-racial discrimination experiences were associated with lower affirmation (Romero & Roberts, 2003; Zeiders et al., 2019) and affirmation mediated the link between discrimination and depressive symptoms (Brittian et al., 2015). Thus, it appears that during adolescence—an important period for the formation of ERI (Umaña-Taylor et al., 2014) when sensitivity to societal messages is high (Cross & Cross, 2012)—negative group evaluations resulting from discriminatory experiences can affect how girls feel about being part of their devalued ethnic-racial ingroup, and these negative perceptions are directly associated with adjustment. The rejection-identification premise, however, would predict that experiences with discrimination would increase one's positive identification with the ingroup, in order to protect one's overall self-worth. Contrary to expectations, but in line with some cross-sectional (Romero & Roberts, 2003) and longitudinal (Zeiders et al., 2019) research, a rejection-*dis*identification process appears to be taking place for the Mexican-origin girls in the current sample. That is, more experiences with discrimination at baseline predicted fewer positive feelings (i.e., disidentifying) toward the Mexican group 2.5 years later, which in turn predicted more depressive symptoms a year later. In this case, adolescent girls appear to go through a rejection-disidentification period in which they may attribute ethnic-racial discrimination to being Mexican, and this diminishes the ability to feel positive about belonging to their ethnic-racial group at a crucial time for girls' ERI (re)formation. It is also possible that when ERI formation is not yet solidified, broader negative societal messages targeted toward one's ethnic-racial group (e.g., border enforcement and anti-immigration sentiments; Torres et al., 2011) can further contribute to the negative association between discrimination experiences and ERI affirmation. Additional longitudinal research is necessary beyond adolescence to determine whether this negative association disappears as youth solidify their sense of affirmation to an ethnic-racial identity.

It is worth noting that this was the only association that reached marginal significance between baseline discrimination and future ERI dimensions. Although past cross-sectional (Brittian et al., 2015) and longitudinal (Pahl & Way, 2006) research showed that discrimination experiences were linked with ERI exploration, the longitudinal link with exploration is not always found (Zeiders et al., 2019). For resolution, Zeiders et al. (2019) did find a longitudinal link from past discrimination to later ERI resolution.

Altogether the individual paths in the full model, alongside past evidence, suggest that earlier experiences with discrimination can prompt youth to report fewer positive views of their ingroup (i.e., ethnic-racial group) at a later time. Importantly, there are concurrent, transactional, and longitudinal processes among different dimensions of ERI at any given time, and how they potentially inform one another across development is important to keep in mind when interpreting current findings (Meca et al., 2020; Umaña-Taylor et al., 2014). For example, it is possible that while discrimination experiences can prompt fewer positive feelings about the ingroup at a given time, exposure and exploration of one's culture through family cultural socialization practices can also be simultaneously at play within an individual, especially for girls who are usually considered the main carriers of culture across generations (Umaña-Taylor & Guimond, 2010). Such exploration, over time, can allow girls to learn about the achievement (despite struggles) of their ethnic-racial group, and this can contribute to feeling more positive about the ingroup and help develop a stronger sense of what it means (i.e., resolution) to be of a specific ethnic-racial background. Future research with longitudinal data should examine transactional processes by which discrimination, ERI, and depressive symptoms are interlocked.

Given the dearth of extant work examining potential differences between U.S.-born and Mexico-born adolescents, we considered potential differences between U.S.-born and foreign-born Mexican-origin adolescents in the model. Overall, results indicated no significant difference in the analytic model between U.S.-born and foreign-born adolescents. However, some coefficients appeared to be substantially different for U.S.-born vs. foreign-born, although the omnibus test of differences was non-significant (see Fig. 13.1). The lack of power to identify potential differences between nativity groups could have resulted from sample size limitations as well as error terms (Cohen et al., 2003). Nonetheless, conducting multigroup modeling suggested that the potential indirect mediating mechanism is driven by U.S.-born girls, as significant and marginally significant individual paths were present only for U.S.-born girls. This suggests that U.S.-born Latina adolescents' ethnic-racial identity may be more impacted by experiences with discrimination, and their ethnic-racial identity affirmation appears to contribute to their later adjustment, perhaps due to their family's longer exposure and awareness of such discriminatory issues (Sirin et al., 2015). It is important that future work continues to consider potential differences across youth who are U.S.- or foreign-born as these groups may experience and interpret culturally related experiences differently.

The current study findings and interpretations must be considered in the presence of some limitations. First, the sample suffered significant attrition across the four years. Baseline data was collected in 2006–2007, whereas the additional two time points took place in 2009 and 2010. In 2008, the United States suffered from a financial crisis (Kochhar, 2008) that could have contributed to our inability to contact families from the original study three years after. Latinos suffered significant job loss during this period, which may have led families to relocate to other parts of the United States or return to Mexico. Despite this limitation, however, it is important to note that we found significant associations longitudinally that provided some support for the hypothesized associations at the bivariate level (i.e., test of mediation did not reach significance). An additional consideration is that the current study focused on girls of Mexican origin, limiting its interpretation beyond this group's demographic. There is evidence that the influence of discrimination on youth adjustment can vary as a function of gender, with some associations being significant only for boys (Delgado et al., 2019). Future research should also examine the proposed model with boys and examine linkages with other Latino groups such as Dominican or Puerto Rican.

Despite the above limitations, our findings suggest an indirect mechanism by which discrimination could hinder adolescent girls' mental health long term via ERI. As adolescents' ERI often develops in a cyclical fashion (Cross & Cross, 2012), they may have experiences, such as discrimination or daily microaggressions, that cumulatively require them to (re)negotiate their identity placing greater or lesser importance on this (i.e., ethnic-racial membership) aspect of themselves and prioritizing other aspects of self. The increasing negative climate that ethnic-racial minoritized groups and Latino immigrants, more specifically, are currently experiencing in the United States has resulted in individuals from such groups expressing increased fear for themselves and families in terms of unfair treatment (Ayón, 2018). For Latina girls who are navigating multiple marginalized statuses (i.e., ethnic-racial membership, gender, nativity), discrimination may be a consequence of either of these identities, which could make them more vulnerable and transform their trajectories. An intersectional approach to identity could further elucidate ways to help Latina girls cope with this lived experience. Further, as intervention programs focused on ERI development are beginning to emerge (Umaña-Taylor et al., 2018), they could become more tailored to incorporate multilayered, transactional learning components to the curriculum to better prepare Latina girls. Taken together, today's increased ethnic-racial diversity along with increased fear of unfair treatment can place minoritized youth, and specifically Latina girls, at an increased risk for poor mental health via the effects of experiences with discrimination on youths' ethnic-racial identity feelings to belong to their ingroup. Continuing to explore ERI as an important cultural developmental process and focus on it as a catalyst for Latina adolescent girls' well-being despite cultural stressors may enrich their future possibilities. Therefore, a greater focus on interventions that promote a healthy development of ethnic-racial identity is paramount as youth navigate adolescence as part of a minoritized group.FundingThis research was supported by a grant from the National Institute on Mental Health (R36MH077425) and a research grant from the Graduate and Professional Student Association at Arizona State University awarded to Mayra Y. Bámaca-Colbert. This research was supported in part by the Prevention and Methodology Training Program (T32 DA017629) with funding from the National Institute on Drug Abuse and by the Institute of Education Sciences, U.S. Department of Education, through grant R305B090007 to the Pennsylvania State University.

References

Alegría, M., Canino, G., Shrout, P. E., Woo, M., Duan, N., Vila, D., Torres, M., Chen, C.-N., & Meng, X.-L. (2008). Prevalence of mental illness in immigrant and non-immigrant U.S. Latino groups. *American Journal of Psychiatry, 165*(3), 359–369. https://doi.org/10.1176/appi.ajp.2007.07040704

Anderson, E. R., & Mayes, L. C. (2010). Race/ethnicity and internalizing disorders in youth: A review. *Clinical psychology review, 30*(3), 338–348. https://doi.org/10.1016/j.cpr.2009.12.008

Arora, P. G., & Wheeler, L. A. (2018). Depressive symptoms in Mexican-origin adolescents: Interrelations between school and family contexts. *Contemporary School Psychology, 22*(3), 266–278. https://doi.org/10.1007/s40688-017-0150-1

Ayón, C. (2018). "vivimos en jaula de oro": The impact of state-level legislation on immigrant Latino families. *Journal of Immigrant & Refugee Studies, 16*(4), 351–371. https://doi.org/10.1080/15562948.2017.1306151

Bámaca-Colbert, M. Y., Greene, K. M., Killoren, S. E., & Noah, A. J. (2014). Contextual and developmental predictors of sexual initiation timing among Mexican-origin girls. *Developmental Psychology, 50*(10), 2353–2359. https://doi.org/10.1037/a0037772

Benner, A. D., Wang, Y., Shen, Y., Boyle, A. E., Polk, R., & Cheng, Y. (2018). Racial/ethnic discrimination and well-being during adolescence: A meta-analytic review. *The American Psychologist, 73*(7), 855–883. https://doi.org/10.1037/amp0000204

Bentler, P. M. (1990). Comparative fit indexes in structural models. *Psychological Bulletin, 107*(2), 238–246. https://doi.org/10.1037//0033-2909.107.2.238

Branscombe, N. R., Schmitt, M. T., & Harvey, R. D. (1999). Perceiving pervasive discrimination among African Americans: Implications for group identification and well-being. *Journal of Personality and Social Psychology, 77*(1), 135–149. https://doi.org/10.1037/0022-3514.77.1.135

Brittian, A. S., Kim, S. Y., Armenta, B. E., Lee, R. M., Umaña-Taylor, A. J., Schwartz, S. J., Villalta, I. K., Zamboanga, B. L., Weisskirch, R. S., Juang, L. P., Castillo, L. G., & Hudson, M. L. (2015). Do dimensions of ethnic identity mediate the association between perceived ethnic group discrimination and depressive symptoms? *Cultural Diversity and Ethnic Minority Psychology, 21*(1), 41–53. https://doi.org/10.1037/a0037531

Brondolo, E., Gallo, L. C., & Myers, H. F. (2009). Race, racism and health: Disparities, mechanisms, and interventions. *Journal of Behavioral Medicine, 32*(1), 1–8. https://doi.org/10.1007/s10865-008-9190-3

Browne, M. W., & Cudeck, R. (1992). Alternative ways of assessing model fit. In K. A. Bollen & J. S. Long (Eds.), *Testing Structural Equation Models* (pp. 136–162). Sage. https://doi.org/10.1177/0049124192021002005

Causadias, J. M. (2013). A roadmap for the integration of culture into developmental psychopathology. *Development and Psychopathology, 25*(4 Pt 2), 1375–1398. https://doi.org/10.1017/S0954579413000679

Causadias, J. M., & Korous, K. M. (2019). Racial discrimination in the United States: A national health crisis that demands a national health solution. *Journal of Adolescent Health, 64*(2), 147–148. https://doi.org/10.1016/j.jadohealth.2018.11.001

Centers for Disease Control and Prevention. (2018). Youth risk behavior surveillance—United States, 2017. *Morbidity and Mortality Weekly Report, 67*(8). https://www.cdc.gov/healthyyouth/data/yrbs/pdf/2017/ss6708.pdf

Cheon, Y. M., & Yip, T. (2019). Longitudinal associations between ethnic/racial identity and discrimination among Asian and Latinx adolescents. *Journal of Youth and Adolescence, 48*(9), 1736–1753. https://doi.org/10.1007/s10964-019-01055-6

Cohen, J., Cohen, P., West, G. S., & Aiken, S. L. (2003). *Applied multiple regression/correlation analysis for the behavioral sciences* (3rd ed.). Erlbaum.

Cross, F. L., Hoffman, A. J., Constante, K., & Rivas-Drake, D. (2018). Ethnic-racial identity content and the development of depressive symptoms among Latino adolescents. *Development and Psychopathology, 30*(5), 1557–1569. https://doi.org/10.1017/S0954579418001086

Cross, W. E. and Cross, T. B. (2008). Theory, research, and models. In S. M. Quintana & C. McKown (Eds.), *Handbook of race, racism, and the developing child* (pp. 154–181). Wiley. https://doi.org/10.1002/9781118269930.ch8

Cross, W. E., & Cross, T. B. (2012). Theory, research, and models. In S. M. Quintana & C. McKown (Eds.), *Handbook of race, racism, and the developing child* (pp. 154–181). Wiley. https://doi.org/10.1002/9781118269930

Delgado, M. Y., Nair, R. L., Zeiders, K. H., & Jones, S. K. (2019). Latino adolescents' experiences with ethnic discrimination: Moderating factors and mediating mechanisms. In H. E. Fitzgerald, D. J. Johnson, D. Baolian Qin, & F. A. Villarruel (Eds.), *Handbook of children and prejudice* (pp. 515–531). Springer. https://doi.org/10.1007/978-3-030-12228-7

Erikson, E. H. (1968). *Identity: Youth and Crisis*. Norton.

Gonzales-Backen, M. A., Bámaca-Colbert, M. Y., & Allen, K. (2016). Ethnic identity trajectories among Mexican-origin girls during early and middle adolescence: Predicting future psychosocial adjustment. *Developmental Psychology, 52*(5), 790–797. https://doi.org/10.1037/a0040193

García Coll, C. G., Lamberty, G., Jenkins, R., McAdoo, H. P., Crnic, K., Wasik, B. H., & García, H. V. (1996). An integrative model for the study of developmental competencies in minority children. *Child Development, 67*(5), 1891–1914. https://doi.org/10.2307/1131600

Hu, L., & Bentler, P. M. (1999). Cutoff criteria for fit indexes in covariance structure analysis: Conventional criteria versus new alternatives. *Structural Equation Modeling: A Multidisciplinary Journal, 6*(1), 1–55. https://doi.org/10.1080/10705519909540118

Jones, M. D., & Galliher, R. V. (2007). Ethnic identity and psychosocial functioning in Navajo adolescents: Ethnic identity and psychosocial functioning. *Journal of Research on Adolescence, 17*(4), 683–696. https://doi.org/10.1111/j.1532-7795.2007.00541.x

Kochhar, R. (2008). *Latino workers in the ongoing recession: 2007 to 2008*. Pew Hispanic Center. https://www.pewhispanic.org/2008/12/15/latino-workers-in-the-ongoing-recession-2007-to-2008/

MacKinnon, D. P., Lockwood, C. M., & Williams, J. (2004). Confidence limits for the indirect effect: Distribution of the product and resampling methods. *Multivariate Behavioral Research, 39*(1), 99–128. https://doi.org/10.1207/s15327906mbr3901_4

McLaughlin, K. A., Hilt, L. M., & Nolen-Hoeksema, S. (2007). Racial/ethnic differences in internalizing and externalizing symptoms in adolescents. *Journal of Abnormal Child Psychology, 35*(5), 801–816. https://doi.org/10.1007/s10802-007-9128-1

Meca, A., Gonzales-Backen, M., Davis, R., Rodil, J., Soto, D., & Unger, J. B. (2020). Discrimination and ethnic identity: Establishing directionality among Latino/a youth. *Developmental Psychology, 56*(5), 982–992. https://doi.org/10.1037/dev0000908

Muthén, L. K., & Muthén, B. O. (1998–2012). Mplus user's guide (7th ed.). Muthén & Muthén. http://www.statmodel.com/download/usersguide/Mplus%20user%20guide%20Ver_7_r3_web.pdf

Nair, R. L., White, R. M. B., Roosa, M. W., & Zeiders, K. H. (2013). Cultural stressors and mental health symptoms among Mexican Americans: A prospective study examining the impact of the family and neighborhood context. *Journal of Youth and Adolescence, 42*(10), 1611–1623. https://doi.org/10.1007/s10964-012-9834-z

Orth, U., Robins, R. W., Widaman, K. F., & Conger, R. D. (2014). Is low self-esteem a risk factor for depression? Findings from a longitudinal study of Mexican-origin youth. *Developmental Psychology, 50*(2), 622–633. https://doi.org/10.1037/a0033817

Pahl, K., & Way, N. (2006). Longitudinal trajectories of ethnic identity among urban black and Latino adolescents. *Child Development, 77*(5), 1403–1415. https://doi.org/10.1111/j.1467-8624.2006.00943.x

Pew Research Center. (2018). *More Latinos have serious concerns about their place in America under Trump* [Report]. https://www.pewresearch.org/hispanic/2018/10/25/more-latinos-have-serious-concerns-about-their-place-in-america-under-trump/

Polo, A. J., & López, S. R. (2009). Culture, context, and the internalizing distress of Mexican American youth. *Journal of Clinical Child & Adolescent Psychology, 38*(2), 273–285. https://doi.org/10.1080/15374410802698370

Radloff, L. S. (1977). The CES-D scale: A self-report depression scale for research in the general population. *Applied Psychological Measurement, 1*(3), 385–401. https://doi.org/10.1177/014662167700100306

Rivas-Drake, D., Seaton, E. K., Markstrom, C., Quintana, S., Syed, M., Lee, R. M., & Ethnic and Racial Identity in the 21st Century Study Group. (2014). Ethnic and racial identity in adolescence: Implications for psychosocial, academic, and health outcomes. *Child Development, 85*(1), 40–57. https://doi.org/10.1111/cdev.12200

Romero, A. J., Edwards, L. M., Fryberg, S. A., & Orduña, M. (2014). Resilience to discrimination stress across ethnic identity stages of development. *Journal of Applied Social Psychology, 44*(1), 1–11. https://doi.org/10.1111/jasp.12192

Romero, A. J., & Roberts, R. E. (2003). The impact of multiple dimensions of ethnic identity on discrimination and adolescents' self-esteem. *Journal of Applied Social Psychology, 33*(11), 2288. https://doi.org/10.1111/j.1559-1816.2003.tb01885.x

Romero-Acosta, K., Penelo, E., Noorian, Z., Ferreira, E., & Domènech-Llaberia, E. (2014). Racial/ethnic differences in the prevalence of internalizing symptoms: Do Latin-American immigrant show more symptomatology than Spanish native-born adolescents? *Journal of Health Psychology, 19*(3), 381–392. https://doi.org/10.1177/1359105312471568

Roosa, M. W. (2000). Some thoughts about resilience versus positive development, main effects versus interactions, and the value of resilience. *Child Development, 71*(3), 567–569. https://doi.org/10.1111/1467-8624.00166

Schafer, J. L., & Graham, J. W. (2002). Missing data: Our view of the state of the art. *Psychological Methods, 7*(2), 147–177. https://doi.org/10.1037//1082-989X.7.2.147

Sirin, S. R., Rogers-Sirin, L., Cressen, J., Gupta, T., Ahmed, S. F., & Novoa, A. D. (2015). Discrimination-related stress effects on the development of internalizing symptoms among Latino adolescents. *Child Development, 86*(3), 709–725. https://doi.org/10.1111/cdev.12343

Stein, G. L., Castro-Schilo, L., Cavanaugh, A. M., Mejia, Y., Christophe, N. K., & Robins, R. (2019). When discrimination hurts: The longitudinal impact of increases in peer discrimination on anxiety and depressive symptoms in Mexican-origin youth. *Journal of Youth and Adolescence, 48*(5), 864–875. https://doi.org/10.1007/s10964-019-01012-3

Stein, G. L., Supple, A. J., Huq, N., Dunbar, A. S., & Prinstein, M. J. (2016). A longitudinal examination of perceived discrimination and depressive symptoms in ethnic minority youth: The roles of attributional style, positive ethnic/racial affect, and emotional reactivity. *Developmental Psychology, 52*(2), 259–271. https://doi.org/10.1037/a0039902

Substance Abuse and Mental Health Services Administration (2018). *2018 national survey on drug use and health* (NSDUH) [Data tables]. https://www.samhsa.gov/data/sites/default/files/cbhsqreports/NSDUHDetailedTabs2018R2/NSDUHDetTabsSect9pe2018.htm.

Tajfel, H. (1981). *Human groups and social categories: Studies in social psychology*. Cambridge University Press.

Toomey, R. B., Umaña-Taylor, A. J., Updegraff, K. A., & Jahromi, L. B. (2013). Ethnic identity development and ethnic discrimination: Examining longitudinal associations with adjustment for Mexican-origin adolescent mothers. *Journal of Adolescence, 36*(5), 825–833. https://doi.org/10.1016/j.adolescence.2013.06.004

Torres, L., Yznaga, S. D., & Moore, K. M. (2011). Discrimination and Latino psychological distress: The moderating role of ethnic identity exploration and commitment. *American Journal of Orthopsychiatry, 81*(4), 526–534. https://doi.org/10.1111/j.1939-0025.2011.01117.x

Tummala-Narra, P. (2015). Ethnic identity, perceived support, and depressive symptoms among racial minority immigrant-origin adolescents. *American Journal of Orthopsychiatry, 85*(1), 23–33. https://doi.org/10.1037/ort0000022

Umaña-Taylor, A. J., Diversi, M., & Fine, M. A. (2002). Ethnic identity and self-esteem of Latino adolescents: Distinctions among the Latino populations. *Journal of Adolescent Research, 17*(3), 303–327. https://doi.org/10.1177/0743558402173005

Umaña-Taylor, A. J., Gonzales-Backen, M. A., & Guimond, A. B. (2009). Latino adolescents' ethnic identity: Is there a developmental progression and does growth in ethnic identity predict growth in self-esteem? *Child Development, 80*(2), 391–405. https://doi.org/10.1111/j.1467-8624.2009.01267.x

Umaña-Taylor, A. J., & Guimond, A. B. (2010). A longitudinal examination of parenting behaviors and perceived discrimination predicting Latino adolescents' ethnic identity. *Developmental Psychology, 46*(3), 636–650. https://doi.org/10.1037/a0019376

Umaña-Taylor, A. J., Kornienko, O., Douglass Bayless, S., & Updegraff, K. A. (2018). A universal intervention program increases ethnic-racial identity exploration and resolution to predict adolescent psychosocial functioning one year later. *Journal of Youth and Adolescence, 47*(1), 1–15. https://doi.org/10.1007/s10964-017-0766-5

Umaña-Taylor, A. J., Quintana, S. M., Lee, R. M., Cross, W. E., Jr., Rivas-Drake, D., Schwartz, S. J., & Ethnic and Racial Identity in the 21st Century Study Group. (2014). Ethnic and racial identity during adolescence and into young adulthood: An integrated conceptualization. *Child Development, 85*(1), 21–39. https://doi.org/10.1111/cdev.12196

Umaña-Taylor, A. J., Yazedjian, A., & Bámaca-Gómez, M. (2004). Developing the ethnic identity scale using Eriksonian and social identity perspectives. *Identity, 4*(1), 9–38. https://doi.org/10.1207/S1532706XID0401_2

Whitbeck, L. B., Hoyt, D. R., McMorris, B. J., Chen, X., & Stubben, J. D. (2001). Perceived discrimination and early substance abuse among American Indian children. *Journal of Health and Social Behavior, 42*(4), 405–424. https://doi.org/10.2307/3090187

Zeiders, K. H., Bayless, S. D., Derlan, C. L., Umaña-Taylor, A. J., Updegraff, K. A., & Jahromi, L. B. (2019). Discrimination and ethnic–racial identity: Understanding direction of effects using within- and between-person analyses. *Child Development, 90*(3), e373–e385. https://doi.org/10.1111/cdev.12978

Zeiders, K. H., Umaña-Taylor, A. J., & Derlan, C. L. (2013). Trajectories of depressive symptoms and self-esteem in Latino youths: Examining the role of gender and perceived discrimination. *Developmental Psychology, 49*(5), 951–963. https://doi.org/10.1037/a0028866

Zeiders, K. H., Umaña-Taylor, A. J., Martinez-Fuentes, S., Updegraff, K. A., Douglass Bayless, S., & Jahromi, L. B. (2018). Latina/o youths' discrimination experiences in the U.S. southwest: Estimates from three studies. *Applied Developmental Science, 25*(1), 1–11. https://doi.org/10.1080/10888691.2018.1527695

Zeiders, K. H., Umaña-Taylor, A. J., Updegraff, K. A., & Jahromi, L. B. (2015). An idiographic and nomothetic approach to the study of Mexican-origin adolescent mothers' socio-cultural stressors and adjustment. *Prevention Science, 16*(3), 386–396. https://doi.org/10.1007/s11121-014-0502-2

Chapter 14
Victimization, Ethnic Identity, and Psychosocial Adjustment During First Year Experiences of Latina/o College Students from Immigrant Families

Meeta Banerjee and Scott W. Plunkett

According to the U.S. Census Bureau (2016), Latinos comprised 17.6% of the population in the United States in 2015 and are expected to increase to 28.6% of the population by 2060. Furthermore, within California, Latinos have surpassed the numbers of their ethnic and White counterparts residing within the state, making it the largest demographic within California (Panzar, 2015). Additionally, immigrant Latino parents often lack educational opportunities in their country of origin, consequently immigrating to areas of opportunity for their children (Guan et al., 2014). Hence, identifying factors that may promote or inhibit psychosocial adjustment in Latino young adults may be of import. Moreover, Los Angeles (LA) County is the most populous county of Latinos in the United States (U.S. Census Bureau, 2016), thus Latinos in LA County are of special importance.

With the demographics showing that Latinos are one of the fastest and largest growing populations within the United States, Latinos are entering college at a higher rate than Whites (Lopez & Fry, 2013). However, according to the Pew Hispanic Center, Latino college students often culminate at a bachelor's degree (Lopez & Fry, 2013). Many incoming Latino college students identify as first-generation college students, given their experience as the first in their families to pursue a higher secondary education. First-generation college students have various challenges (e.g., lower financial resources, family pressures) that may hinder their success within the university setting (Vazquez-Salgado et al., 2015). Moreover, research on ethnic minority populations has found that factors such as discrimination and peer victimization may be directly associated with psychosocial well-being (Schmitt et al., 2014).

M. Banerjee (✉)
University of South Carolina, Columbia, SC, USA
e-mail: MEETA@mailbox.sc.edu

S. W. Plunkett
California State University Northridge, Northridge, CA, USA
e-mail: scott.plunkett@csun.edu

D. J. Johnson et al. (eds.), *Re/Formation and Identity*, Advances in Immigrant Family Research, https://doi.org/10.1007/978-3-030-86426-2_14

Research has shown that psychosocial indicators such as self-regulation may be linked to motivation and success in the college/university setting (Pintrich, 2004). Additionally, studies have found that other psychosocial indicators (e.g., attachment) may be associated with either reducing or exacerbating stress in college students (Kenny, 1987; Mason, 2016). Within the college setting, these factors may also be directly or indirectly related to indicators of psychosocial adjustment. The purpose of this study was to investigate the relationship between ethnic discrimination and peer victimization on various indicators of psychosocial adjustment in a sample of Latino college students from immigrant families. Furthermore, we examined how ethnic identity affirmation may provide a buffer for Latino college students from immigrant families.

Theoretical Foundations

For this study, we drew from two different theoretical frameworks to provide a better understanding of how these processes may work for Latino first-generation college students from immigrant families. First, this study utilized García Coll et al.'s (1996) integrative model for developmental competencies in ethnic minority children. In particular, when examining how social stratification mechanisms such as discrimination play a role in the outcome of ethnic minority young adults, exploring these pathways is key to developing an understanding of how they may be related to damaging effects. Moreover, this framework highlights how factors such as ethnic discrimination can influence individuals' experiences in everyday spaces. The second theoretical framework was informed by a risk and resilience paradigm; in particular, research has indicated that ethnic identity can be considered a component of self-concept that may protect individuals from risks such as discrimination or peer victimization (Phinney, 2003). In the trajectory of development, as youth reach emerging adulthood, their ethnic identities may have strengthened, and they have likely moved from exploration to commitment (Marcia, 1980). Hence, this study will utilize both frameworks to provide deeper insight into these processes within emerging adulthood.

Latinos and the College Experience

For Latino college students, enrollment in undergraduate programs has more than doubled in the past 14 years (Rischall & Meyers, 2017). Kena et al. (2016) noted the initial enrollment grew from 1.3 million in 2000 to 3 million by 2014. Many students who entered into the college sphere were first-generation college students, where parents have not attended college and their children are the first in their families to attend. Hence, more work needs to be done exploring the Latino college experience.

Many studies focusing on the experiences of Latino college students have been largely qualitative (Martinez, 2018; Vega, 2016). Martinez (2018) found that in 15 first-generation Latino college students, one of the themes that emerged from the qualitative data was that their academic interests were linked to their ethnic identity. Additionally, Vega (2016) found that having supportive networks was a factor that influenced high achieving college students. These supportive networks were composed of school personnel who provided positive experiences, family members who instilled the importance of secondary education, and friends who could provide guidance with issues such as enrollment into courses.

Challenges and Barriers Within the Latino College Experience

Within the undergraduate experience, many ethnic minority college students report certain barriers and challenges within the system. Soria and Stebleton (2012) in their multiethnic sample of undergraduates at public universities found that first-generation college students reported more obstacles that impacted their academic success such as responsibilities to job and family, perceptions of weak math and English skills, and experiencing mental health issues. Additionally, some students have had challenges with feelings of belongingness. It is important to note that while first-generation college students are experiencing these challenges and barriers within the college system, they may also be experiencing other challenges in everyday life. Particularly, they may be experiencing the effects of ethnic/racial discrimination, which has been found to have detrimental effects on mental health (see meta-analyses by Pascoe & Smart Richman, 2009 and Schmitt et al., 2014). Hence, it is important to examine how negative factors such as discrimination and peer victimization might be impacting the mental health of Latino college students.

Experiences of Peer Victimization

According to Graham (2006), peer victimization is defined by the physical, verbal, or psychological abuse that takes place in settings such as school. Individuals experiencing frequent victimization from peers will generally experience elevated emotional frustration, brooding rumination, and emotion dysregulation leading to internalizing symptoms (see meta-analysis by Reijntjes et al., 2010b), externalizing problems (e.g., aggression; see meta-analysis by Reijntjes et al., 2010a), and psychosomatic issues (see meta-analysis by Gini & Pozzoli, 2013). Within the research on peer victimization, most research focuses on ethnic minority children and adolescents. Also, most studies on children and adolescents indicate that peer victimization is related to poorer psychosocial outcomes in ethnic minority youth (Graham et al., 2006; Seaton et al., 2013). There is little research that examines peer victimization within the college context. This study will examine whether peer victimization is related to psychosocial outcomes to expand the field on the literature within college student populations.

Perceived Ethnic Discrimination and Its Negative Effects

Perceived ethnic discrimination is characterized as a salient experience of unfair treatment due to ethnic background. According to Schmitt and Branscombe (2002), perceived discrimination suggests that each individual has a certain place in society due to their race/ethnicity, which can negatively affect their well-being. These actions are often systematic and result in inequitable and negative effects on the minority group (Williams et al., 2003). In their meta-analytic review of the effects of perceived discrimination, Pascoe and Smart Richman (2009) found that perceived discrimination was related to both psychological and physical health in ethnic minority populations. Additionally, another meta-analysis by Schmitt et al. (2014) illustrated the deleterious effects of discrimination on the psychological well-being of individuals. The deleterious relationship between discrimination and mental health has also been demonstrated within ethnic minority youth (Cobb et al., 2018). For example, racial discrimination was related to depressive symptomatology (Seaton & Douglass, 2014) and psychosocial adjustment (Neblett et al., 2016) within African American youth. The work on Latino youth has found similar effects. Specifically, Huq et al. (2016) found that ethnic discrimination was related to depressive symptomatology in a sample of Latino youth. The effects are also prevalent in Latino young adults who are attending college. Hwang and Goto (2008) found that experiences of discrimination were related to Latino and Asian college students' anxiety and psychological distress. Similarly, Brittian et al. (2015) reported that perceived ethnic discrimination was related to depressive symptomatology in a sample of ethnic minority college students. In addition, research has shown there is a significant relationship between discrimination and ethnic identity affirmation (Baldwin-White et al., 2017).

These studies underscore the importance of examining the harmful effects of ethnic discrimination within Latino college populations. Additionally, this work also emphasizes the relation between perceived ethnic discrimination and ethnic identity formation. Although these previous studies have shown direct linkages between perceived ethnic/racial discrimination and certain mental health factors, we know less about how perceived ethnic discrimination is related to outcomes such as anxious romantic attachment in young adults. Further, outcomes such as brooding and verbal hostility have not been explored in relation to perceived ethnic discrimination. These mental health outcomes might be important to shed light as they could be informative to the college student experience and be factors of psychosocial adjustment that are important to student success.

Ethnic Identity as a Buffer Among Populations of Color

In a seminal meta-analysis conducted by Rivas-Drake et al. (2014), a distinction between ethnic identity versus racial identity has been identified in that ethnic identity encompasses individuals' feelings of belongingness with their cultural background (Cokley, 2007), while racial identity can refer to identities that are formulated as a response to social stratification factors (Sellers et al., 1998). Research has found aspects of ethnic identity to be a protective factor for positive adjustment and academic achievement among ethnic minorities (Ong et al., 2006; Rivas-Drake et al., 2014). Furthermore, Phinney (2003) emphasized that ethnic identity may be a protective source for Latino populations, specifically in the face of discrimination.

Many previous studies on ethnic identity have used composite indices of ethnic identity dimensions (Umaña-Taylor, 2011) which can obscure how each ethnic identity dimension may contribute to psychosocial adjustment. When examining the dimensions, much of the attention has focused on ethnic commitment (i.e., sense of belonging to one's ethnic group) and ethnic exploration (i.e., seeking out knowledge about one's ethnic group), with less attention on ethnic affirmation (i.e., positive feelings associated with being a member of one's ethnic group). This is surprising since ethnic affirmation can be an important part of one's internal representation of self-worth that can facilitate well-being. For example, Brittian et al. (2013) found that ethnic identity affirmation was negatively related to depression and anxiety in Asian American, African American, and Latino college students. Also, Jones et al. (2017) found that ethnic identity affirmation was related to worse depressive symptoms and greater self-esteem of Mexican American college students.

Studies have frequently examined how various dimensions of ethnic identity are mediators between cultural stressors (e.g., discrimination) and mental health outcomes (e.g., Brittian et al., 2015; Jones et al., 2017; Umaña-Taylor & Updegraff, 2007). However, there is conceptual support for the idea that ethnic affirmation could moderate the relationship between stressors and adaptation. Additionally, Tajfel and Turner (1986) indicated that when individuals experience discrimination, ethnic identity may be heightened. For example, individuals who experience ethnic discrimination may feel their self-worth is attacked, yet a strong internal ethnic affirmation can potentially alter the reaction to discrimination so that its harm is buffered. More specifically, a person with strong ethnic affirmation may be more likely to attribute the discrimination to the actions of the perpetrator instead of internalizing diminished feelings toward one's ethnicity and self. However, for those with low ethnic affirmation, perceptions of ethnic discrimination or victimization could potentially result in even worse psychosocial outcomes.

Recently, research has examined the protective nature of ethnic identity within ethnic minority populations (Brittian et al., 2015). In one particular study, Brittian et al. (2015) found that in a sample of 2315 Latino and African American college students, perceived group discrimination was mediated via two paths of ethnic identity dimensions. Specifically, they learned that the pathway between ethnic group discrimination and depression was mediated by ethnic identity affirmation for

Latino college students. Moreover, Umaña-Taylor and Updegraff (2007) found that ethnic identity mediated the relationship between perceived ethnic discrimination and psychosocial outcomes in a sample of Latino adolescents. Additionally, Jones et al. (2017) found that in a sample of Latino college students, there was a negative relationship between ethnic identity affirmation and ethnic discrimination.

As stated above, previous literature has shown that ethnic identity relates to certain mental health outcomes. However, we do not know if ethnic identity, and in particular, the dimension of ethnic identity affirmation, relates directly to outcomes such as brooding, verbal hostility, anger regulation, or romantic attachment. Moreover, we do not know if the dimension of ethnic identity affirmation is a moderator between discrimination or victimization and these mental health outcomes. Given the importance of establishing and maintaining relationships during emerging adulthood, these psychosocial outcomes may be especially relevant.

Addressing the Gaps: The Present Study

Although previous work has examined the effects of ethnic discrimination on Latino college populations (Huq et al., 2016), little is known about how both peer victimization and ethnic discrimination may influence the psychosocial outcomes of Latino college students. In particular, we know less when examining outcomes, such as anxious attachment, brooding, and anger regulation coping, that can often be important in the psychosocial adjustment of college youth while they are navigating the college environment. Additionally, many Latino college students from immigrant families are first-generation students. In the current sample, most are in the first year of their college experience. Encompassing a strong ethnic identity, specifically around the sphere of affirmation, may help these students navigate through the first year successfully as they are experiencing the effects of ethnic discrimination or victimization by peers.

The present study furthers the field by exploring the potential direct and moderating effects of ethnic identity affirmation within Latino students from immigrant families. This study specifically examines the relation between discrimination, victimization, and ethnic affirmation on the psychosocial outcomes of Latino college students (see Fig. 14.1). The following three research questions guided the study. First, how do negative factors (e.g., victimization or discrimination) relate to Latino college students' psychosocial outcomes? We believe that higher reports of peer victimization or ethnic discrimination will be related to reports of more brooding, more anxious attachment, more verbal hostility, and less anger regulation coping. Second, is ethnic identity, specifically affirmation, associated with positive psychosocial outcomes? In particular, we hypothesize that Latino college students who have greater ethnic identity affirmation would have less brooding, less anxious attachment, less verbal hostility, and more anger regulation coping. Finally, does ethnic identity moderate the relationship between either discrimination or victimization and the psychosocial outcomes? We posit that higher levels of ethnic identity

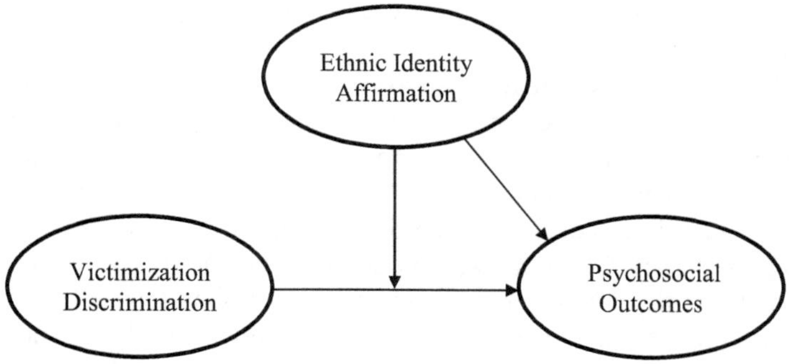

Fig. 14.1 Conceptual model

affirmation will buffer the effects of discrimination and victimization on their psychosocial outcomes. These research questions will be addressed utilizing data from approximately 1727 Latino students at a Hispanic Serving Institution (HSI).

Method

Sample

Data were used from 1727 undergraduate Latina/o college students from immigrant families (67.4% women, 32.4% men, and 0.2% other). Their ages ranged from 18 to 24 years (M = 19.2 years; SD = 1.36 years). Their classifications were 73.6% first year, 7.9% sophomore, 12.0% junior, and 6.5% senior. Generation statuses were 12.0% first generation (participant and both parents born outside the United States), 79.4% second generation (participant U.S.-born and both parents born outside the United States), and 8.6% 2.5 generation (participant U.S.-born and one parent born outside the United States). Most parents were born in various Latin American countries: Mexico (64.0% fathers, 62.1% mothers), El Salvador (14.4% fathers, 15.0% mothers), Guatemala (9.4% fathers, 9.1% mothers), and other (12.2% fathers, 13.8% mothers).

Procedures

This study was approved by the university institutional review board prior to data collection. The data were collected from students at one comprehensive university in southern California designated as an Hispanic serving institution. Participants were recruited using a research subject pool for lower-division psychology classes. Given that the research subject pool is primarily made up of lower-division students in a general education course, the subject pool was composed of students from a variety of majors.

Students in the classes were required to either participate as subjects in research studies or complete an alternative assignment. The participants signed up through the Sona system which protects their anonymity, yet still allows researchers to assign credit for completion of the study. The participants completed an online survey through Qualtrics. Students had the right to leave questions unanswered or to stop taking the survey. Questions were inserted throughout the survey to ascertain whether participants were carefully reading the questions (e.g., "Please answer 'strongly disagree' to show you are reading the survey carefully"). Students who missed more than one of these questions were not included in the data. The data were collected in 2012, 2014, 2015, and 2016. Each year had slightly different foci, so the samples sizes vary depending on the mental health variable. Students who identified as Latino, Hispanic, or a related category (e.g., Mexican American, Chicano, Salvadoran) were included in the analyses.

Measures

For all the scales, items were averaged to create mean scores. Means, standard deviations, and ranges are shown in Table 14.1.

Ethnic Identity Affirmation

A 7-item subscale of the Multi-group Ethnic Identity Measure (MEIM; Phinney, 1992) was used to assess ethnic identity affirmation (i.e., an affective component measuring affirmation, belonging, and commitment to one's ethnic identity).

Table 14.1 Intercorrelations and descriptives on study variables

	1	2	3	4	5	6	*M* (*SD*)	Range
1. Ethnic identity affirmation	–						4.70 (0.58)	1.00–5.00
2. Peer victimization	−0.10**	–					1.51 (0.64)	1.00–5.00
3. Ethnic discrimination	0.05*	0.53**	–				2.01 (0.88)	1.00–5.56
4. Anxious attachment	−0.10**	0.20**	0.26**	–			3.52 (0.99)	1.00–6.83
5. Brooding rumination	−0.00	0.32**	0.22**	0.34**	–		2.27 (0.76)	1.00–4.00
6. Anger regulation	0.06*	−0.19**	−0.09**	−0.09*	−0.11**	–	2.26 (0.46)	1.00–3.00
7. Verbal hostility	−0.01	0.31**	0.31**	0.23**	0.32**	−0.37**	0.91 (0.77)	0.00–4.00

*$p < 0.05$, **$p < 0.01$

Sample items were "I have a strong sense of belonging to my own ethnic group" and "I have a lot of pride in my ethnic group." Response choices were: 1 = *strongly disagree*, 2 = *disagree*, 3 = *neutral*, 4 = *agree*, and 5 = *strongly agree*. The affirmation subscale has been used with diverse samples, demonstrating good reliability with Cronbach's alphas above 0.80 (Phinney, 1992; Roberts et al., 1999). Roberts et al. (1999) reported positive correlations between ethnic identity and various indicators of psychological well-being and negative correlations between ethnic identity and depression/loneliness, demonstrating construct validity. In the current data, the Cronbach's alpha was 0.89.

Peer Victimization

A 10-item modification of the Bullying Scale (Champion & Clay, 2007) was used to measure frequency of perceived peer victimization in the last 6 months. Three items about physical aggression (e.g., "hit of kicked to hurt me") were dropped, but four other items were added: (1) "Excluded me from social events," (2) "Harassed me," (3) "Intentionally hurt my feelings," and (4) "Shared private information about me with other people." The response choices were 1 = *never*, 2 = *once or twice*, 3 = *sometimes*, 4 = *once a week*, and 5 = *more than once per week*. In the current data, the Cronbach's alpha was 0.91.

Ethnic Discrimination

The 9-item Everyday Discrimination Scale was used to measure frequency of perceiving routine, minor acts of unfair treatment (Essed, 1991). The stem is "How often have any of the following happened to you because of your RACE, CULTURE, LANGUAGE, and/or SKIN COLOR." Two sample items include "You are treated with less respect than other people" and "You are called names or insulted." The response options were 1 = *never,* 2 = *less than once a year,* 3 = *a few times a year,* 4 = *a few times a month,* 5 = *at least once a week,* and 6 = *almost every day*. In the current sample, the Cronbach's alpha was 0.92. Using a sample of ethnically diverse university students in immigrant families, Bakhtiari et al. (2018) found a Cronbach's alpha of 0.92. In the current data, the Cronbach's alpha was 0.92.

Anger Regulation Coping

A 4-item anger regulation coping subscale was used from the Children's Sadness and Anger Management Scale (Zeman et al., 2001; Zeman et al., 2002). A sample item included, "When I am feeling mad, I control my temper." The response choices follow: 1 = *hardly ever,* 2 = *sometimes*, and 3 = *often*. In the current data, a Cronbach's alpha of 0.63 was found.

Anxious Attachment

The 6-item anxious romantic attachment subscale was used from the 12-item Experiences in Close Relationship Scale (ECR)-Short Form (Wei et al., 2007). One item was reverse coded. Two sample items are "I need a lot of reassurance that I am loved by my partner" and "I rarely worry about being abandoned" (reverse coded). Response choices were 1 = *strongly disagree,* 2 = *disagree*, 3 = *somewhat disagree*, 4 = *neutral*, 5 = *somewhat agree*, 6 = *agree*, and 7 = *strongly agree.* In the current data, a Cronbach's alpha of 0.64 was found.

Verbal Hostility

Verbal hostility was measured with a 5-item scale adapted from a hostility scale by Derogatis, Rickels, and Rock (1976). The items ask the participants how often they yell, call names, curse when angry, have temper outbursts, and get into frequent arguments. Response choices were 0 = *never*, 1 = *sometimes*, 2 = *frequently*, 3 = *very frequently*, and 4 = *always*. In the current data, a Cronbach's alpha of 0.84 was found.

Brooding

A 5-item scale was used to measure brooding rumination (Treynor et al., 2003). The stem for the items was "When you feel down, sad, or depressed, please indicate how frequently you generally do the following (not what you think you should do)." Two sample items were (1) "Think 'Why can't I handle things better?'" and (2) "Think 'What am I doing to deserve this?'" Response choices follow: 1 = *almost never*, 2 = *sometimes*, 3 = *often*, and 4 = *almost always*. In the current data, the Cronbach's alpha was 0.83.

Results

Type of Analysis

Preliminary analyses were conducted to test the relationships between the study variables. This included running descriptives and correlations on the variables of interests. To test the study hypotheses, a set of hierarchical regressions were conducted (Aiken & West, 1991) on the following dependent variables: anxious romantic attachment, brooding rumination, anger regulation, and verbal hostility. The first set of hierarchical regressions tested the main effects of ethnic discrimination and ethnic identity affirmation in steps 1 and 2, respectively. The third step examined the effects of the interaction between ethnic discrimination and ethnic affirmation. The second set of hierarchical regressions tested the main effects of peer victimization and ethnic affirmation in steps 1 and 2, respectively. The third step examined the effects of the interaction between peer victimization and ethnic affirmation.

Preliminary Analyses

Intercorrelations were conducted on the study variables (see Table 14.1). Reports of peer victimization and ethnic discrimination were significantly and positively related to anxious attachment, brooding rumination, and verbal hostility. Ethnic identity affirmation was significantly and positively correlated to anger regulation, negatively correlated to anxious attachment, but not significantly correlated to brooding rumination or verbal hostility.

Hierarchical Regressions

To examine the relationships between ethnic identity, peer victimization, and ethnic discrimination on the outcome variables, a series of hierarchical regressions was conducted. The first set investigated the associations between ethnic identity affirmation and ethnic discrimination (see Table 14.2). Participants who experienced higher levels of ethnic discrimination also reported participating in more verbal hostility, brooding rumination, and verbal hostility, and they reported lower anger regulation. In the second step, ethnic identity affirmation was negatively related to anxious romantic attachment, suggesting that those with higher levels of ethnic affirmation had lower levels of anxious attachment. However, ethnic identity affirmation was not significantly related to the other outcomes after controlling for ethnic discrimination. In the final step, the interaction between ethnic discrimination and ethnic identity affirmation was significantly related to anxious attachment. As shown in Fig. 14.2, ethnic identity affirmation buffered the relationship between discrimination and anxious romantic attachment, especially at high levels of discrimination. Anxious romantic attachment was highest when Latino students perceived high discrimination and low ethnic identity affirmation.

The second set of hierarchical regressions were conducted where peer victimization and ethnic identity affirmation were the designated independent variables (see Table 14.3). In the first step, peer victimization was significantly and positively related to anxious romantic attachment, verbal hostility, and brooding rumination, and peer victimization was significantly and negatively related to anger regulation. In the second step, ethnic identity affirmation was not significantly related to the outcome variables after controlling for peer victimization. In the last step, a significant interaction was found between ethnic identity affirmation and victimization for anxious romantic attachment, anger regulation, and verbal hostility. As shown in Fig. 14.3, ethnic identity affirmation buffered the relationship between peer

Table 14.2 Hierarchical multiple regressions examining perceived discrimination, ethnic identity, and Latino adolescents' psychosocial adjustment

	Anxious romantic attachment	Brooding rumination	Anger regulation	Verbal hostility
N	1084	1717	795	1157
Step 1				
Discrimination	0.257**	0.222**	−0.093**	0.098**
r^2 change	0.066**	0.049**	0.009**	0.313**
Step 2				
Discrimination	0.261**	0.223**	−0.096**	0.313**
Ethnic identity	−0.107**	−0.015	0.067	−0.018
r^2 change	0.012**	0.000	0.004	0.000
Step 3				
Discrimination	0.254**	−0.015	−0.095**	0.316**
Ethnic identity	−0.101**	0.223**	0.066	−0.021
Discrimination × ethnic identity	−0.064*	−0.014	0.011	0.026
r^2 change	0.004*	0.002	0.000	0.001

*$p < 0.05$, **$p < 0.01$

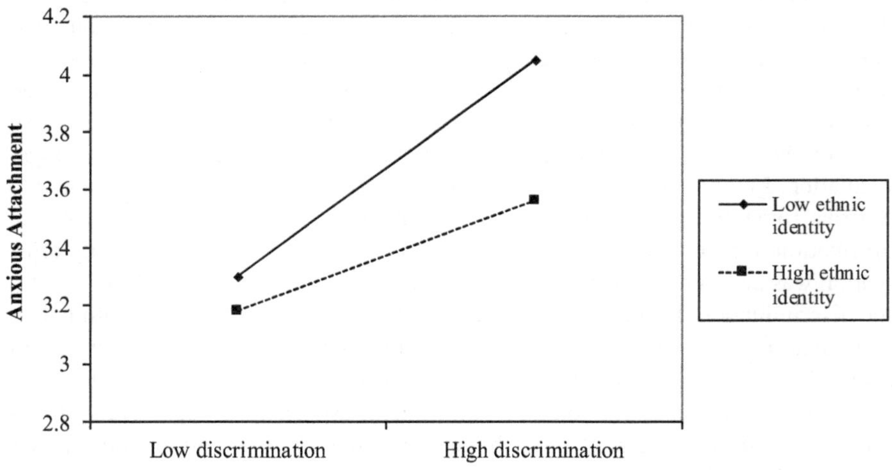

Fig. 14.2 Ethnic identity moderating ethnic discrimination on anxious romantic attachment

victimization and anxious romantic attachment, especially at high levels of victimization. Anxious romantic attachment was highest when victimization was high and ethnic affirmation was low. As shown in Fig. 14.4, ethnic identity affirmation moderated the relationship between peer victimization and anger regulation, especially at high levels of victimization. Anger regulation was lowest at high levels of

Table 14.3 Hierarchical multiple regressions examining perceived peer victimization, ethnic identity, and Latino adolescents' psychosocial adjustment

	Anxious romantic attachment	Brooding rumination	Anger regulation	Verbal hostility
N	1084	1194	793	1155
Step 1				
Victimization	0.201**	0.322**	−0.187**	0.308**
r^2 change	0.040**	0.104**	0.035**	0.095**
Step 2				
Victimization	0.193**	0.323**	−0.183**	0.310**
Ethnic identity	−0.079**	0.003	0.056	0.019
r^2 change	0.006**	0.000	0.003	0.000
Step 3				
Victimization	0.171**	0.326**	−0.168**	0.299**
Ethnic identity	−0.059	0.000	0.044	0.029
Victimization × ethnic identity	−0.132**	0.019	0.074*	−0.063*
r^2 change	0.016**	0.000	0.005*	0.004*

*$p < 0.05$, **$p < 0.01$

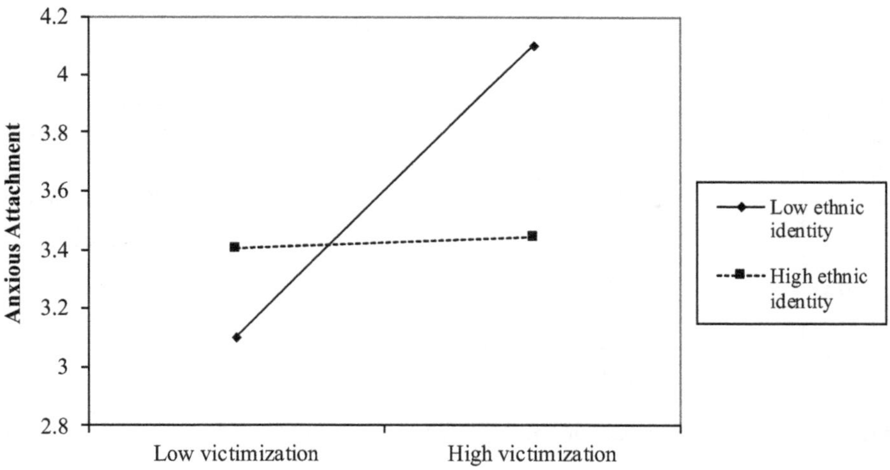

Fig. 14.3 Ethnic identity moderating peer victimization on anxious romantic attachment

victimization and low levels of ethnic affirmation. As shown in Fig. 14.5, ethnic identity affirmation moderated the relationship between peer victimization and verbal hostility, especially at low levels of victimization. Verbal hostility was highest when victimization was high and ethnic identity affirmation was low; however, verbal hostility was lowest when ethnic identity affirmation and victimization were lowest.

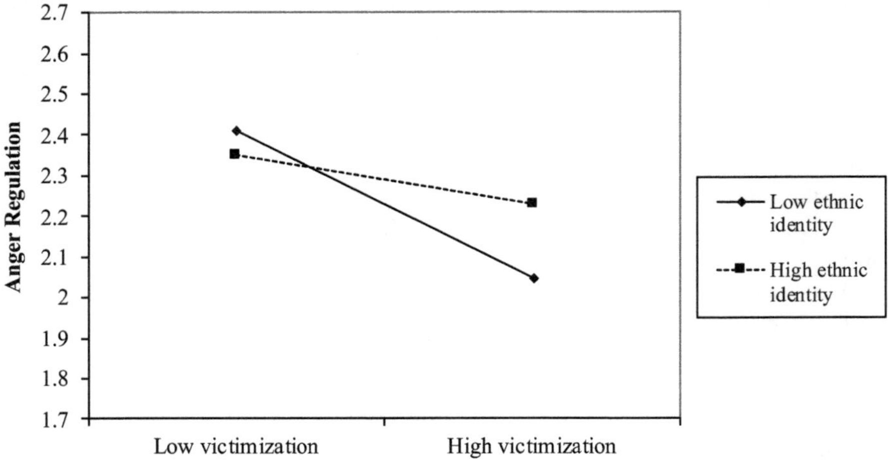

Fig. 14.4 Ethnic identity moderating peer victimization on anger emotion regulation

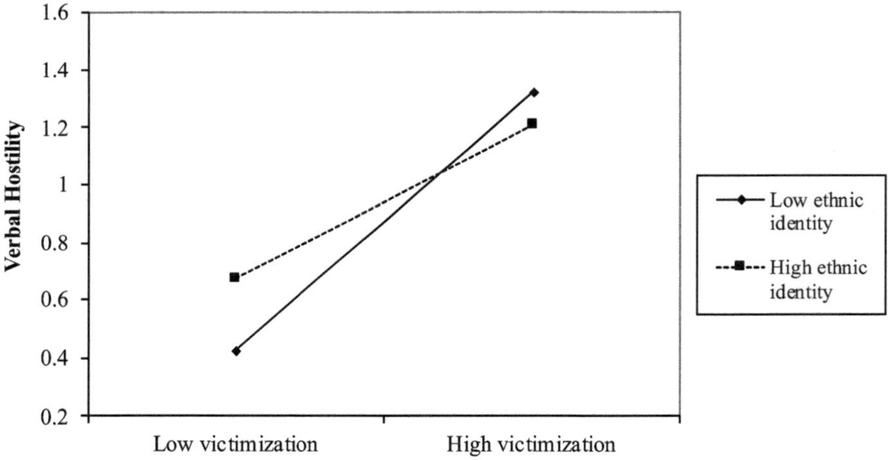

Fig. 14.5 Ethnic identity moderating peer victimization on verbal hostility

Discussion

The current study explored: (1) the effects of peer victimization and discrimination on four psychosocial outcomes, and (2) whether ethnic identity affirmation would buffer the effects of ethnic discrimination or peer victimization on the psychosocial outcomes of Latino college students. This study found that higher reports of peer victimization were associated with less anger regulation coping and anxious romantic attachment and also more brooding rumination and verbal hostility. These findings are consistent with a meta-analysis that found that peer victimization was related to internalizing issues (e.g., depressive symptoms, anxiety; see Reijntjes et al., 2010b) and externalizing behaviors (e.g., aggression; see Reijntes et al., 2010a). Victimization by peers (e.g., threats, name-calling, harassment, intentional exclusion) can heighten emotional reactivity and threaten self-worth. Thus, Latino college students who feel victimized by peers may brood about why this happens to them, have difficulty controlling their anger and verbal outbursts in response to the victimization, and feel more anxious in relationships.

Additionally, this study indicated that ethnic discrimination was also related to the four psychosocial outcomes. These findings are in line with Pascoe and Smart Richman's (2009) meta-analysis that showed that ethnic discrimination in ethnic minority youth was linked to both internalizing and externalizing behaviors. Similar to peer victimization, when Latino college students experience ethnic discrimination, they might ruminate on why people discriminate against them just because of their ethnicity, skin color, accent, etc. These experiences can provoke feelings of anger, which can undermine anger regulation and result in more verbal hostility. Also, when Latino college students experience ethnic discrimination, they may question their own value in the dating scene leading to anxious romantic attachment.

Support was found for the hypotheses regarding the direct relationships between ethnic affirmation and the psychosocial outcomes. Specifically, ethnic affirmation was related to greater anger regulation and less anxious romantic attachment. Thus, when Latino college students have more positive feelings toward their ethnicity, they may have more internal value (i.e., self-worth) which can help regulate anger emotions and provide more security in romantic relationships. Ethnic affirmation was not significantly related to brooding or verbal hostility in this sample. Brooding (i.e., negative rumination) often requires an activating stressor. This may be the same for verbal hostility. Thus, experiences of discrimination and victimization are more likely to elicit brooding and verbal hostility than positive feelings toward one's ethnicity.

Additionally, support was found for the moderating effect of ethnic affirmation. First, ethnic affirmation buffered the relationship between both discrimination and victimization on anxious romantic attachment. Specifically, anxious romantic attachment was highest when Latino students perceived high discrimination or victimization and low ethnic affirmation. Thus, low ethnic affirmation seems to be an additional risk factor for Latino college students' anxious attachment. Specifically, when Latino college students experience discrimination or victimization, they may feel anxious regarding romantic relationships. If they have higher ethnic affirmation, it can help minimize those anxieties because it enhances feelings of self-worth, but low ethnic affirmation might result in more confirmation bias resulting from the discrimination or victimization. Ethnic affirmation also moderated the relationship between peer victimization and anger regulation. Thus, anger regulation was lowest at high levels of victimization and low levels of ethnic affirmation. Once again, low levels of ethnic affirmation might be another risk factor when combined with victimization by peers. When Latino college students experience bullying from peers, it can aggravate negative emotions, and if they feel worse about their ethnicity, their anger dysregulation is aggravated.

Although this study adds to the literature on the experiences of Latino college students, there are some limitations. First, the study relies on the self-report of Latino students, this underscores the subjective nature of perceived ethnic discrimination, peer victimization, and ethnic identity. Additionally, future studies could include reports from those individuals surrounding the student, such as peers, parents, or sibling to gather a more complete picture of what the individual is truly experiencing. Furthermore, this research focuses on cross-sectional data from Latino college students. To understand when these factors of ethnic discrimination, peer victimization, and ethnic identity become key factors in development, it might be essential to conduct longitudinal studies on these constructs.

Conclusions and Implications

Latino college students are an important population to focus on, given the increases of Latino college students and the large number of Latino students in immigrant families. Understanding how various risk factors (e.g., discrimination, peer victimization) impact students' experience may help Latino emerging adults to persist and achieve within the college experience. Moreover, this study highlights the protective nature of ethnic affirmation on some psychosocial outcomes, which may be especially pertinent to the identity process and formation for Latino college students from immigrant families, many of whom are first-generation college students. Hence, universities and colleges may want to consider implementing interventions, preventions, or curriculum modifications that: (1) try to diminish discrimination and victimization, (2) teach coping strategies to those who experience discrimination or victimization, and (3) encourage positive views of one's ethnicity. Each of these efforts can potentially enhance the psychosocial outcomes of Latino college students from immigrant families, as well as other students from marginalized groups. As previous research has shown, positive mental health may be impactful to gains in academic achievement (Keyes et al., 2012). Therefore, it is imperative that we stress the importance of personal cultural factors such as ethnic affirmation within ethnic minority college populations.

References

Aiken, L. S., & West, S. G. (1991). *Multiple regression: Testing and interpreting interactions*. Sage.

Bakhtiari, F., Benner, A. D., & Plunkett, S. W. (2018). Life quality of university students from immigrant families in the United States. *Family and Consumer Sciences Research Journal, 46*(4), 331–346. https://doi.org/10.1111/fcsr.12260

Baldwin-White, A. J. M., Kiehne, E., Umana-Taylor, A., & Marsiglia, F. F. (2017). In pursuit of belonging: Acculturation, perceived discrimination and ethnic-racial identity among Latino youths. *Social Work Research, 41*(1), 43–52. https://doi.org/10.1093/swr/svw029

Brittian, A. S., Kim, S. Y., Armenta, B. E., Lee, R. M., Umaña-Taylor, A. J., Schwartz, S. J., Villalta, I. K., Zamboanga, B. L., Weisskirch, R. S., Juang, L. P., Castillo, L. G., & Hudson, M. L. (2015). Do dimensions of ethnic identity mediate the association between perceived ethnic group discrimination and depressive symptoms? *Cultural Diversity and Ethnic Minority Psychology, 21*(1), 41–53. https://doi.org/10.1037/a0037531

Brittian, A. S., Umaña-Taylor, A. J., Lee, R. M., Zamboanga, B. L., Kim, S. Y., Weisskirch, R. S., Castillo, L. G., Whitbourne, S. K., Hurley, E. A., Huynh, Q.-L., Brown, E. J., & Caraway, S. J. (2013). The moderating role of centrality on associations between ethnic identity affirmation and ethnic minority college students' mental health. *Journal of American College Health, 61*(3), 133–140. https://doi.org/10.1080/07448481.2013.773904

Champion, K. M., & Clay, D. L. (2007). Individual differences in responses to provocation and frequent victimization by peers. *Child Psychiatry & Human Development, 37*, 205–220. https://doi.org/10.1007/s10578-006-0030-9

Cobb, C. L., Meca, A., Branscombe, N. R., Schwartz, S. J., Xie, D., Zea, M. C., Fernandez, C. A., & Sanders, G. L. (2018). Perceived discrimination and well-being among unauthorized Hispanic immigrants: The moderating role of ethnic/racial group identity centrality. *Cultural Diversity and Ethnic Minority Psychology, 25*(2), 280–287. https://doi.org/10.1037/cdp0000227

Cokley, K. (2007). Critical issues in the measurement of ethnic and racial identity: A referendum on the state of the field. *Journal of Counseling Psychology, 54*(3), 224–234. https://doi.org/10.1037/0022-0167.54.3.224

Derogatis, L. R., Rickels, K., & Rock, A. F. (1976). SCL-90 and the MMPI: A step in the validation of a new self-report scale. *British Journal of Psychiatry, 128*(3), 280–289. https://doi.org/10.1192/bjp.128.3.280

Essed, P. (1991). *Understanding everyday racism*. Sage. https://doi.org/10.4135/9781483345239

García Coll, C. G., Lamberty, G., Jenkins, R., McAdoo, H. P., Crnic, K., Wasik, B. H., & Garcia, H. V. (1996). An integrative model for the study of the developmental competencies in minority children. *Child Development, 67*(5), 1891–1914. https://doi.org/10.2307/1131600

Gini, G., & Pozzoli, T. (2013). Bullied children and psychosomatic problems: A meta-analysis. *Pediatrics, 132*(4), 720–729. https://doi.org/10.1542/peds.2013-0614

Graham, S. (2006). Peer victimization in school: Exploring the ethnic context. *Current Directions in Psychological Science, 15*(6), 317–321. https://doi.org/10.1111/j.1467-8721.2006.00460.x

Graham, S., Bellmore, A. D., & Mize, J. (2006). Peer victimization, aggression, and their co-occurrence in middle school: Pathways to adjustment problems. *Journal of Abnormal Child Psychology, 34*(3), 363–378. https://doi.org/10.1007/s10802-006-9030-2

Guan, S. S. A., Greenfield, P. M., & Orellana, M. F. (2014). Translating into understanding: Language brokering and prosocial development in emerging adults from immigrant families. *Journal of Adolescent Research, 29*(3), 331–355. https://doi.org/10.1177/0743558413520223

Huq, N., Stein, G. L., & Gonzalez, L. M. (2016). Acculturation conflict among Latino youth: Discrimination, ethnic identity and depressive symptoms. *Cultural Diversity and Ethnic Minority Psychology, 22*(3), 377–385. https://doi.org/10.1037/cdp0000070

Hwang, W. H., & Goto, S. (2008). The impact of perceived racial discrimination on the mental health of Asian American and Latino college students. *Cultural Diversity and Ethnic Minority Psychology, 14*(4), 326–335. https://doi.org/10.1037/1099-9809.14.4.326

Jones, S. K., Killoren, S. E., Alfaro, E. C., & Gonzales-Backen, M. A. (2017). Mexican American college students' perceived experiences of discrimination, ethnic identity, and adjustment: The protective role of sibling support. *Emerging Adulthood, 6*(4), 280–287. https://doi.org/10.1177/2167696817711606

Kena, G., Hussar, W., McFarland, J., de Brey, C., Musu-Gillette, L., Wang, X., Zhang, J., Rathbun, A., Wilkinson-Flicker, S., Diliberti, M., Barmer, A., Bullock Mann, F., & Dunlap Velez, E. (2016). *The condition of education 2016 (NCES 2016–144)*. National Center for Education Statistics. http://files.eric.ed.gov/fulltext/ED565888.pdf

Kenny, M. E. (1987). The extent and function of parental attachment among first-year college students. *Journal of Youth and Adolescence, 16*(1), 17–29. https://doi.org/10.1007/bf02141544

Keyes, C. L. M., Eisenberg, D., Perry, G. S., Dube, S. R., Kroenke, K., & Dhingra, S. S. (2012). The relationship of level of positive mental health with current mental disorders in predicting suicidal behavior and academic impairment in college students. *Journal of American College Health, 60*(2), 126–133. https://doi.org/10.1080/07448481.2011.608393

Lopez, H. M., & Fry, R. (2013). *Among recent high school grads, Hispanic college enrollment rates surpass that of Whites.* Pew Research Center. http://www.pewresearch.org/fact-tank/2013/09/04/hispanic-college-enrollment-rate-surpasses-whites-for-the-first-time/

Marcia, J. E. (1980). Identity in adolescence. In J. Addison (Ed.), *Handbook of adolescent psychology* (pp. 159–187). Wiley.

Martinez, E. (2018). College choice processes and experiences of Latino/a/x student-athletes in the community college context. *Journal of the Study of Sports and Athletes in Education, 12*(1), 14–34. https://doi.org/10.1080/19357397.2018.1444700

Mason, H. O. (2016). Multiple measures of family and social support as predictors of psychological well-being: An additive approach. *Journal of Educational and Developmental Psychology, 6*(2), 97–112. https://doi.org/10.5539/jedp.v6n2p97

Neblett, E. W., Bernard, D. L., & Banks, K. H. (2016). The moderating roles of gender and socioeconomic status in the association between racial discrimination and psychological adjustment. *Cognitive and Behavioral Practice, 23*(3), 385–397. https://doi.org/10.1016/j.cbpra.2016.05.002

Ong, A., Phinney, J. S., & Dennis, J. (2006). Competence under challenge: Exploring the protective influence of parental support and ethnic identity in Latino college students. *Journal of Adolescence, 29*(6), 961–979. https://doi.org/10.1016/j.adolescence.2006.04.010

Panzar, J. (2015). *It's official: Latinos now outnumber Whites in California.* Los Angeles Times. https://www.latimes.com/local/california/la-me-census-latinos-20150708-story.html

Pascoe, E. A., & Smart Richman, L. (2009). Perceived discrimination and health: A meta-analytic review. *Psychological Bulletin, 135*(4), 531–554. https://doi.org/10.1037/a0016059

Phinney, J. S. (1992). The multi-group ethnic identity measure: A new scale for use with adolescents and young adults from diverse groups. *Journal of Adolescent Research, 7*(2), 156–176. https://doi.org/10.1177/074355489272003

Phinney, J. S. (2003). Ethic identity and acculturation. In K. M. Chun, P. B. Organista, & G. Marín (Eds.), *Acculturation: Advances in theory, measurement, and applied research* (pp. 63–81). American Psychological Association. https://doi.org/10.1037/10472-006

Pintrich, P. R. (2004). A conceptual framework for assessing motivation and self-regulated learning in college students. *Educational Psychology Review, 16*(4), 385–407. https://doi.org/10.1007/s10648-004-0006-x

Reijntjes, A., Kamphuis, J. H., Prinzie, P., Boelen, P. A., van der Schoot, M., & Telch, M. J. (2010a). Prospective linkages between peer victimization and externalizing problems in children: A meta-analysis. *Aggressive Behavior, 37*(3), 215–222. https://doi.org/10.1002/ab.20374

Reijntjes, A., Kamphuis, J. H., Prinzie, P., & Telch, M. J. (2010b). Peer victimization and internalizing problems in children: A meta-analysis of longitudinal studies. *Child Abuse & Neglect, 34*(4), 244–252. https://doi.org/10.1016/j.chiabu.2009.07.009

Rischall, E., & Meyers, S. A. (2017). Conflicting cultural pressures and the adjustment of Latino college students. *Journal of Hispanic Higher Education, 18*(4), 1–14. https://doi.org/10.1177/1538192717747948

Rivas-Drake, D., Syed, M., Umaña-Taylor, A., Markstrom, C., French, S., Schwartz, S. J., Lee, R., & Ethnic and Racial Identity in the 21st Century Study Group. (2014). Feeling good, happy, and proud: A meta-analysis of positive ethnic–racial affect and adjustment. *Child Development, 85*(1), 77–102. https://doi.org/10.1111/cdev.12175

Roberts, R., Phinney, J., Masse, L., Chen, Y., Roberts, C., & Romero, A. (1999). The structure of ethnic identity of young adolescents from diverse ethnocultural groups. *The Journal of Early Adolescence, 19*(3), 301–322. https://doi.org/10.1177/0272431699019003001

Schmitt, M. T., & Branscombe, N. R. (2002). The meaning and consequences of perceived discrimination in disadvantaged and privileged social groups. *European Review of Social Psychology, 12*(1), 167–199. https://doi.org/10.1080/14792772143000058

Schmitt, M. T., Branscombe, N. R., Postmes, T., & Garcia, A. (2014). The consequences of perceived discrimination for psychological well-being: A meta-analytic review. *Psychological Bulletin, 140*(4), 921–948. https://doi.org/10.1037/a0035754

Seaton, E. K., & Douglass, S. (2014). School diversity and racial discrimination among African-American adolescents. *Cultural Diversity and Ethnic Minority Psychology, 20*(2), 156–165. https://doi.org/10.1037/a0035322

Seaton, E. K., Neblett, E. W., Cole, D. J., & Prinstein, M. J. (2013). Perceived discrimination and peer victimization in African American and Latino youth. *Journal of Youth and Adolescence, 42*(3), 342–350. https://doi.org/10.1007/s10964-012-9848-6

Sellers, R. M., Smith, M. A., Shelton, J. N., Rowley, S. A. J., & Chavous, T. M. (1998). Multidimensional model of racial identity: A reconceptualization of African American racial identity. *Personality and Social Psychology Review, 2*(3), 18–39. https://doi.org/10.1207/s15327957pspr0201_2

Soria, K. M., & Stebleton, M. J. (2012). First-generation students' academic engagement and retention. *Teaching in Higher Education, 17*(6), 673–685. https://doi.org/10.1080/13562517.2012.666735

Tajfel, H., & Turner, J. C. (1986). The social identity theory of intergroup behavior. In S. Worchel & W. G. Austin (Eds.), *Psychology of intergroup relations* (pp. 7–24). Nelson Hall. https://doi.org/10.4324/9780203505984-16

Treynor, W., Gonzalez, R., & Nolen-Hoeksema, S. (2003). Rumination reconsidered: A psychometric analysis. *Cognitive Therapy and Research, 27*, 247–259. https://doi.org/10.1023/A:1023910315561

U.S. Census Bureau. (2016). *Facts for features: Hispanic heritage month 2016 (report # CB16-FF.16)*. Author. https://www.census.gov/content/dam/Census/newsroom/facts-for-features/2016/cb16-ff16.pdf

Umaña-Taylor, A. J. (2011). Ethnic identity. In S. J. Schwartz, K. Luyckx, & V. L. Vignoles (Eds.), *Handbook of identity theory and research* (pp. 791–809). Springer.

Umaña-Taylor, A. J., & Updegraff, K. A. (2007). Latino adolescents' mental health: Exploring the interrelations among discrimination, ethnic identity, cultural orientation, self-esteem, and depressive symptoms. *Journal of Adolescence, 30*(4), 549–567. https://doi.org/10.1016/j.adolescence.2006.08.002

Vazquez-Salgado, Y., Greenfield, P. M., & Burgos-Cienfuegos, R. (2015). Exploring home-school value conflicts: Implications of academic achievement and well-being among Latino first-generation college students. *Journal of Adolescent Research, 30*(3), 271–305. https://doi.org/10.1177/0743558414561297

Vega, D. (2016). "Why not me?": College enrollment and persistence in high-achieving first-generation Latino college students. *School Psychology Forum, 10*(3), 307–320.

Wei, M., Russell, D. W., Malinckrodt, B., & Vogel, D. L. (2007). The Experiences in Close Relationship Scale (ECR)-Short Form: Reliability, validity, and factor structure. *Journal of Personality Assessment, 88*(2), 187–204. https://doi.org/10.1080/00223890701268041

Williams, D. R., Neighbors, H. W., & Jackson, J. S. (2003). Racial/ethnic discrimination and health: Findings from community studies. *American Journal of Public Health, 93*(2), 200–208. https://doi.org/10.2105/ajph.93.2.200

Zeman, J., Shipman, K., & Penza-Clyve, S. (2001). Development and initial validation of the children's sadness management scale. *Journal of Nonverbal Behavior, 25*, 187–205. https://doi.org/10.1023/A:1010623226626

Zeman, J., Shipman, K., & Suveg, C. (2002). Anger and sadness regulation: Predictions to internalizing and externalizing symptoms in children. *Journal of Clinical Child and Adolescent Psychology, 31*(3), 393–398. https://doi.org/10.1207/s15374424jccp3103_11

Chapter 15
Parent-Adolescent Dissimilarity in Ethnic-Racial Identity Exploration: Implications for Adolescent Family Cohesion and Depressive Symptoms

Jessica P. Montoro, Stephanie Miller-Tejada, Fernanda L. Cross, and Deborah Rivas-Drake

Ethnic-racial identity formation is a normative developmental task influenced by our social contexts (Phinney, 1990). Primarily established during adolescence, ethnic-racial identity formation consists of the exploration and development of strong, clear feelings of what it means to be a part of one's ethnic-racial group (Umaña-Taylor et al., 2014). Multiple systematic reviews of the literature report that, among Latinx youth, *ethnic-racial identity exploration*—engagement in activities to learn about the heritage culture—has been linked to positive mental health (e.g., Rivas-Drake et al., 2014a, 2014b; Smith & Silva, 2011). Exploration for immigrants often requires navigating unfamiliar social terrain and acculturation to a foreign host culture (Berry, 1997; Portes & Rumbaut, 1990). Ultimately, ethnic-racial identity exploration may involve an identity *reformation* process as immigrants reevaluate what it means to be a part of their ethnic-racial group within their new context (Suárez-Orozco, 2001). Previous categorizations of who is in their "ingroup" and "outgroup" may fluctuate as they adapt to the social stratification of their host culture, generating reexaminations of one's own identity (Schwartz et al., 2006). Solidifying their ethnic-racial identity can serve as an anchor through the acculturation process, as they restructure their understanding of who they are, reexamine their commitments and obligations to others in their ethnic-racial group, and find a balance in maintaining values and behaviors from their heritage culture while adopting other values and behaviors from the host culture (Schwartz et al., 2006).

Among immigrant families, the pace at which ethnic-racial identity development unfolds for both parents and adolescents could be the result of an acculturation discordance. As immigrant-origin youth are beginning to form their ethnic-racial identities in the host culture, their parents are reforming their own. Compared with their parents,

J. P. Montoro (✉) · S. Miller-Tejada · F. L. Cross · D. Rivas-Drake
University of Michigan, Ann Arbor, MI, USA
e-mail: jmontoro@umich.edu; stepmm@umich.edu; flcross@umich.edu; drivas@umich.edu

D. J. Johnson et al. (eds.), *Re/Formation and Identity*, Advances in Immigrant Family Research, https://doi.org/10.1007/978-3-030-86426-2_15

children of immigrants are often more immersed in the host culture through their interactions at school and with peers, prompting their exposure to and adoption of host culture values and lifestyle preferences (Lau et al., 2005). Further, immigrant-origin adolescents are entering their key identity formation period while engaging in a different environment than their immigrant parents, who reached maturity in their heritage culture (Telzer, 2010). This may create tension between immigrant-origin youth and their parents, who spend less time engaging in the host culture as they often work multiple jobs in settings that are populated by fellow immigrants (Suárez-Orozco, 2001). Differences in context and developmental stages filter the lens through which the immigrant parent and youth acculturate to the host culture and re/form their ethnic-racial identities, setting the stage for potential cultural clashes as the families navigate two worlds (García Coll & Magnuson, 1997).

Although many scholars have investigated the implications of acculturation gaps between immigrant parents and their children, the bulk of this literature has examined differences in the domains of values, behaviors, and belief systems (Lau et al., 2005; Telzer, 2010). Less is known about the familial functioning implications and well-being outcomes of immigrant families who are dissimilar in the domain of identity re/formation. In this chapter, we contribute to the literature by examining ethnic-racial identity exploration dissimilarity among Latinx parent-adolescent dyads residing in the midwestern United States. Specifically, we examined the longitudinal associations between parent-youth dissimilarity in ethnic-racial identity exploration, adolescents' perceptions of family cohesion, and adolescent depressive symptoms. Following suggestions raised by Telzer (2010), we examined these relationships through three different methods of computing parent-youth dissonance: (1) the interaction approach, (2) a difference score, and (3) a gap value.

Theoretical Framework

To examine how ethnic-racial identity dissimilarity within Latinx parent-adolescent dyads and their family cohesion relate to the youth's depressive symptoms, we draw on the integrative model of ethnic minority child development (García Coll et al., 1996) and the acculturation gap-distress model (Szapocznik & Kurtines, 1993). The integrative model proposes links between social position factors (e.g., ethnicity, race), family factors (e.g., family cohesion), and youth's mental health outcomes. In addition, the model acknowledges how social stratification informs the interactions between the child and their environment and, in turn, helps or hinders their development (García Coll et al., 1996). Among the contextual factors, the model accounts for *adaptive culture* and family dynamics that influence developmental trajectories. Adaptive culture is defined as the values, goals, and attitudes that differ from those of the dominant culture and includes the acculturation process (García Coll et al., 1996; Perez-Brena et al., 2018). Importantly, the integrative model acknowledges the ever-changing social environment that influences ethnic-racial identity development and, in turn, youth's psychological well-being (Perez-Brena et al., 2018).

To frame the identity dissimilarity context more specifically, we draw on the acculturation gap-distress model (Szapocznik & Kurtines, 1993). This model postulates that immigrant children acculturate to their new context more quickly than their parents, creating an acculturation dissonance that adds to the stress of normative intergenerational conflict between parents and adolescents. The model also posits that acculturative stress leads to strained family functioning and negative psychosocial outcomes for youth. Importantly, the acculturative gap-distress model's emphasis on parent-youth acculturation differences and family functioning complements the integrative model's link between ethnicity, familial factors, and youth mental health.

Parent-Adolescent Dissimilarity in Ethnic-Racial Identity Exploration

When immigrants arrive to the host country, they undergo a process of acculturation as they adapt to the new culture, shifting their beliefs, values, and behaviors to varying degrees (Berry, 1997). Ethnic-racial identity is one acculturation domain that focuses on the subjective sense of belonging to a group or culture, as well as the processes by which these attitudes develop (Phinney, 1990; Umaña-Taylor et al., 2014). Ethnic-racial identity exploration is a key component of the ethnic-racial identity process that immigrants experience as they navigate being a part of their ethnic-racial group in the new host context. This involves behaviors that promote learning about one's heritage culture and includes activities such as reading about one's culture, attending cultural events, or speaking to others about their ethnic-racial group (Phinney, 1990). Studies have demonstrated that experiences of prejudice and discrimination often serve as a catalyst for heritage and identity exploration (Rivas-Drake et al., 2014a, 2014b). As Latinx immigrants come into contact with social institutions in the United States, they may experience anti-immigrant and anti-Latinx rhetoric in a hostile sociopolitical climate (Roehling et al., 2010), which may spark an introspection of how one relates to their ethnic-racial identity.

In addition, as individuals grapple with their cultural identities between their host and home contexts, they may also experience disconnections among family members who are also experiencing acculturation at potentially different rates (Buckingham & Brodsky, 2015). For instance, whereas Latinx immigrant parents reached maturity in their home culture, their children are simultaneously developing their values, beliefs, and identities in the host culture and maintaining ties to the heritage culture (Telzer, 2010). Latinx youth are also gaining more autonomy from their parents as they transition into adolescence, learning daily about the host culture in the midst of an ethnic-racial identity exploration period that involves not only heritage culture messages but also those of the host culture. Meanwhile, some immigrant parents may hesitate for their child to participate in cultural attitudes and behaviors embedded in American youth culture (Bacallao & Smokowski, 2007). Some immigrant parents may consider the host culture's values and behaviors as anti-education, anti-family, and glorifying of violence—in contrast with their heritage culture values (Suárez-Orozco, 2001). More individualistic cultural orientations promoted in the host culture may be in direct opposition to highly endorsed Latinx cultural values such as *familismo* that center around the family's rather than the individual's well-being (Stein et al., 2014). Furthermore, this potential misalignment in acculturation experiences within immigrant-origin families is thought to be detrimental to youth and family well-being.

Acculturation Gap-Distress and Youth and Family Well-Being: A Mixed Picture

Extant research has examined the implications of parent-youth differences in navigating and maintaining connections to their culture (Lau et al., 2005; Portes & Rumbaut, 1990; Telzer, 2010), and a good deal of this work supports the idea that acculturation gaps are less optimal for youth and familial well-being. For example, cultural discrepancies between Latinx parents and adolescents have been associated with more substance abuse (Martinez, 2006; Unger et al., 2009) and increased depressive symptoms (Huq et al., 2016; Hwang & Wood, 2009). In addition, cultural dissonance within parent-youth relationships has been linked to family discord in Latinx families (Gonzales et al., 2006; Sciarra & Ponterotto, 1991; Smokowski et al., 2008). In a study of Latinx adolescents, cultural orientation differences between father-child dyads were related to more father-child conflict in families with lower father-child relationship quality (Schofield, et al., 2008). Findings from a separate study showed parent-adolescent acculturation dissonance to be negatively associated with family cohesion among Latinx families (Smokowski et al., 2008). Intergenerational differences surrounding language preference, ethnic and racial makeup of peer networks, and endorsement of host culture values may contribute to greater family conflict among Latinx families (Sciarra & Ponterotto, 1991).

Moreover, some research suggests that parent-adolescent acculturation discordance may be linked with youth outcomes via its impact on family functioning. For example, among families of Mexican background, discordance in mother-adolescent language preference was linked to increased family conflict, thus diminishing family cohesion and amplifying youth internalizing and externalizing behaviors (Gonzales et al., 2006). Nair et al. (2018) found that acculturative family distancing was related to more parent-youth conflict and less family cohesion among Latinx adolescents; greater family conflict was also associated with more alcohol intake among adolescents.

Yet, although much research indicates that less acculturation toward one's heritage culture is detrimental to family functioning and youth well-being, others report conflicting or null findings. For example, Lau et al. (2005) found that among Mexican-American families, acculturation discrepancies were not associated with more adolescent conflict or conduct issues. Rather, youth who were more acculturated toward their *heritage* culture than their parents also reported more conduct issues (Lau et al., 2005). These findings are in direct contrast to the acculturation gap-distress model, which posits that youth who acculturate to the host culture more rapidly than their parents experience more distress (Szapocznik & Kurtines, 1993). Telzer and colleagues (2016) found no associations between Mexican-American adolescent-parent acculturation gaps and negative adolescent psychological outcomes. Instead, they found that independent of parental acculturation levels, adolescents who showed more Mexican behavioral practices, higher Spanish proficiency, and greater family obligation exhibited more positive outcomes, including more family support as well as fewer internalizing and externalizing behaviors (Telzer et al., 2016). Similarly, other studies have found no relation between acculturative discordance and parent-adolescent relationship quality, youth externalizing behaviors, or family conflict (Nieri et al., 2016; Pasch et al., 2006).

In sum, the inconclusive findings regarding parent-adolescent acculturative gaps illustrate the ever-changing, complicated nature of acculturation and its impact on familial dynamics and youth adjustment (Telzer, 2010). Contextual and developmental factors involved in acculturation must also be acknowledged when investigating how ethnic-racial identity differences may or may not disrupt familial processes and youth well-being (Quintana, 2007; Telzer, 2010). Although previous work underscores the mediating role of both positive and negative familial factors, such as family cohesion or family conflict (Miranda et al., 2006), few scholars have empirically examined the ways in which parent-youth dissimilarity in ethnic-racial identity may influence youth mental health via familial dynamics (Lau et al., 2005; Telzer, 2010).

Identity Dissimilarity and Youth Mental Health: Family Functioning as a Mediator

Examining ethnic-racial identity dissimilarity is a novel way to consider acculturation among parents and youth, as the majority of empirical research has focused on acculturation gap dissonance in language use or cultural values (Oh & Fuligni, 2010; Telzer, 2010). In a review of over 20 studies that explored the acculturation gap-distress model, family conflict, and youth well-being, only seven looked at identity as a cultural domain of acculturation (Telzer, 2010). Of these seven, only one study (Lau et al., 2005) assessed ethnic-racial identity among Latinx families. Lau et al. (2005) examined parent-youth acculturation differences in ethnic-racial identity among Mexican-origin families but only assessed whether ethnicity reflected an affinity toward American or their heritage culture. They found no relations among acculturation discrepancies and youth behavioral problems as well as perceived conflict (Lau et al., 2005). In our review of the literature, we did not find a study that explicitly looked at parent-youth dissimilarity in ethnic-racial identity exploration, in particular. Consequently, it is unclear whether the pathways between such dissimilarity and youth mental health would manifest as they do for other acculturation processes. Given the acculturative gap-distress model's focus on family harmony, parent-youth ethnic-racial identity dissimilarity may be an important contributor to family functioning and youth mental health to explore.

The Current Study

The current longitudinal study was guided by two research objectives. First, we examined whether parent-adolescent dissimilarity in ethnic-racial identity exploration was related to adolescent depressive symptoms concurrently and longitudinally. Second, we examined whether family cohesion helped explain the relation between parent-adolescent ERI exploration discordance and depressive symptoms. Drawing from previous empirical research and theory (García Coll et al., 1996; Szapocznik & Kurtines, 1993), we hypothesized that having greater ethnic-racial identity exploration dissimilarity would lead to higher depressive symptoms among adolescents. Given that we were capturing parent-youth exploration dissimilarity with three different approaches, we developed hypotheses specific to each approach. Specifically, for the tests of interaction approach: We expected that *similarity*, where both parent and adolescent were high in exploration, would result in the most optimal outcomes, that is, reporting the greatest levels of family cohesion and fewest depressive symptoms. We also expected that a *dissimilarity* where the parent was high and the adolescent was low in exploration would result in the least optimal outcomes, with the lowest family cohesion and the most depressive symptoms. In terms of the difference score approach: We expected that any parent-youth dissimilarity (i.e., difference) in ethnic-racial identity exploration, regardless of the direction of the difference, would be associated with lower family cohesion and more youth depressive symptoms; we hypothesized that family cohesion would account for the link between dissimilarity and depressive symptoms. Finally, with regard to the gap score approach: We expected that dyads exhibiting an identity exploration gap where parents' levels were higher than adolescents' would exhibit less family cohesion and more adolescent depressive symptoms; as with the difference score approach, we expected that family cohesion would mediate the association of having such an identity exploration gap with youths' depressive symptoms. By examining how ethnic-racial identity development dissimilarity among Latinx families may influence family cohesion, this study sheds light on a novel aspect of acculturation discordance and its implications for the family ecology and, in turn, youth mental health.

Method

Participants

Data were drawn from Waves 1 and 2 of a longitudinal study of Latinx families living in the midwestern region of the United States. One hundred and forty-eight parent-adolescent dyads participated at Wave 1. Among the parents, 78.7% were mothers, with ages ranging between 33 and 56 years of age (M = 39.86 years, SD = 5.47). The large majority, 90.5%, were born abroad and 63% reported being of Mexican origin. The remaining came from several other Latin American countries (e.g., Dominican Republic, El Salvador, Guatemala, Perú, Honduras, Argentina). Of those born abroad, their average length of U.S. residence was 17.21 years (SD = 6.16). Sixty-five percent of parents reported household incomes below $30,000, 26% between $30,000 and $60,000, and 9% above $60,000. Forty-five percent of parents reported not completing high school, 31% obtained a high school diploma, and 24% reported schooling beyond high school. The overwhelming majority (84.5%) of parents reported that their child was eligible to receive free or reduced-fee lunch at school. Seventy-six percent of parents were married, and 9.9% reported not being married, but living together with a partner.

At Wave 1, adolescents were 13 (45.9%) and 14 (54.1%) years of age (M = 13.54 years; SD = 0.5), with 59.5% in the eighth and 41.5% in the ninth grade. Eighty-two percent of the adolescents were born in the United States. One hundred and forty adolescents were surveyed again about one year later. At Wave 2, adolescents were 13 (0.7%), 14 (41.7%), 15 (51.3%), and 16 (2.2%) years of age, and the remaining did not report their age (M = 14.59 years, SD = 0.55). Fifty-eight percent were in 9th grade, 42% in 10th, and the rest did not report a grade. Fifty-four percent of the adolescents were females, and 46% were male.

Procedure

To be eligible for participating in the study, both the adolescents and parents had to self-identify as Latinx or Hispanic. In addition, the adolescents had to be either 13 or 14 years of age and enrolled in the eighth or ninth grade. Flyers in English and Spanish were distributed throughout southeast Michigan, in areas of heavy Latinx presence, such as in community centers, churches, international grocery stores, recreational areas, and youth centers. The flyers included information regarding the purpose of the study and phone numbers for those interested to contact the research lab. Recruiting was also supported through the use of a community liaison who is fully fluent in Spanish and had years of experience in community outreach in the region. A brief pre-screening protocol was administered in person, through telephone, or email to determine interested participants' eligibility for the study.

Once deemed eligible, an appointment for completion of informed consent and survey was scheduled at a location of the parents' choosing to protect confidentiality and maximize participants' comfort. Informed consent was obtained by parents for their own participation as well as the participation of their child. Informed assent was obtained from adolescents. Ninety percent of parents completed the survey in Spanish and all adolescents completed their survey in English. Data from parents who preferred to complete the survey in Spanish were collected by the community liaison or Spanish-speaking research assistants. The survey completion took approximately 45 minutes. Wave 1 (W1) data collection happened between 2014 and 2015, and Wave 2 (W2) between 2015 and 2016. Parents participated only in the first wave, whereas the adolescents participated in both. Parents received a $30 gift card for their participation. The adolescents were compensated with a $20 gift card at W1 and a $25 gift card at W2.

Measures

Parent and Adolescent Ethnic-Racial Identity Exploration. Parents and adolescents reported on their ethnic-racial identity exploration (7 items; W1 parent $\alpha = 0.77$; W1 adolescent $\alpha = 0.77$) using the Ethnic Identity Scale (Umaña-Taylor et al., 2004). A sample item is "I have participated in activities that have taught me about my ethnicity." Responses were measured on a 4-point scale, from 1 (*does not describe me at all*) to 4 (*describes me very well*), and means were calculated where a higher value reflected greater levels of exploration.

Family Cohesion. The Family Cohesion Scale (Alegría & Takeuchi, 2007) was employed to assess adolescents' feelings of family closeness, togetherness, time spent together, and communication (10 items; W1 $\alpha = 0.93$; W2 $\alpha = 0.92$). A sample item is "Family members like to spend free time with each other." Response options ranged from 1 (*Strongly disagree*) to 5 (*Strongly agree*) with higher values indicating greater family cohesion.

Depressive Symptoms. Adolescents reported on their levels of depressive symptoms (20 items; W1 $\alpha = 0.94$; W2 $\alpha = 0.95$) with the Center for Epidemiological Studies-Depression Scale (CES-D; Eaton et al., 2004) by rating how often they had felt a list of 20 feelings and behaviors. A sample item is "I could not shake off the blues." Responses ranged from 1 (*not at all or less than once a day*) to 5 (*nearly every day for two weeks*), with higher values indicating greater frequency of depressive symptoms.

Covariates

Adolescent self-reported gender at Wave 1 was coded as 1 (male) and 2 (female). Parents reported information on adolescent eligibility for free and reduced-fee lunch was coded as 0 (full price) and 1 (free or reduced-fee).

Analytic Plan

Concurrent and longitudinal associations between parent-adolescent ethnic-racial identity exploration dissimilarity, adolescent-reported family cohesion, and depressive symptoms were examined in the same model. Our first aim was to examine whether the dissimilarity in ethnic-racial identity exploration between parent-adolescent dyads was related to youth's depressive symptoms. Second, we were interested in the role of family cohesion as a mediator in the relationship between ethnic-racial identity discordance and youth's depressive symptoms. Family cohesion levels at W1 and W2 were included as mediators of the associations between dissimilarity and adolescent depressive symptoms at W1 and W2. Adolescent gender and enrollment in free or reduced-fee school lunch at W1 were included as covariates in all models. Mplus 7.2 (Muthén & Muthén, 2012) was used for analyses. Recommended thresholds for model fit were followed, including a CFI and TLI of >0.90 and an RMSEA of <0.05 with a 90% confidence interval <0.08 and a standardized root mean square residual (SRMR) of <0.08 (Hu & Bentler, 1999; Kline, 2015). We examined bias-corrected bootstrap 95% confidence intervals (CIs) to assess the significance of indirect associations; the indirect effect is significant when these CIs do not contain zero.

To examine ethnic-racial identity dissimilarity, three approaches were taken: (a) an interaction term, (b) a difference score, and (c) a gap value. Each is described in greater detail below. In addition, it should be noted that each dissimilarity approach was examined in a separate model.

Interaction of Parent and Adolescent Ethnic-Racial Identity Exploration

An interaction of both parent and adolescent ethnic-racial identity exploration at W1 was computed by calculating the product of the two. This allowed for examination of specific types of similarity and dissimilarity scenarios between parent and youth (Telzer, 2010). This method of examining parent-youth dissimilarity allowed us to identify whether parent ethnic-racial identity exploration, adolescent ethnic-racial identity exploration, or the interaction of the two predicted family cohesion and, in turn, adolescent depressive symptoms. A separate strength of creating an interaction is that, unlike a difference score, it accounts for mean levels of parent and youth's acculturation levels (Laird & De Los Reyes, 2013; Telzer et al., 2016). The interaction also indicates whether one type of exploration gap or a general gap is associated with family cohesion and, in turn, youth depressive symptoms (Telzer, 2010). When contrasted with a difference score, the interaction approach captures both the unique effects of parent and adolescents' ethnic-racial identity exploration, and the effect of the interaction, as well as the direction of the ethnic-racial identity dissimilarity effect (Birman, 2006).

Parent-Adolescent Exploration Difference Score

A difference score between parents' and adolescents' exploration score was computed by subtracting the adolescent ethnic-racial identity exploration score from the parent ethnic-racial identity exploration score. We then took the absolute value of that difference score, such that a larger value, regardless of valence, was indicative of a larger discrepancy between their levels of exploration. The difference score method is the most commonly used approach to calculating differences in parent-youth informant scores (Hou et al., 2020; Telzer, 2010). An advantage of taking the absolute value of the difference score is that it indicates the magnitude or "distance" of the dissimilarity between parent and adolescent ethnic-racial identity levels (Atzaba-Poria & Pike, 2007; Telzer, 2010). However, scholars have distinguished limitations of this approach, including measurement and variance differences between parent and adolescent scores that complicate interpretation of the difference score (Hou et al., 2020). Despite the limitations of this method, we thought it important to identify whether *any* "distance" between parent and adolescent ethnic-racial identity exploration influenced family cohesion and adolescent mental health outcomes.

Parent-Adolescent Exploration Gap Value

Per Stein and Polo's (2014) recommendation, we computed a dummy code using the ethnic-racial identity exploration difference score to create a gap value. A score of 0 indicated the parent's ethnic-racial identity exploration score at W1 was 0.5 standard deviations less than or equal to that of the adolescent's score at W1. A score of 1 indicated the parent's ethnic-racial identity exploration score was 0.5 standard deviations greater than that of the adolescent's score. To capture the most commonly explored acculturation gap type, the gap variable allowed us to examine the effect of a dissimilarity type in which the parent is more connected to their heritage culture (as measured by a higher ethnic-racial identity exploration) than the youth. This type of discordance is the most commonly explored gap in the acculturation gap-distress model literature, and it is the gap most often linked to lower family cohesion (Telzer, 2010). Unlike the difference score, the gap thus allows us to identify whether a specific type of dissimilarity rather than *any* difference (regardless of direction) in parent and adolescent exploration contributes to family cohesion and youth depressive symptoms.

Results

Preliminary Analyses

Means, standard deviations, and bivariate correlations for all variables of interest are presented in Table 15.1. Adolescents reported experiencing depressive symptoms on average 1 to 2 days or less within the last 2 weeks. Adolescent ethnic-racial

Table 15.1 Descriptive statistics and correlations for study variables

Variable	*M*	*SD*	1	2	3	4	5	6
1. W1 parent ERI exploration	3.05	0.72		0.03	−0.01	−0.08	−0.04	−0.01
2. W1 adolescent ERI exploration	2.75	0.64			0.34***	0.23**	−0.17*	−0.12
3. W1 family cohesion	40.31	7.27				0.39***	−0.29***	−0.08
4. W2 family cohesion	40.88	6.68					−0.21*	−0.31***
5. W1 depressive symptoms	1.67	0.71						0.44*
6. W2 depressive symptoms	1.81	0.75						

Note. ERI Ethnic-racial identity
*$p < 0.05$, **$p < 0.01$, ***$p < 0.001$

identity exploration was correlated with family cohesion at both waves and with W1 depressive symptoms. In addition, family cohesion at W1 and W2 was negatively correlated with depressive symptoms at W1, respectively. Finally, family cohesion at W2 was negatively correlated with depressive symptoms at W2.

Primary Analyses

Interaction Approach. The first model included W1 parent ethnic-racial identity, W1 adolescent ethnic-racial identity, and their interaction as predictors, following guidelines proposed by Telzer (2010) to identify the unique effects of both parent and adolescent exploration. This model fits the data well $\chi^2(11) = 15.17, p = 0.18$ (CFI = 0.95, TLI = 0.89, RMSEA = 0.05, 90% CI [0.00, 0.11], SRMR = 0.05). There were no significant effects of the interaction term on adolescent depressive symptoms or family cohesion. Additionally, neither parent nor adolescent exploration score directly influenced family cohesion on its own. Next, in order to assess improvement in model fit, we removed the interaction and ran the same model examining only the main effects. A chi-square difference test was not significant, χ^2 (11, $N = 146$) = 15.17, $p = 0.95$, indicating that removing the interaction did not improve model fit (CFI = 0.95, RMSEA = 0.06, 90% CI [0.00, 0.12], SRMR = 0.05). Thus, the models with and without the interaction term fit the data equally well, and the more parsimonious main effects only model could be preferred.

Difference Score Approach. This model included the ethnic-racial identity exploration difference score (absolute value) as a predictor to assess whether *any* parent-adolescent difference in exploration influenced family cohesion and youth depressive symptoms. Path analysis results for the parent-adolescent exploration difference indicated that the model fits the data somewhat well $\chi^2(5) = 7.93$, $p = 0.16$ (CFI = 0.96, TLI = 0.87, RMSEA = 0.06, 90% CI [0.00, 0.14], SRMR = 0.05). There was no direct effect of parent-adolescent exploration difference on adolescent depressive symptoms at either wave. However, there was a direct negative association of parent-adolescent exploration difference on W1 family cohesion, such that a greater ethnic-racial identity exploration parent-adolescent difference (irrespective of valence) was linked to less adolescent-reported family cohesion at W1.

In addition, W1 family cohesion was positively associated with W2 family cohesion and negatively associated with W1 depressive symptoms. There was also a direct negative association of W2 family cohesion on W2 depressive symptoms, and a positive association of W1 depressive symptoms with W2 depressive symptoms. In addition, there was a significant indirect effect of W1 exploration difference on W1 adolescent depressive symptoms through W1 family cohesion, $b = 0.06$, 95% CI [0.01, 0.16]. This indirect association of W1 parent-adolescent exploration difference on W2 adolescent depressive symptoms via family cohesion at both time points was also present, $b = 0.02$, 95% CI [0.01, 0.08]. This suggests that when there is any parent-adolescent dissimilarity in ethnic-racial identity exploration, sustained family cohesion may have a protective role for adolescents' depressive symptoms. Full model results are provided in Table 15.2.

Gap Value Approach. The final approach considered the particular type of dissimilarity in which parents' exploration levels were greater than those of their adolescents, as indicated by a gap value. Compared to both the interaction and difference score approach, this model fits the data best $\chi^2(5) = 6.44$, $p = 0.27$ (CFI = 0.98, TLI = 0.93 RMSEA = 0.04, 90% CI [0.00, 0.13], SRMR = 0.05). Ethnic-racial identity exploration gap was not directly related to adolescent depressive symptoms at either wave. However, an ethnic-racial identity exploration gap was negatively associated with family cohesion at Wave 1. Family cohesion at Wave 1 was negatively associated with adolescent depressive symptoms concurrently, but not longitudinally. In addition, family cohesion at W2 was negatively associated with W2 adolescent depressive symptoms and positively associated with W1 cohesion. Further, adolescent depressive symptoms at W1 was positively associated with W2 adolescent depressive symptoms. Lastly, the indirect association of the W1 gap on W1 adolescent depressive symptoms via family cohesion was significant, $b = 0.08$,

Table 15.2 Path model for ERI exploration difference score, family cohesion, and depressive symptoms

Paths	*b* (*SE*)	*b*/*SE*	β (*SE*)
W1 gender → W1 depressive symptoms	0.31 (0.12)	2.67**	0.22 (0.07)
W1 free/reduced lunch → W1 depressive symptoms	0.14 (0.15)	0.92	0.07 (0.07)
W1 exploration difference → W1 depressive symptoms	0.05 (0.10)	0.53	0.04 (0.08)
W1 family cohesion → W1 depressive symptoms	−0.03 (0.01)	−2.51**	−0.30 (0.11)
W1 gender → W2 depressive symptoms	−0.01 (0.11)	−0.02	−0.01 (0.08)
W1 free/reduced lunch → W2 depressive symptoms	−0.09 (0.15)	−0.63	−0.05 (0.07)
W1 exploration difference → W2 depressive symptoms	0.05 (0.10)	0.44	0.03 (0.08)
W1 family cohesion → W2 depressive symptoms	0.02 (0.01)	1.65	0.16 (0.08)
W2 family cohesion → W2 depressive symptoms	−0.03 (0.01)	−2.49*	−0.28 (0.09)
W1 depressive symptoms → W2 depressive symptoms	0.43 (0.10)	4.39***	0.42 (0.09)
W1 exploration difference → W1 family cohesion	−2.05 (1.01)	−2.02*	−0.16 (0.08)
W1 exploration difference → W2 family cohesion	0.58 (0.94)	0.62	0.05 (0.08)
W1 family cohesion → W2 family cohesion	0.36 (0.10)	3.52***	0.40 (0.09)

Note. ERI Ethnic-racial identity
*$p < 0.05$, **$p < 0.01$, ***$p < 0.001$

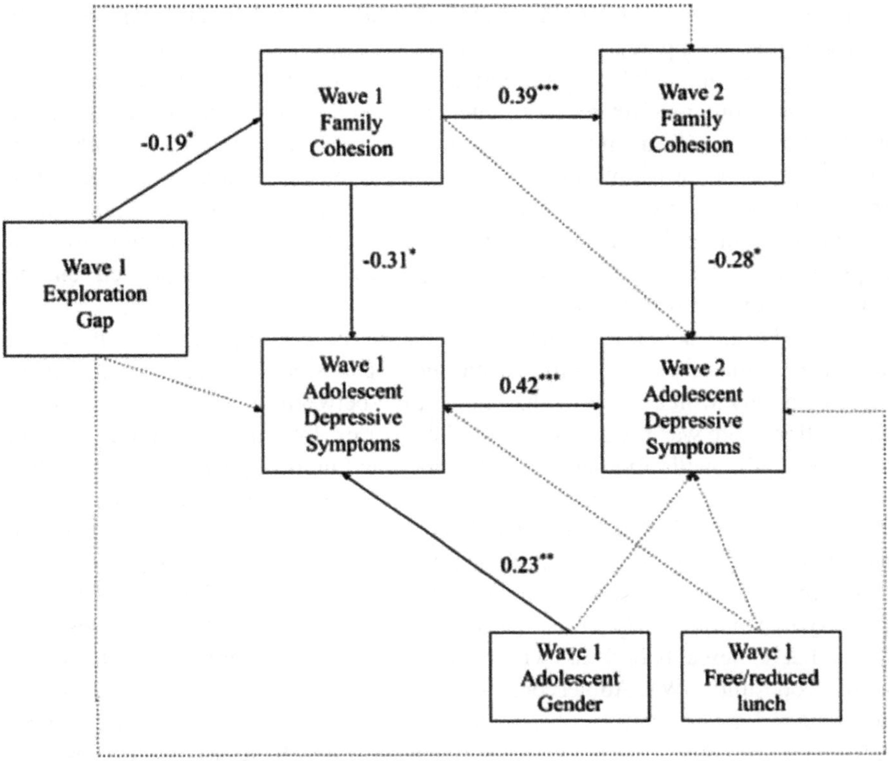

Fig. 15.1 *Note.* Standardized coefficients for path analysis ($n = 146$) linking ethnic-racial identity exploration gap (1 = parent > adolescent, 0 = other) with adolescent-reported family cohesion and adolescent depression at Waves 1 and 2. *Dashed lines* represent nonsignificant paths. $^{*}p < 0.05$, $^{**}p < 0.01$, $^{***}p < 0.001$

95% CI [0.02, 0.22]. This association was also present longitudinally across both waves through W1 and W2 family cohesion $b = 0.03$, 95% CI [0.01, 0.09], demonstrating the beneficial effect of sustained family cohesion for adolescent mental health. See Fig. 15.1 for full model.

Discussion

Our study provides empirical insight into how ethnic-racial identity exploration dissimilarity among Latinx parent-adolescent dyads may impact youths' outcomes. We had two research objectives. First, we aimed to examine the concurrent and longitudinal associations between parent-adolescent discordance in ethnic-racial identity exploration and youth depressive symptoms. The results of our study supported the hypothesis that parent-adolescent exploration dissimilarity (as assessed by a difference score and a gap value) would be indirectly related to more depressive symptoms in youth. With both the difference score approach and the gap approach, we found that when parents and adolescents had greater differences in ethnic-racial identity exploration, adolescents reported lower family cohesion and, in turn, more depressive symptoms. However, our interaction model did not fit the data well, and we found no significant interaction effect of parent and adolescent exploration. Therefore, we were unable to probe the interaction to examine which gap type resulted in most or least optimal outcomes for youths' well-being. Thus, the extent to which parent-adolescent dissimilarity in ethnic-racial identity exploration had implications for family cohesion and depressive symptoms depended on the approach taken to operationalize dissimilarity.

When the interaction approach was used, we did not find an effect of parent-adolescent similarity or dissimilarity on family cohesion or on adolescent depressive symptoms, which suggests that the relation between family cohesion and one exploration score was not dependent on the other exploration score. Findings from this model align with prior studies that found acculturation gaps in Latinx families were not associated with greater family conflict (Lau et al., 2005; Pasch et al., 2006; Telzer et al., 2016). Instead, we found that the magnitude of the difference between parent and adolescent exploration negatively influenced family cohesion.

For the ethnic-racial identity exploration difference, we found no direct effect of parent-youth dissimilarity on adolescent depressive symptoms. Nonetheless, we found that adolescents in dyads with a larger exploration difference reported lower family cohesion at W1. Adolescents who reported greater family cohesion at W1 also reported greater family cohesion at W2. At both time points, adolescents who reported more family cohesion also experienced fewer depressive symptoms. The same pattern of findings was found when the exploration dissimilarity was examined with the exploration gap approach. Stated differently, when the parent engaged in more ethnic-racial identity exploration than the adolescent, adolescents also reported lower family cohesion and, in turn, more depressive symptoms. Findings from both the difference and gap models were consistent with prior research on the links between parent-youth acculturation dissonance, familial factors such as family cohesion, and youth mental health (Nair et al., 2018; Schofield et al., 2008). Furthermore, of all three approaches, the gap score showed the best model fit and captured the specific type of exploration dissimilarity between parent and youth when compared to the difference and interaction approach.

Our findings indicate that differences in how immigrant parents and U.S.-born youth engage with their culture may be particularly detrimental to family cohesion. The vast majority of our parent sample were immigrants, whereas the majority of our adolescent sample were U.S.-born. Immigrant parents may have learned about their heritage culture in their home countries and may be currently engaging less in exploring their heritage culture in adulthood. Additionally, immigrant Latinx parents may be more exposed to discriminatory and stigmatizing experiences than their U.S.-born children and may be less interested in exploring their culture in the host country. Torres et al. (2011) found that Latinx adults who engaged in more ethnic-racial identity exploration also experienced more distress when discriminated against in work, academic, or public settings. Contrastingly, youth may be seeking out more knowledge about their heritage culture in adolescence as a time of heightened identity exploration (Erikson, 1968). Additionally, some items from the ethnic-racial identity scale may be more salient exploration strategies for adolescents than for parents. For example, adolescents may be more inclined than their parents to learn about their heritage culture through media exposure or from the internet (Te'Neil Lloyd, 2002). If parents and adolescents are connecting to their cultures in distinct ways, there may be less opportunities to promote parent-youth communication. For ethnic-racial minority adolescents, decreased parent-child communication has been shown to hinder family cohesion and may be particularly damaging to adolescent mental health (Ackard et al., 2006).

Second, we were interested in examining whether family cohesion accounted for the relation between ethnic-racial identity exploration and youth depressive symptoms. As expected, we found two indirect effects of exploration dissimilarity on depressive symptoms. In both the difference and gap models, we found indirect effects of parent-youth exploration dissimilarity on adolescent depressive symptoms through family cohesion. Through family cohesion at W1, ethnic-racial identity exploration difference was related to more adolescent depressive symptoms at W1. Exploration difference was also related to increased depressive symptoms at W2 via family cohesion and both W1 and W2. This is consistent with prior studies that also found mediating effects of familial factors in the associations between parent-youth acculturation dissonance and youth psychological well-being (Gonzales et al., 2006; Miranda et al., 2006). The family context is one of the most salient ecologies in which youth learn about their heritage culture (Umaña-Taylor et al., 2009). Parents aiming to inform their children about their heritage culture may be more inclined to engage in ethnic-racial identity exploration alongside their children. Behaviors such as eating ethnic food, listening to music, or watching movies that reflect the heritage culture are all activities that Latinx families highly endorse and engage in together (Umaña-Taylor et al., 2009). Moreover, family members who are jointly exploring their ethnicity may have more similar exploration levels and, in turn, enjoy greater family cohesion.

Our study also provides some support for the acculturation gap-distress model, contributing to the literature that has found mixed results for how intergenerational differences influence family dynamics and youth outcomes. As discussed in previous studies, family cohesion is often times not studied as a mediator in the acculturation gap-distress model (Nair et al., 2018). Our study findings parallel those of Nair et al. (2018), which also demonstrated that familial differences in acculturation impacted family cohesion and family conflict. However, unlike previous studies, we found that parent-youth ethnic-racial identity gaps are indirectly associated with youths' depressive symptoms via family cohesion. Further, our results contribute to the dialogue around familial dynamics and youth well-being. Much of the previous literature examines the acculturation gap-distress model using the domains of cultural practices and values, yet fewer have examined the implications of intergenerational differences during the ethnic-racial identity re/formation *process* (Telzer, 2010). Rather than observing differences in outcomes of acculturation, we observed differences in the acculturation process by investigating an aspect of ethnic-racial identity development among immigrant parents and youth.

Our findings indicate that acculturation dissimilarity in the domain of ethnic-racial identity exploration may have implications for youth adjustment. Such dissimilarity captures how immigrant parents and youth may be undergoing transformations of their ethnic-racial identities in the host country at different rates. The context in which we collected our data also offers an important look into how immigrant families are acculturating in a new immigrant destination. Demographic shifts in the Midwest suggest Latinx immigrant families residing in the area are undergoing ethnic-racial identity re/formation and acculturation in locations that have few established immigrant communities and resources, as compared to Latinx immigrant families acculturating to the host country in established ethnic enclaves (Witherspoon et al., 2020). Because the families reside in areas with less ethnic, racial, and cultural diversity, there may be fewer opportunities to learn about one's ethnic-racial identity outside of the home. Youth may be exploring more of the host culture through their exposure at school, with peers, and through the media. Meanwhile, parents may be more inclined to work toward maintaining their connection to their heritage culture, further accentuating acculturation differences within immigrant families (Gibson, 2001). Given the lack of established cultural supports in new immigrant areas, it is important to further investigate how parents and youths' identities are impacted by their community context (Morales et al., 2012).

Limitations and Future Directions

Although these findings contribute to the current literature on this topic, there are limitations that should be noted. First, while parents reported on their own ethnic-racial identity, we relied mainly on adolescent self-report of all other constructs included in the study, which presents a source for bias. Future studies should include multiple reporters by assessing parents' or relative's account of family cohesion as well as of adolescent depressive symptoms to help reduce self-report bias (De Los Reyes et al., 2015). This is especially important to determine whether parent-youth discordance in parenting reports are due to issues of family functioning or to normative priority shifts that occur in adolescence. While adolescents' autonomy-seeking behavior may be developmentally appropriate, it may also influence differences in reports of family dynamics between parent and adolescent (Hou et al., 2020). Additionally, only dyads in which *both* the parent and adolescent self-identified as Latinx or Hispanic were included in our sample. Future research should explore ethnic-racial identity dissimilarity when only one dyad member identifies as Latinx, as ethnic-racial identity discordance may be more salient within these parent-youth dyads.

Our findings also have limited generalizability in terms of geographic location and socioeconomic status. The present data were collected in a constrained geographic region in which there are small pockets of ethnic enclaves in areas where the Latinx population accounts for less than 8% of the population (Warikoo, 2018). Youth growing up with more access to others from their same ethnic background and more readily exposed to their native language might have more opportunities to explore their ethnic-racial identity than those raised in areas of lower Latinx immigrant concentration (Umaña-Taylor et al., 2002). Thus, future studies should examine parent-child ethnic-racial identity dissimilarity, family cohesion, and their longitudinal associations to adolescent mental health outcome in a variety of contexts. In addition, future studies should aim for more economically diverse samples since research has found greater discordance in parent and adolescent-reported parenting among families of lower socioeconomic status (Hou et al., 2020). Our participants tended to be mainly from lower socioeconomic backgrounds, which could have potentially impacted our findings considering the added stress that living under economic stress presents to families (García Coll et al., 1996). In terms of stressors, another potential limitation of our study was not considering parents' and adolescents' documentation status (which was not available for this sample). Mixed-status families dealing with the restrictions of undocumented status and various types of visas have an additional layer of vulnerability, which potentially impacts several aspects of their family functioning, including their levels of engagement with and exploration of their ethnic-racial identity (Stein et al., 2016).

While our study benefited from a longitudinal design that allowed us to examine the influence of parent-adolescent ethnic-racial identity gap on youth's mental health at different points in time, we did not examine variations of such dissimilarity over time and how these possible changes were related to youth's outcome. Considering past findings indicating decreases in ethnic-racial identity discrepancies throughout adolescence, future studies should account for these possible oscillations (Nieri et al., 2016; Schwartz et al., 2016). Furthermore, early adulthood is an important, but understudied period of ethnic-racial identity re/formation. Thus, studies would benefit from employing longitudinal designs spanning larger periods of time accounting for shifts and reexaminations of ethnic-racial identity in both parents and children (Stein et al., 2017).

Finally, it is important to continue examining possible different pathways that link parent-child ethnic-racial identity gap and youth's depressive symptoms. It is likely that a different mediating mechanism could be involved in this relationship. We examined family cohesion given the knowledge that the family environment is of particular importance for youths' ethnic-racial identity development (Rivas-Drake et al., 2014a, 2014b). However, it is likely that other factors—relationship quality or *familismo*, for example—could help also explain this connection. Additionally, such discordances may inform the individual ethnic-racial identity development or acculturation process of either parent or youth. Both parent and youth's individual ethnic-racial identity exploration may influence the rate at which each acculturate to the native or host culture. Given the importance of ethnic-racial identity re/formation for Latinx immigrant children and its implications for youths' outcomes, our study clarifies yet another piece of the puzzle by demonstrating some of the pathways through which discordances in parent-adolescent ethnic-racial identity re/formation impact family and youth outcome (Neblett et al., 2012; Rivas-Drake et al., 2014a, 2014b).

Implications

The current study provides important implications for practice. Clinicians working with youth facing increased depressive symptomatology should consider acculturation discordances, such as ethnic-racial identity differences between parents and adolescents, as a possible source of risk for the youth. Practitioners should expect and acknowledge differences in how Latinx immigrant parents and their children relate to and explore their ethnicity. Interventions at the family level should be culturally sensitive and provide opportunities for intergenerational discussions regarding family members' different experiences and engagement with their native and host culture (Ayon et al., 2018). These opportunities could expose and explain differences in thinking that could not only impact family members' ethnic-racial identity exploration, but also their feelings and perceptions of family cohesion.

Studies on ethnic-racial identity have demonstrated the value of parental conversations of culture and ethnicity for children's own identity development (Miranda et al., 2006). Thus, practitioners should also utilize Latinx immigrants' strong family orientation, connections to their culture, and value placed on transmitting it to their children by building onto these strengths. Services should be provided in both Spanish and English and focus on celebrating their cultural heritage and maintaining ethnic pride in the family (Ayón, et al., 2018). In addition, practitioners working with immigrants should consider ways to help families adjust to and explore their identities in their new context by focusing on practices that support and sustain family cohesion, such as improved communication skills (Leidy et al., 2010). Future studies should investigate how Latinx immigrant families navigate cultural clashes and identity re/formation processes, while maintaining a positive family ecology and promoting youth adjustment. Lastly, both scholars and families should remain cognizant of the ways in which discriminatory and acculturation experiences, psychological well-being, and familial dynamics are continuously reshaping Latinx immigrants' identity process.

Acknowledgments Special thanks to the families who participated in this project. We also thank the CASA Lab at the University of Michigan and all research assistants, research staff, and collaborators for their contributions to the research process.

References

Ackard, D. M., Neumark-Sztainer, D., Story, M., & Perry, C. (2006). Parent–child connectedness and behavioral and emotional health among adolescents. *American Journal of Preventive Medicine, 30*(1), 59–66. https://doi.org/10.1016/j.amepre.2005.09.013

Alegria, M., Jackson, J. S., Kessler, R. C., & Takeuchi, D. (2007). *Collaborative psychiatric epidemiology surveys (CPES), 2001–2003 [United States]*. Inter-university Consortium for Political and Social Research.

Atzaba-Poria, N., & Pike, A. (2007). Are ethnic minority adolescents at risk for problem behaviour? Acculturation and intergenerational acculturation discrepancies in early adolescence. *British Journal of Developmental Psychology, 25*(4), 527–541. https://doi.org/10.1348/026151006X173053

Ayón, C., Ojeda, I., & Ruano, E. (2018). Cultural socialization practices among Latino immigrant families within a restrictive immigration socio-political context. *Children and Youth Services Review, 88*, 57–65. https://doi.org/10.1016/j.childyouth.2018.02.042

Bacallao, M. L., & Smokowski, P. R. (2007). The costs of getting ahead: Mexican family systems after immigration. *Family Relations, 56*, 52–66. https://doi.org/10.1111/j.1741-3729.2007.00439.x

Berry, J. (1997). Immigration, acculturation, and adaptation. *Applied Psychology: An International Review, 46*, 5–34. https://doi.org/10.1111/j.1464-0597.1997.tb01087.x

Birman, D. (2006). Measurement of the "acculturation gap" in immigrant families and implications for parent-child relationships. In M. H. Bornstein & L. R. Cote (Eds.), *Acculturation and parent-child relationships: Measurement and development* (pp. 113–134). Erlbaum.

Buckingham, S. L., & Brodsky, A. E. (2015). "Our differences don't separate us": Immigrant families navigate intrafamilial acculturation gaps through diverse resilience processes. *Journal of Latina/o Psychology, 3*(3), 143–159. https://doi.org/10.1037/lat0000042

De Los Reyes, A., Augenstein, T. M., Wang, M., Thomas, S. A., Drabick, D. A., Burgers, D. E., & Rabinowitz, J. (2015). The validity of the multi-informant approach to assessing child and adolescent mental health. *Psychological Bulletin, 141*(4), 858–900. https://doi.org/10.1037/a0038498

Eaton, W. W., Muntaner, C., Smith, C., Tien, A., & Ybarra, M. (2004). Center for Epidemiologic Studies Depression Scale: Review and revision (CESD and CESD-R). In M. E. Maruish (Ed.), *The use of psychological testing for treatment planning and outcomes assessment* (Instruments for Adults) (Vol. 3, 3rd ed., pp. 363–377). Erlbaum.

Erikson, E. H. (1968). *Identity: Youth and crisis* (7th ed.) WW Norton & Company.

García Coll, C., Crnic, K., Lamberty, G., Wasik, B. H., Jenkins, R., Garcia, H. V., & McAdoo, H. P. (1996). An integrative model for the study of developmental competencies in minority children. *Child Development, 67*(5), 1891–1914. https://doi.org/10.1111/j.1467-8624.1996.tb01834.x

García Coll, C., & Magnuson, K. (1997). The psychological experience of immigration: A developmental perspective. In A. Booth, A. C. Crouter, & N. Landale (Eds.), *A developmental perspective* (*Immigration and the family: Research and policy on US immigrants*) (pp. 91–131). Routledge.

Gibson, M. A. (2001). Immigrant adaptation and patterns of acculturation. *Human Development, 44*(1), 19–23. https://doi.org/10.2307/26763493

Gonzales, N. A., Deardorff, J., Formoso, D., Barr, A., & Barrera, M., Jr. (2006). Family mediators of the relation between acculturation and adolescent mental health. *Family Relations, 55*(3), 318–330. https://doi.org/10.1111/j.1741-3729.2006.00405.x

Hou, Y., Benner, A. D., Kim, S. Y., Chen, S., Spitz, S., Shi, Y., & Beretvas, T. (2020). Discordance in parents' and adolescents' reports of parenting: A meta-analysis and qualitative review. *American Psychologist, 75*(3), 329. https://doi.org/10.1037/amp0000463

Hu, L. T., & Bentler, P. M. (1999). Cutoff criteria for fit indexes in covariance structure analysis: Conventional criteria versus new alternatives. *Structural Equation Modeling: A Multidisciplinary Journal, 6*(1), 1–55. https://doi.org/10.1080/10705519909540118

Huq, N., Stein, G. L., & Gonzalez, L. M. (2016). Acculturation conflict among Latino youth: Discrimination, ethnic identity, and depressive symptoms. *Cultural Diversity and Ethnic Minority Psychology, 22*(3), 377–387. https://doi.org/10.1037/cdp0000070

Hwang, W. C., & Wood, J. J. (2009). Acculturative family distancing: Links with self-reported symptomatology among Asian Americans and Latinos. *Child Psychiatry and Human Development, 40*(1), 123–138. https://doi.org/10.1007/s10578-008-0115-8

Kline, R. B. (2015). *Principles and practice of structural equation modeling*. Guilford Publications.
Laird, R. D., & De Los Reyes, A. (2013). Testing informant discrepancies as predictors of early adolescent psychopathology: Why difference scores cannot tell you what you want to know and how polynomial regression may. *Journal of Abnormal Child Psychology, 41*(1), 1–14. https://doi.org/10.1007/s10802-012-9659-y
Lau, A. S., McCabe, K. M., Yeh, M., Garland, A. F., Wood, P. A., & Hough, R. L. (2005). The acculturation gap-distress hypothesis among high-risk Mexican-American families. *Journal of Family Psychology, 19*(3), 367–375. https://doi.org/10.1037/0893-3200.19.3.367
Leidy, M. S., Guerra, N. G., & Toro, R. I. (2010). Positive parenting, family cohesion, and child social competence among immigrant Latino families. *Journal of Family Psychology, 24*(3), 252–260. https://doi.org/10.1037/a0019407
Martinez Jr, C. R. (2006). Effects of differential family acculturation on Latino adolescent substance use. *Family Relations, 55*(3), 306–317. https://doi.org/10.1111/j.1741-3729.2006.00404.x
Miranda, A. O., Bilot, J. M., Peluso, P. R., Berman, K., & Van Meek, L. G. (2006). Latino families: The relevance of the connection among acculturation, family dynamics, and health for family counseling research and practice. *The Family Journal, 14*(3), 268–273. https://doi.org/10.1177/1066480706287805
Morales, A., Yakushko, O. F., & Castro, A. J. (2012). Language brokering among Mexican-immigrant families in the Midwest: A multiple case study. *The Counseling Psychologist, 40*(4), 520–553. https://doi.org/10.1177/0011000011417312
MPLUS (Version 7.2). Computer software. Muthén & Muthén (2012).
Nair, R. L., Roche, K. M., & White, R. M. (2018). Acculturation gap distress among Latino youth: Prospective links to family processes and youth depressive symptoms, alcohol use, and academic performance. *Journal of Youth and Adolescence, 47*(1), 105–120. https://doi.org/10.1007/s10964-017-0753-x
Neblett, E. W., Jr., Rivas-Drake, D., & Umaña-Taylor, A. J. (2012). The promise of racial and ethnic protective factors in promoting ethnic minority youth development. *Child Development Perspectives, 6*(3), 295–303. https://doi.org/10.1111/j.1750-8606.2012.00239.x
Nieri, T., Grindal, M., Adams, M. A., Cookston, J. T., Fabricius, W. V., Parke, R. D., & Saenz, D. S. (2016). Reconsidering the "acculturation gap" narrative through an analysis of parent–adolescent acculturation differences in Mexican-American families. *Journal of Family Issues, 37*(14), 1919–1944. https://doi.org/10.1177/0192513X14551175
Oh, J. S., & Fuligni, A. J. (2010). The role of heritage language development in the ethnic identity and family relationships of adolescents from immigrant backgrounds. *Social Development, 19*(1), 202–220. https://doi.org/10.1111/j.1467-9507.2008.00530.x
Pasch, L. A., Deardorff, J., Tschann, J. M., Flores, E., Penilla, C., & Pantoja, P. (2006). Acculturation, parent-adolescent conflict, and adolescent adjustment in Mexican American families. *Family Process, 45*(1), 75–86. https://doi.org/10.1111/j.1545-5300.2006.00081.x
Perez-Brena, N. J., Rivas-Drake, D., Toomey, R. B., & Umaña-Taylor, A. J. (2018). Contributions of the integrative model for the study of developmental competencies in minority children: What have we learned about adaptive culture? *American Psychologist, 73*(6), 713–726. https://doi.org/10.1037/amp0000292
Phinney, J. S. (1990). Ethnic identity in adolescents and adults: Review of research. *Psychological Bulletin, 108*(3), 499–514. https://doi.org/10.1037/0033-2909.108.3.499
Portes, A., & Rumbaut, R. G. (1990). *Immigrant America: A portrait*. University of California Press.
Quintana, S. M. (2007). Racial and ethnic identity: Developmental perspectives and research. *Journal of Counseling Psychology, 54*(3), 259–270. https://doi.org/10.1037/0022-0167.54.3.259
Rivas-Drake, D., Seaton, E. K., Markstrom, C., Quintana, S., Syed, M., Lee, R. M., & Ethnic and Racial Identity in the 21st Century Study Group. (2014b). Ethnic and racial identity in adolescence: Implications for psychosocial, academic, and health outcomes. *Child Development, 85*(1), 40–57. https://doi.org/10.1111/cdev.12200

Rivas-Drake, D., Syed, M., Umaña-Taylor, A., Markstrom, C., French, S., Schwartz, S. J., & Ethnic and Racial Identity in the 21st Century Study Group. (2014a). Feeling good, happy, and proud: A meta-analysis of positive ethnic–racial affect and adjustment. *Child Development, 85*(1), 77–102. https://doi.org/10.1111/cdev.12175

Roehling, P. V., Hernandez Jarvis, L., Sprik, J. M., & Campbell, P. H. (2010). The immigration debate and its relationship to the ethnic identity development and well-being of Latino and White youth. *Hispanic Journal of Behavioral Sciences, 32*(2), 292–308. https://doi.org/10.1177/0739986310366458

Schofield, T. J., Parke, R. D., Kim, Y., & Coltrane, S. (2008). Bridging the acculturation gap: Parent-child relationship quality as a moderator in Mexican-American families. *Developmental Psychology, 44*(4), 1190–1194. https://doi.org/10.1037/a0012529

Schwartz, S. J., Montgomery, M. J., & Briones, E. (2006). The role of identity in acculturation among immigrant people: Theoretical propositions, empirical questions, and applied recommendations. *Human Development, 49*(1), 1–30. https://doi.org/10.1159/000090300

Schwartz, S. J., Unger, J. B., Baezconde-Garbanati, L., Zamboanga, B. L., Córdova, D., Lorenzo-Blanco, E. I., Huang, S., Des Rosiers, S. E., Soto, D. W., Lizzi, K. M., Villamar, J. A., Pattarroyo, M., & Szapocznik, J. (2016). Testing the parent–adolescent acculturation discrepancy hypothesis: A five-wave longitudinal study. *Journal of Research on Adolescence, 26*(3), 567–586. https://doi.org/10.1111/jora.12214

Sciarra, D. T., & Ponterotto, J. G. (1991). Counseling the Hispanic bilingual family: Challenges to the therapeutic process. *Psychotherapy: Theory, Research, Practice, Training, 28*(3), 473. https://doi.org/10.1037/0033-3204.28.3.473

Smith, T. B., & Silva, L. (2011). Ethnic identity and personal well-being of people of color: A meta-analysis. *Journal of Counseling Psychology, 58*(1), 42–60. https://doi.org/10.1037/a0021528

Smokowski, P. R., Rose, R., & Bacallao, M. L. (2008). Acculturation and Latino family processes: How cultural involvement, biculturalism, and acculturation gaps influence family dynamics. *Family Relations, 57*(3), 295–308. https://doi.org/10.1111/j.1741-3729.2008.00501.x\

Stein, G. L., Cupito, A. M., Mendez, J. L., Prandoni, J., Huq, N., & Westerberg, D. (2014). Familism through a developmental lens. *Journal of Latina/o Psychology, 2*(4), 224–250. https://doi.org/10.1037/lat0000025

Stein, G. L., Gonzales, R. G., García, C. C., & Prandoni, J. I. (2016). Latinos in rural, new immigrant destinations: A modification of the integrative model of child development. In L. Crockett & G. Carlo (Eds.), *Rural ethnic minority youth and families in the United States* (*Advancing Responsible Adolescent Development*). Springer. https://doi.org/10.1007/978-3-319-20976-0_3

Stein, G. L., & Polo, A. J. (2014). Parent–child cultural value gaps and depressive symptoms among Mexican American youth. *Journal of Child and Family Studies, 23*(2), 189–199. https://doi.org/10.1007/s10826-013-9724-3

Stein, G. L., Rivas-Drake, D., & Camacho, T. C. (2017). Ethnic identity and familism among Latino college students: A test of prospective associations. *Emerging Adulthood, 5*(2), 106–115. https://doi.org/10.1177/2167696816657234

Suárez-Orozco, C. (2001). Identities under siege: Immigration stress and social mirroring among the children of immigrants. In A. Robben & M. Suárez-Orozco (Eds.), *Cultures under siege: Social violence and trauma* (pp. 194–206). Cambridge University Press.

Szapocznik, J., & Kurtines, W. M. (1993). Family psychology and cultural diversity: Opportunities for theory, research, and application. *American Psychologist, 48*(4), 400–407. https://doi.org/10.1037/0003-066X.48.4.400

Telzer, E. H. (2010). Expanding the acculturation gap-distress model: An integrative review of research. *Human Development, 53*(6), 313–340. https://doi.org/10.1159/000322476

Telzer, E. H., Yuen, C., Gonzales, N., & Fuligni, A. J. (2016). Filling gaps in the acculturation gap-distress model: Heritage cultural maintenance and adjustment in Mexican–American families. *Journal of Youth and Adolescence, 45*(7), 1412–1425. https://doi.org/10.1007/s10964-015-0408-8

Te'Neil Lloyd, B. (2002). A conceptual framework for examining adolescent identity, media influence, and social development. *Review of General Psychology, 6*(1), 73–91. https://doi.org/10.1037/1089-2680.6.1.73

Torres, L., Yznaga, S. D., & Moore, K. M. (2011). Discrimination and Latino psychological distress: The moderating role of ethnic identity exploration and commitment. *American Journal of Orthopsychiatry, 81*(4), 526. https://doi.org/10.1111/j.1939-0025.2011.01117.x

Umaña-Taylor, A. J., Diversi, M., & Fine, M. A. (2002). Ethnic identity and self-esteem of Latino adolescents: Distinctions among the Latino populations. *Journal of Adolescent Research, 17*(3), 303–327. https://doi.org/10.1177/0743558402173005

Umaña-Taylor, A. J., Yazedjian, A., & Bámaca-Gómez, M. Y. (2004). Developing the ethnic identity scale using Eriksonian and social identity perspectives. *Identity: An International Journal of Theory and Research, 4*(1), 9–38. https://doi.org/10.1207/S1532706XID0401_2

Umaña-Taylor, A. J., Alfaro, E. C., Bámaca, M. Y., & Guimond, A. B. (2009). The central role of familial ethnic socialization in Latino adolescents' cultural orientation. *Journal of Marriage and Family, 71*(1), 46–60. https://doi.org/10.1111/j.1741-3737.2008.00579.x

Umaña-Taylor, A. J., Quintana, S. M., Lee, R. M., Cross Jr, W. E., Rivas-Drake, D., Schwartz, S. J., Syed, M., Yip, T., Seaton, E., & Ethnic and Racial Identity in the 21st Century Study Group. (2014). Ethnic and racial identity during adolescence and into young adulthood: An integrated conceptualization. *Child development, 85*(1), 21–39. https://doi.org/10.1111/cdev.12196

Unger, J. B., Ritt-Olson, A., Soto, D. W., & Baezconde-Garbanati, L. (2009). Parent-child acculturation discrepancies as a risk factor for substance use among Hispanic adolescents in Southern California. *Journal of Immigrant and Minority Health, 11*(3), 149–157. https://doi.org/10.1007/s10903-007-9083-5

Warikoo, N. (2018). *Michigan non-criminal immigrant arrests, deportations soar under Trump*. https://www.freep.com/story/news/local/michigan/2018/03/20/non-criminal-immigrants-arrests-deportations/424066002/

Witherspoon, D. P., Bámaca-Colbert, M. Y., Stein, G. L., & Rivas-Drake, D. (2020). Hidden populations: Uncovering the developmental experiences of minoritized populations across contexts. *Developmental Psychology, 56*(8), 1425-1430. https://doi.org/10.1037/dev0001055

Part V
Positive Youth and Emerging Adult Development

Chapter 16
Positive Identity Development as an Integration Measure: Reframing Alternative Conceptual Findings for Academically Resilient, Low Income Urban Youth

Sophie Fajardo, Bronwyn Nichols Lodato, Margaret B. Spencer, and Christopher G. Graziul

As an identity-focused cultural ecological perspective and systems theoretical framework, the phenomenological variant of ecological systems theory (PVEST) acknowledges the universality of human vulnerability and potential for resilience. Consistent with Eriksonian (1950) identity formation and life course process, the framework is helpful for studying human diversity and as such holds valuable promise for the study of increasingly diverse migration experiences to the United States. PVEST highlights the processual nature of identity development across the life course and considers the unavoidable need to both cope with challenges and garner benefit from accessible supports in multiple contexts. This framework envisions identity as a dynamic, context-linked response to an individual's balanced or imbalanced *vulnerability status*. Given exposures and access to both challenges and supports across diverse contexts, some level of vulnerability is unavoidable for all humans. Incorporating a PVEST perspective, this chapter suggests that theories of assimilation consider identity as an important and developmentally sensitive measure of an individual's experiences within a particular context. Focusing on identity centers immigrants' experiences of core coping processes required within their receiving contexts. How immigrants perceive and experience a receiving context over time corresponds to changes to their identities, thus offering important windows into the dynamic relationship between individuals' meanings made of interpersonal relationships and diverse contexts. We therefore suggest that assessing immigrants' identification relative to their receiving contexts is an important measure of their integration into those contexts.

S. Fajardo (✉) · B. N. Lodato · M. B. Spencer · C. G. Graziul
University of Chicago, Chicago, IL, USA
e-mail: fajardos@uchicago.edu; bronwyn@uchicago.edu; mbspencer@uchicago.edu; graziul@uchicago.edu

D. J. Johnson et al. (eds.), *Re/Formation and Identity*, Advances in Immigrant Family Research, https://doi.org/10.1007/978-3-030-86426-2_16

This chapter provides an empirical demonstration of a developmentally sensitive and theoretically grounded approach to immigrant integration. Specifically, drawing on a surveyed sample of 333 high performing secondary school students in a large northeastern U.S. city (referred to in this chapter as "Philtren"), we explore the impact of adolescent immigrants' perspectives of their receiving contexts, colorism, and ethnic identity on their positive self-identification. First, we discuss the results in the context of assimilation theory in the social sciences and its relationship to a white middle-class standard. Second, PVEST is presented as an alternative conceptual framework for its potential to illuminate the integrative experiences of immigrants beyond classic and current assimilation theories. Third, we use our present study and the lessons learned as an empirical demonstration to explore positive identity development of first-generation immigrant adolescents. Finally, we discuss the implications of the combined effort for future research on identity development and immigrant integration.

Perspectives on Migration, Self, and Society

The ways in which immigrants relate to their host countries change over time. Children and families migrate as outsiders to a host country's context, prompting policy makers, citizens, and academics alike to ask how (and if) immigrants ultimately assimilate into their host societies. Thus, it is important to first examine the explanations of immigrants' economic and sociocultural assimilation into their host societies according to classic and segmented assimilation theories. Another theoretical framework, PVEST, with its vulnerability-resilience emphasis, is presented to address these issues by centering identity as a measure of self-reported immigrant integration within specific receiving contexts.

Classic Theories: Assimilation to the White Middle Class

Classical sociological theories consider immigrant integration and assimilation as a "straight line" trend of upward mobility in which immigrants become part of the U.S. middle class while simultaneously becoming ethnically indistinct from the American mainstream (Gans, 1979). According to Gans (1979), "Acculturation and assimilation are viewed as secular trends that culminate in the eventual absorption of the ethnic group into a single host society" (p. 2). Although later generations of immigrants may retain some ethnic distinctiveness, Gans (1979) argues that these expressions are mostly symbolic and independent of country of origin cultures and domestic ethnic organizations (unlike the ethnic expressions of earlier generations of immigrants). However, as Gans (1992) notes in later work, the downward mobility of many immigrants post-1965 challenges this original theory. According to

Gans, changing economic conditions primarily account for this shift in mobility, although race plays a secondary role.

Critics of this original theory of assimilation contend that it relies on a narrow definition of economic assimilation. According to Alba and Nee (2003), immigrant incorporation and assimilation should be considered not only as an upwardly mobile path that destitute immigrants take to join the American middle class, but as a wide set of trajectories that socioeconomically diverse immigrants with different sets of skills and resources may experience. These experiences can lead to upward, lateral, or downward trending paths that lead to incorporation at various levels of the American socioeconomic hierarchy. Furthermore, Alba and Nee (2003) complicate Gans' macroeconomic emphasis and attempt to introduce more agency into the explanatory model of assimilation by pointing to immigrants' purposive action, network mechanisms, and forms of capital (economic and non-economic) as important factors that shape integrative outcomes.

Similarly, Rumbaut (1999) destabilizes the "straight line" theory of immigration by questioning the singularity with which past theorists have considered the socioeconomic position of typical immigrants, and also the stability of the typical "American" experience to which the immigrant supposedly assimilates. Rumbaut expands the notion of assimilation beyond the economic scope, pointing to positive and negative aspects of health outcomes and behaviors, criminal activity, academic achievement, language, and ethnic self-identification as other realms of assimilation that immigrants experience with upward, lateral, or downward returns.

Finally, Bean and Stevens (2003) argue that classical immigration theory suffers from the assumption that economic assimilation is closely followed by sociocultural assimilation. According to Bean and Stevens (2003), classical immigration theory cannot explain current immigration trends particularly because sociocultural and economic assimilation have become increasingly separated. Bean and Stevens discuss this separation particularly for racial identification as a marker of sociocultural assimilation. The authors note that for some groups strong ethnic identities form after the immigrant group has gained a measure of relative economic success, while for others a strong ethnic identity can be used as a strategy for achieving economic incorporation (Bean & Stevens, 2003). In this view, assimilation theory should pay due attention to the ways that economic and sociocultural assimilation interact temporally and causally.

Segmented Assimilation: Greater Diversity and a Vanishing White Middle Class

After the Immigration and Naturalization Act of 1965, immigration to the United States became much more diverse. At the same time, the domestic economy produced greater inequality throughout the last half of the twentieth century up to today. Thus, immigrants to the United States post-1965 experienced far less economic mobility than immigrants in the early twentieth century and, as the domestic middle class also shrank over this period, "straight-line" theories of assimilation became increasingly irrelevant. Portes and Zhou (1993) introduced segmented assimilation theory to address this issue. By connecting segmented labor market theory and immigrants' experiences of racism and colorism, they theorize that immigrants of less racially marginalized status in the United States are more likely to experience upward mobility and assimilate economically and culturally with the American middle class than immigrants of color, who are more likely to experience lateral or downward mobility and assimilate economically and culturally with the American lower classes. A third path was also conceptualized, in which immigrants deliberately retain their cultures and communities while experiencing upward economic mobility, as in the example of Cuban refugees in the late 1980s in Miami. This theory highlights the importance of receiving context conditions—such as local hostility to particular races and ethnicities—for the integration of immigrant populations in the United States and adds a degree of autonomy for immigrants in recognizing efforts to resist and combat racial discrimination and its resulting social and economic disadvantages in the U.S. context.

In addition to race and color, Portes and Zhou (1993) point to location and the absence or presence of economic mobility "ladders" as organizing factors leading to each of these paths. In this conception, racially marginalized immigrants in poor American communities with few structural opportunities for economic advancement are most likely to experience cultural and economic assimilation into the American lower classes, while immigrants of the same background living in primarily ethnic neighborhoods with more community embedded resources and opportunities for advancement are more likely to take the third, ethnically and economically resilient path.

Mary Waters' (1994) findings concerning second-generation Black immigrants' racial identification support this theory, but also reveal the intentional strategies that adolescents attach to each of these paths. Among second-generation Black immigrants living in poorer neighborhoods and attending public schools in New York's inner city, assimilating culturally with Black Americans was the most popular choice, while second-generation Black immigrants living in neighborhoods with a stronger ethnic presence and attending schools with more white students tended to hang on to their ethnic self-identifications (Waters, 1994). According to Waters (1994), ethnically identifying youth were more likely to do so because their contacts with white people indicated to them that they were treated better when they revealed their immigrant backgrounds than when they did not resist assumptions that they were Black Americans. Youth identifying with Black Americans, on the other hand, had fewer contacts with white people and did not perceive a difference in the treatment of Black Americans and Black immigrants by whites. Instead, they identified with Black Americans out of a desire to avoid ethnic teasing and out of social benefits of American values relative to their parents' immigrant values (i.e., more perceived gender egalitarianism).

Waters' ethnographic findings point to the importance of how immigrants, and in particular adolescent immigrants, experience the receiving contexts with which they interact. Importantly, these meaning making processes have important consequences for the ways that adolescent immigrants identify ethnically and racially and are informed by developmentally linked race socialization experiences in families, communities, and school contexts. Thus, identification is associated with developmental change and experiences in multiple social contexts and as such may hold promise as an individual-level measure of immigrant integration. In what follows, the authors demonstrate how the phenomenological variant of ecological systems theory (PVEST) provides a mechanism for considering this potential integration measure (see Spencer 1995, 2006, 2008; Spencer et al., 1997, 2019).

The Present Study: Development in an Urban Context

This study makes use of a longitudinal data set collected in a large, urban, northeast center of the United States referred to as "Philtren." The data set contained significant individual and context level data describing discrepancies and relationships within and among home and school neighborhood characteristics and how those impacted the sample of high performing youths' coping strategies and education outcomes (Spencer et al., 2005). Utilizing a "windshield" observational strategy and the PVEST framework described as the underlying conceptual framing device, copious neighborhood level data were collected. The researchers mapped Philtren's neighborhoods and aligned them with 1990 Census data and Municipal data.

These findings were used to create a Neighborhood Assessment of Community Characteristics (NACC) tool, which captures key information on the characteristics of the residential and school neighborhoods of the 40 high schools of Philtren. With a driver and an observational protocol trained observer, neighborhood data collection afforded "Windshield" observations (i.e., data captured through navigating neighborhoods by car and collecting observations). The overall Nested Context Assessment Strategy described below guided the analytic and operational components of the NACC (see Spencer et al., 2012). The strategy allowed for the incorporation of the broader contextual factors that inform the developmental and educational trajectories and outcomes of the students under study.

Over a 2-year period, a team of researchers conducted a total of 483 observations of neighborhood characteristics within a 1-mile radius of the high schools of Philtren to ascertain which characteristics correlated with high academic achievement (Spencer et al., 2005).

The neighborhood data analysis was informed by a careful determination of an appropriate definition of a neighborhood. Combining this observational information with student survey data provided for a deeper understanding of the psychological effects of the neighborhood and school contexts; in this illustration it meant linking with the project measure: Perceived and actual risk, referred to as "Fear of Calamity"

(Riechard & McGarrity, 1994). The assessment of the assets in a given community, centering the analysis of the utility of these assets in mitigating or exacerbating risk or "fear of calamity" for high school students of Philtren, most salient, expands insights about perceptions for improving understanding of adolescents' phenomenological processes in context.

Significantly, students clearly understood their neighborhood boundaries and the key safety considerations for navigating those areas. Review of the data from a group of 94 high performing and low performing students determined that they resided in the same neighborhoods (Spencer, 2001). Additionally, the majority of the students who participated in a second-year pilot study did not attend their neighborhood school. This suggests that these students and their families were negotiating experiences in broader spaces outside their neighborhood, thus, providing exposure to different resources as well as risks in different environments. In terms of friendships, high achieving students reported more close friends outside of their home neighborhood than their moderately achieving counterparts.

Sociohistorical Considerations

In 1999, when these data were collected, public opinion toward immigration in the United States was at an all-time high. Among respondents to Gallup, Princeton Survey Research Associates, and CBS/The New York Times surveys in 1999, 41% of respondents preferred for immigration levels to stay the same (the highest percentage since 46% in 1964), 10% preferred that immigration to the United States increase (the highest recorded percentage at the time), and 44% preferred that immigration decrease (the lowest percentage since 42% in 1977) (Buck et al., 2004). However, these attitudes varied according to immigrants' regions of origin. Gallup, Princeton Survey Research Associates, and CBS/The New York Times surveys in 1995 reported that respondents thought there were "too many" immigrants of Latin American (56%), Asian (44%), African (33%), and European origin (31%).

Since 1999, the immigrant receiving context of the United States has gone through several dramatic changes. Public contempt for Middle Eastern immigrants increased dramatically in the aftermath of September 11, 2001 (Schildkraut, 2002). Public opinion of Latin American immigrants also declined, as these immigrants became the target of increased deportations, peaking at a single year record of 438,421 under the Obama administration in 2013 (Gonzalez-Barrera & Krogstad, 2014). In a Gallup Poll conducted just one month after September 11, 2001, 30% of respondents thought immigration to the United States should remain at present levels, 8% thought immigration should increase, and 58% were in favor of decreasing immigration (Gallup, 2007). This is compared to 42% (same), 14% (increase), and 41% (decrease) just 4 months prior; public opinion toward general immigration to the United States has not been this positive since. Given that the data utilized in the present investigation were collected before this lasting decrease in positive public opinion toward immigrants, no inferences about the experiences of immigrants in the twenty-first century should be made from this empirical demonstration. Readers can only imagine that any indication of negative receiving context experiences reported here would likely be augmented for today's immigrants.

PVEST as an Integration Measure: Reframing the Assimilation Question

Both classic and segmented theories of assimilation lack a basic theorization of the ways individuals interact with their lived environments. To address this issue, we utilize PVEST (Spencer 1995, 2006, 2008), which accounts for the processual development of identity vis-a-vis an individual's experiences of their lived environments. According to PVEST, as the individual is confronted with various life course challenges as well as supports to confront challenges in their lived environments, they respond with various coping mechanisms. As an individual becomes more familiar with these challenges, they retain some of these coping mechanisms while discarding others through trial and error. As a function of basic redundancy, those coping mechanisms that stabilize across trials become part of the *individual's emergent identity*. Thus, the PVEST perspective sees identity as both: (1) a process born out of an individual's experiences within their specific environments, and (2) an agentic tool with which an individual navigates environments where developmental processes and experiences are had.

As a theory of resilience that considers identity development a process of give and take between individuals and their lived environments, PVEST complements theories of acculturation. Acculturation refers to cultural changes that occur when people of different cultures come into contact via migration (Motti-Stefanidi, 2018). Resilient developmental outcomes are achieved on an individual level when immigrants combine some aspects of their original cultures with aspects of their host cultures. The process of acculturation, therefore, is conceptualized as an integration strategy (Sam & Berry, 2016).

PVEST can further the study of acculturation in its consideration of identity as a dynamic, contextually informed process. Previous studies of acculturation consider identity a more static set of cultural attributes that individual immigrants may or may not retain (Berry et al., 2006). With PVEST's conceptualization of identity, we center immigrants' individual meanings of their experiences with challenges (like racism and colorism) and supports (like immigrant mobility ladders) in various receiving contexts. In doing so, we address the issues of defining economic and sociocultural assimilation by taking seriously immigrant's own dynamic perspectives, meanings accrued, and definitions of their economic and social relationships. We contend that how immigrants experience a receiving context over time and corresponding changes in their identities offer important windows into the dynamic relationship between individuals and their contexts. More specifically, conceptualized as active coping with situations endured in their migration process, the redundancy of strategies enacted as reactive coping responses matter for stable emergent identities manifested in their new contexts. Further implicated, we submit, is that assessing immigrants' identification relative to their receiving contexts is important for understanding their strategies and integrative outcomes across time.

Hypotheses

The primary research question that this study attempts to answer is: How do adolescent immigrants' perceptions of their receiving contexts impact positive self-identification? To answer this question, we tested three hypotheses:

Hypothesis 1: Contexts of Reception. Perceptions of both contexts of reception (school and residential neighborhood) will be directly related to positive self-identification.

Hypothesis 2: Color and Colorism. Darker skin color and negative skin color perceptions will be inversely related to positive self-identification.

Hypothesis 3: Ethnic Identity. Strong ethnic identity will be directly related to positive self-identification.

Method

The data utilized in this study were originally collected for an evaluation of a program that granted monetary stipends to high academic achieving, low-income high school students in Philtren in 1999. The academic achievement component required students to have earned A and B grades in major subjects to be eligible for the program. For guaranteeing the economic status of low income, the program under evaluation also required that participants provided documentation for ascertaining that the family met the guidelines for the Federal Free Lunch program (130% of the poverty line). Families of applicants were required to provide proof of income and information about family household composition at the time of application. Eligibility was determined by comparing family income levels and household sizes with the Federal Free Lunch program guidelines (see Spencer, Noll & Cassidy, 2005).

Description of Sample

A total of 172 ninth graders, 198 tenth graders, and 164 eleventh graders participated in the study. Females comprised 69% (n = 367) of the study participants and 31% (n = 167) were male. Respondents identifying as Black (n = 242, 45%) and Asian (n = 120, 23%) comprised the largest ethnic groups, followed by Hispanic

(n = 65, 12%), White (n = 56, 11%), and students who identified themselves as Other (9%; n = 50). One student's ethnicity was not reported.

Ninety-three percent of participating students (n = 502) completed the baseline survey, which comprises this study. There were no significant differences in survey response by assigned treatment group, gender, or grade level. With respect to race/ethnicity, however, Black, Hispanic, and "Other" students were significantly more likely to respond than Asian or white students, $F(4, 537) = 3.98$, $p = 0.003$.

Although both age and grade level information were captured in data collection, we used age as a developmental proxy, and excluded grade. This is because most of the first-generation immigrants represented in the sample are of either Asian or Latin American origin, and at the time of data collection Vietnam and Cuba represented the fourth and fifth highest sending countries of refugees to the United States. In refugee populations in particular, adolescents often experience disruptions to their education that cause them to have to enroll in lower grades than most of their peers. Furthermore, for both refugees and traditional first-generation immigrants limited English language skills also often cause students to be placed at lower grade levels than their peers. Thus, age was deemed to be a more appropriate proxy of adolescent development than grade level.

Ultimately, due to missing data, our final sample consisted of 333 urban adolescent high school students that were both at least 130% of the poverty line and earned all As and Bs in the previous academic term. Table 16.1 presents descriptive statistics for individual characteristics among the sample of respondents ultimately analyzed.

Intersection of Ethnic Heritage and First-Generation Status

Due to both collinearity issues and a desire to compare the unique experiences of first-generation immigrants and native-born U.S. citizens of the same ethnic heritage, we collapsed these variables as reflected in Table 16.2. Students reported country of origin in their surveys. This data was used to generate an "ethnic heritage" variable, which included Asian, African, Latin American, European, and Other. For students who were born in the United States, parents' country of origin and self-reported ethnicity were collapsed into these categories, such that every

Table 16.1 Descriptive statistics for individual characteristics

Variable	*N*	Mean	Std. dev.	Minimum	Maximum
Age	333	15.0	1.2	12	21
Gender (1 = female)	333	0.71	–	0	1
Skin color	333	4.01	2.03	0	8
Skin color perception	333	3.11	0.94	0.81	5.56
MEIM scale	333	3.08	0.44	1.25	4
Perceived school climate	333	2.85	0.30	1.96	3.72
Perceived neighborhood safety	333	50.3	8.9	35	69

Table 16.2 Intersection of ethnic heritage and first-generation status

Category	Ethnic heritage	Migration status	*N*	Percentage
1	Asian	Non-immigrant	24	6.6
2	Asian	First-generation	61	16.8
3	African	Non-immigrant	160	44.0
4	African	First-generation	4	1.1
5	Latin American	Non-immigrant	24	6.6
6	Latin American	First-generation	19	5.2
7	European	Non-immigrant	19	5.2
8	European	First-generation	15	4.1
9	Other	Non-immigrant	23	6.3
10	Other	First-generation	15	4.1

student who identified as "white" was included in the "European" ethnic heritage, "Hispanic" was included in "Latin American," "Asian" in "Asian," and "Black" in "African." We recognize that there are several limitations to these groupings but opted for their use in order to generate large enough groups of domestic and immigrant students that could feasibly be compared. These limitations are discussed in detail later to contextualize reported results.

Measures

General Positive Attitude Measure

The Hare/Funder/Block Ego-Esteem Resiliency Scale HFBEERS is a 23-item scale assessing positive self-concept and identity among adolescents and young adults (Hare 1977; Shoemaker 1980; McDermott & Spencer, 1997). The HFBEERS instrument asks respondents to assess their agreement with statements like "I am usually proud of my report card," "My friends would elect me to a leadership position," and "I am an important person to my family/in my classes." For the purposes of this study, we conceptualize this scale as a measure of contextualized positive self-identification. This is because each item on the scale connects a positive quality of the student to contextualized social validation. For instance, "I am usually proud of my report card" links an assessment from the school context (the report card) to self-assessment (pride). Similarly, "My friends would elect me to a leadership position" highlights the status of the self within a peer context. We therefore see this measure as particularly useful for assessing the integration of immigrant students, such that it indicates students' general positive sentiment about the position of the self in various contexts of their lived environments.

Students' responses to the 23 items comprising the scale were submitted to exploratory common factor analyses using squared multiple correlations as initial communality estimates and using varimax and promax rotations. Ultimately, a one-factor solution representing positive self-concept/identity was the most reliable.

Skin Color Opinions and Perceptions Evaluation (SCOPE) Measure

Skin color and skin color perceptions were measured using the 17-item Skin Color Opinions and Perceptions Evaluation (SCOPE) instrument (Fegley et al., 2008). For the first item, students self-reported their own skin color using a 10-color array of skin tones ordered from lightest to darkest. A scale of skin color perceptions was constructed from the remaining 16 items, which solicited respondents' own skin color preference (i.e., skin color they would most/least like to have) as well as the perceived value of skin color by others (e.g., teachers, peers, etc.). Thus, the skin color perception measure represents students' beliefs about which skin color is more/less desirable given their life experiences. We contend that these are important measures for understanding these adolescents' challenges to integration, as colorism continues to be a persistent problem in Asian, Latino, Black, and white communities (Hunter, 2007).

Multigroup Ethnic Identity Measure

To create a valid measure of adolescent ethnic identity, Phinney (1992) invented the Multi-Ethnic Identity Measure (MEIM) containing a total of 20 items. Items used Likert scale responses from 1 to 4 (1 = *strongly disagree*, 4 = *strongly agree*). For this study, we removed four items ("I sometimes feel it would be better if different ethnic groups didn't try to mix together"; "I am not very clear about the role of my ethnicity in my life"; "I really have not spent much time trying to learn more about the culture and history of my ethnic group"; and "I don't try to become friends with people from other ethnic groups") that were found to be uninformative within this particular sample relative to other items. These revisions are not likely to have impacted the psychometric validity of the MEIM given that a revised 12-item version that also discounts these items was found to have desirable psychometric properties (Roberts et al., 1999). We constructed a scale using the remaining MEIM items to obtain a measure of how strongly respondents identified with their ethnic background.

Receiving Context Variables

Following Spencer et al. (2003), tools and metrics like Cattell's scree plot, eigenvalues, and hyperplane counts were employed to establish the factor structure of two measures described below. For example, Cronbach's alpha greater than 0.70 was considered adequate for internal consistency. All raw scale scores were transformed using z-scores to standardize the measures to allow for comparison of effect sizes.

Residential Neighborhood. As a measure of the residential neighborhood immigrant receiving context, we used a 9-item version of the "fear of calamity" scale developed by Riechard and McGarrity (1994) to capture students' perceptions of their neighborhoods' safety. Items in this scale asked students to indicate their level of concern regarding a series of perceived neighborhood-based risks, such as threat of violence.

School. A 25-item survey designed to capture students' general feelings about their specific school contexts were used as a measure of perceptions of the school receiving context. These items used a 4-point Likert-like scale (1 = *strongly disagree*, 4 = *strongly agree*) and asked students to respond to statements like "I have learned a lot this year," "If I did something wrong at school, I know I would get a second chance," and "Students often disrupt class." A measure was constructed from these items representing students' perceived positive/negative (e.g., supportive or challenging) school climate.

Results

Regression Model

Our multivariate model consisted of an ordinary least squares regression with indicator variables for each class of respondent and positive self-identification as the dependent variable (see Table 16.3). The age measure was centered (i.e., mean value of zero) while measures of perceived school/neighborhood climate, skin color perceptions, and the MEIM were all standardized (mean = 0, $SD = 1$) prior to modeling.

Parameter estimates revealed that first-generation African immigrant adolescents were far more likely to report higher positive self-identification ($b = 12.178$, $t = 2.33$, $p = 0.021$) than their Asian counterparts, as were European first-generation immigrant adolescents, although to a lesser extent ($b = 5.718$, $t = 2.49$, $p = 0.013$). Latinx first-generation immigrant adolescents did not report significantly different positive self-identification (i.e., than their Asian counterparts). Both Black ($b = 7.867$, $t = 5.82$, $p < 0.001$) and White ($b = 6.548$, $t = 2.89$, $p = 0.004$) native-born students reported significantly higher positive self-identification than first-generation Asian immigrant students.

These findings remain notable and suggestive even though the limited number of African first-generation immigrant adolescents included in analysis ($n = 4$) makes it impossible to draw robust conclusions about their relative positive/negative self-identification compared to other groups. One traditionally desirable possibility is that this effect is so large that it is detectable within even a very small sample of respondents.

One traditionally undesirable possibility is that this effect is due to an outlier response by one of the four respondents and thus our estimate is biased. Of these four respondents, all scored above the median on positive self-identification and one reported the most positive self-identification observed among all respondents. Otherwise, these respondents were unremarkable in terms of other characteristics (i.e., evenly split by gender, about the same age as other respondents, and no more than ~1.5 standard deviations from the mean on all other scales). It seems that these four respondents were typical in terms of other characteristics but tended to exhibit

Table 16.3 OLS model predicting positive self-identification

				95% *CI*		
Variable	*b*	Std. Err.	t-statistic	LL	UL	*p* value
Non-immigrant Asian	3.592	2.173	1.65	−0.683	7.867	0.099
First-generation Asian (reference)	–	–	–	–	–	–
Non-immigrant African	7.867	1.351	5.82	5.210	10.525	0.000
First-generation African	12.178	5.230	2.33	1.888	22.469	0.021
Non-immigrant Latin American	6.915	1.931	3.58	3.116	10.713	0.000
First-generation Latin American	3.541	2.086	1.70	−0.563	7.644	0.091
Non-immigrant European	6.548	2.267	2.89	2.087	11.009	0.004
First-generation European	5.718	2.294	2.49	1.205	10.231	0.013
Non-immigrant other	6.911	1.771	3.90	3.426	10.396	0.000
First-generation other	1.835	2.255	0.81	−2.603	6.272	0.417
Age	0.366	0.354	1.03	−0.330	1.062	0.302
Gender	0.265	0.909	0.29	−1.523	2.053	0.771
Skin color	−0.224	0.250	−0.90	−0.715	0.267	0.370
Skin color perception	0.241	0.482	0.50	−0.708	1.189	0.618
MEIM scale	2.018	0.513	3.93	1.008	3.028	0.000
Perceived school climate	3.948	0.479	8.23	3.005	4.892	0.000
Perceived neighborhood safety	0.470	0.462	1.02	−0.439	1.379	0.310
(constant)	44.538	1.234	36.10	42.111	46.966	0.000

Note. $N = 333$; $F(16, 316) = 16.95$; R-squared = 0.406

more positive self-identification than other respondents, with one respondent a clear outlier in this sense.

However, social scientists studying the marginal benefit of "big data" are beginning to suggest outlier cases are informative and can indicate areas for future research. They note that regression modeling and related methods for predicting outcomes of interest to social scientists, especially life course outcomes, can have unexpectedly limited value (i.e., low accuracy), but that outliers and cases where the predicted outcome diverges substantially from the observed outcome provide useful guidance regarding where to focus future research efforts (Garip, 2020; Salganik, 2019; Salganik et al., 2020; Small, 2019). Such cases have often been viewed as problematic when drawing inferences from quantitative models, yet these concerns apply only insofar as we view model fit to be the overriding goal of analysis.

We are not claiming the ability to predict self-identification, only that a relationship exists and that our model estimates may highlight areas for further research. This caveat is especially important to make given that our data were collected to support an unrelated research project but happen to speak to subjects raised in this volume. Keeping these issues in mind, model estimates suggest a penalty in positive self-identification associated with the first-generation immigrant experience for almost all groups except first-generation African immigrant students, who reported exceptionally high levels of positive self-identification. In this regard, our findings suggest the value of further studying first-generation African immigrant students to understand their specific experiences that might support such high levels of positive self-identification and whether these experiences are unique or might inform strategies for improving positive self-identification among other groups.

Robustness Checks

To assess the overall effect of immigrant status and ethnic heritage on positive self-identification, we estimated a hierarchical linear model (HLM) identical to the OLS model except including the combination of immigrant status and ethnic heritage as a random effect rather than as separate indicator variables (Raudenbush & Bryk, 2002). This HLM model included random intercepts where average levels of self-identification for each class of respondent were estimated. With an intra-class correlation (ICC) coefficient of 0.088 (95% CI [0.026, 0.256]), this model suggests that nearly 9% of the variance in self-identification was attributable to the intersection of immigrant status and ethnic heritage. Parameter estimates for other variables included in the OLS model reported earlier did not change substantially in magnitude or statistical significance, leaving little reason to believe that a nested model provides a different (e.g., improved) picture of the relationships in question. This model supports the salience of our original research question, which assumed that such intersectionality played a significant role in shaping self-identification independent of other factors.

Additional robustness checks included running multiple permutations of the OLS model presented earlier using different strategies for classifying respondents according to immigrant status and ethnic heritage. The taxonomy presented here proved most intuitive compared to alternative taxonomies. These efforts also involved the inclusion of multiplicative interaction terms involving, for example, immigrant status and ethnic heritage with skin color perceptions. Though some interaction terms were statistically significant, we found no evidence that such interaction effects were prominent and assume that what effects we found were likely due to chance (i.e., overfitting) since no consistent patterns emerged across (classes of) respondents.

Hypotheses Testing: Mixed Results

We tested Hypothesis 1 that perceptions of both contexts of reception, school and residential neighborhood, will be associated with positive self-identification. The results of the fixed effects model indicate that perceptions of school climate were positively and significantly ($b = 3.948$, $t = 8.23$, $p < 0.001$) associated with positive self-identification, but perceptions of neighborhood safety were not ($b = 0.470$, $t = 1.02$, $\text{p} = 0.310$). This may be because neighborhood safety is too narrow of a focus for assessing the residential neighborhood context. Future studies should consider adopting a broader instrument to capture neighborhood climate (e.g., accessible supports).

For Hypothesis 2, darker skin color and negative skin color perceptions will be inversely related to positive self-identification, neither skin color perceptions ($b = 0.241$, $t = 0.50$, $p = 0.618$) nor skin color ($b = -0.224$, $t = -0.90$, $p = 0.370$) were significantly associated with positive self-identification. This could be because in this particular context the majority of students are Black, as reflected by the data. Thus, negative perceptions associated with darker skin tones in other parts of the country may not be as pronounced in this context. Very salient, as well, for adolescent youth, ethnic identification simply matters more than perceptions of social attitudes toward skin color for positive self-identification.

Although the color and colorism effects anticipated in Hypothesis 2 were not significant, ethnic identity ($b = 2.018$, $t = 3.93$, $p < 0.001$) was significantly and positively associated with positive self-identification. This supports Hypothesis 3, pointing to the importance of ethnic identity as a potential protective factor for the ego resiliency of immigrant adolescents.

Discussion and Future Directions

This research investigated the question of whether adolescent immigrants' perspectives of their receiving contexts impact positive self-identification, in order to determine whether identity could potentially be used as an individual level indicator of immigrant integration. Unique to the present study was the exploration of the effects of school and residential neighborhood contexts, the interactive effects of students' skin color and colorism within these contexts, and the mediating effects of student ethnic identity, on the development of positive identity among first-generation adolescent immigrants.

The strongest support for the case that identity may usefully indicate individual level integration is in the significant relationship between positive self-identification and perception of school climate. Thus, adolescent positive self-identification is strongly linked to a favorable perception of the school context and thus provides evidence for a significant link between perceptions of a receiving context and self-concept. Negative self-concepts among immigrant adolescents may therefore be related to negative experiences of receiving contexts (like schools). Future researchers should therefore explore negative self-concepts as potentially indicating dissonance between an individual immigrant and their receiving contexts, and positive self-concepts as possibly pointing to harmonious integrative experiences in relevant receiving contexts.

The link between dissonant and consonant experiences in receiving contexts on immigrant adolescents' negative and positive self-identification, respectively, suggests an intuitive narrative regarding their process of integration. Perceptions about receiving contexts likely reflect real experiences of social inclusion or exclusion which, in turn, signal whether others value one's identity. Thus, while our findings indicate that African immigrant adolescents reported high levels of positive self-identification within these majority Black schools, an interesting future direction of research could be to investigate whether African immigrant adolescents' positive self-identification changes significantly in school contexts with different racial compositions.

Our findings indicate that immigrant adolescents' perceptions of their school contexts are positively correlated with positive self-identification. For those students who perceive school context negatively, adopting a negative self-identification may be psychologically harmful but less stressful than the cognitive dissonance of maintaining a positive self-identification in the face of experiences that challenge this self-concept. However, this hypothesis requires further testing since it cannot explain situations where, for example, first-generation African immigrants may be less susceptible to the effects of experiences within receiving contexts due to other coping mechanisms (e.g., associated with sending context), unrecognized external supports for coping with stress (e.g., members of the immigrant community), or even reduced risk of stressful experiences (e.g., a distinction between perceptions of African immigrants versus other Black students). The PVEST framework thus helps to suggest specific mechanisms for future study based on the idea that phenomenological experiences of ecological context must be considered in conjunction with the features of these contexts.

Although in this demonstration first-generation immigrants experienced lower positive self-identification than their non-immigrant peers, avenues for supporting positive identity development among immigrant students are indicated in our findings in school climate and ethnic identity. For example, educational policy researchers seeking to support immigrant adolescents should consider focusing attention on the cultivation of welcoming school environments and effective ethnic identity development programming for immigrant youth. This strategy should cultivate positive and supportive experiences and thus more positive perceptions of these receiving contexts.

Limitations

There are several important limitations to the findings presented here that are inherent to the data. First, the data were not collected with the intent of representing adolescent immigrant experiences. Thus, the number of immigrants in our sample is very small and difficult to compare meaningfully. Second, these data were collected at a time when the national U.S. receiving context was favorable toward immigrants at levels unseen since September 11, 2001, and thus provide no insight into the experiences of immigrants of today's U.S. context. Third, the data represent only those adolescent students who earned A and B grades in core classes. Thus, the data cannot speak to the experiences of students who earned lower grades.

These limitations are particularly important to highlight because in current times information has commonly been distorted or misrepresented to be politically weaponized against immigrants. Thus, we wish to emphasize that the patterns indicated here are in no way intended to inform either immigration or macro education policies. Our emphasis on resiliency is intended to highlight the unique strengths of these youth when many academic conversations focus on negative outcomes. While often well-meaning, this emphasis can pathologize immigrants and other marginalized adolescents. Presenting a PVEST-based understanding of immigrant adolescent experiences serves to illustrate a non-pathologizing strategy for studying immigrant identity and integration that also presents multiple future research directions focused on the mechanisms behind observed findings.

While the findings cannot indicate macro policy prescriptions, they do point to avenues for future research that seek to improve individual level interventions. However, those seeking to improve individual level interventions should also consider that the academic resilience of students who earn A and B grades should not be overstated, as there are likely many students who earn lower grades who could be considered academically resilient by virtue of remaining enrolled in school despite untoward challenges. Unfortunately, our data were unable to speak to this form of resiliency and thus suggests yet another future research strategy based on the inclusion and comparison of immigrant adolescents' self-identification as linked to classroom performance.

Conclusion

The results of this inquiry indicate that identity may be an important indicator of the ways immigrant adolescents (re)form and maintain a positive sense of identity in relation to their receiving contexts. In this way, context specific identity how dissonant or harmonious relationships with these contexts may indicate a process of integration at the individual level needs to be examined. This focus on identity allows researchers to avoid the pitfalls of relying on the white middle class standard of comparison, as well as consider integration both from a developmentally and contextually sensitive perspective.

References

Alba, R., & Nee, V. (2003). *Remaking the American mainstream: Assimilation and contemporary immigration*. Harvard University Press.

Allen, W. R., Spencer, M. B., & Brookins, G. K. (1985). Synthesis: Black children keep on growing. In M. B. Spencer, G. K. Brookins, & W. R. Allen (Eds.), *Beginnings: The social and affective development of Black children* (pp. 301–314). Erlbaum.

Bean, F. D., & Stevens, G. (2003). *America's newcomers: Immigrant incorporation and the dynamics of diversity*. Sage.

Berry, J. W., Phinney, J. S., Sam, D. L., & Vedder, P. (2006). Immigrant youth: Acculturation, identity, and adaptation. *Applied Psychology, 55*(3), 303–332. https://doi.org/10.1111/j.1464-0597.2006.00256.x

Buck, B., Hobbs, M., Kaiser, A., Lang, S., Montero, D., Romines, K., & Scott, T. (2004). Immigration and immigrants: Trends in American public opinion, 1964–1999. *Journal of Ethnic and Cultural Diversity in Social Work, 12*(3), 73–90. https://doi.org/10.1300/J051v12n03_04

Erickson, E. (1950). *Childhood and society*. W.W. Norton.

Fegley, S. G., Spencer, M. B., Goss, T. N., Harpalani, V., & Charles, N. (2008). Colorism embodied: Skin tone and psychosocial well-being in adolescence. In W. F. Overton, U. Müller, & J. L. Newman (Eds.), *Developmental perspectives on embodiment and consciousness* (pp. 281–311). Taylor & Francis.

Fisher, C. B., Hoagwood, K., Boyce, C., Duster, T., Frank, D. A., Grisso, T., Levine, R. J., Macklin, R., Spencer, M. B., Takanishi, R., Trimble, J. E., & Zayas, L. H. (2002). Research ethics for mental health science involving ethnic minority children and youths. *American Psychologist, 57*(12), 1024–1040. https://doi.org/10.1037/0003-066X.57.12.1024

Gallup. (2007). *Immigration*. Gallup.Com. https://news.gallup.com/poll/1660/Immigration.aspx

Gans, H. J. (1979). Symbolic ethnicity: The future of ethnic groups and cultures in America. *Ethnic and Racial Studies, 2*(1), 1–20. https://doi.org/10.1080/01419870.1979.9993248

Gans, H. J. (1992). Second-generation decline: Scenarios for the economic and ethnic futures of the post-1965 American immigrants. *Ethnic and Racial Studies, 15*(2), 173–192. https://doi.org/10.1080/01419870.1992.9993740

Garip, F. (2020). What failure to predict life outcomes can teach us [Review of *What failure to predict life outcomes can teach us*]. *Proceedings of the National Academy of Sciences of the United States of America, 117*(15), 8234–8235. https://doi.org/10.1073/pnas.2003390117

Gonzalez-Barrera, A., & Krogstad, J. M. (2014). *US deportations of immigrants reach record high in 2013*. Pew Research Center.

Gutherie, R. (1976). *Even the rat was white: A historical view of psychology*. Harper & Row.

Hare, B. R. (1977). Racial and socioeconomic variations in preadolescent area-specific and general self-esteem. *International Journal of Intercultural Relations, 1*(3), 31–51. https://doi.org/10.1016/0147-1767(77)90018-9

Hunter, M. (2007). The persistent problem of colorism: Skin tone, status, and inequality. *Sociology Compass, 1*(1), 237–254. https://doi.org/10.1111/j.1751-9020.2007.00006.x

McDermott, P. A., & Spencer, M. B. (1997). Racial and social class prevalence of psychopathology among school-age youth in the United States. *Youth and Society, 28*(4), 387–414. https://doi.org/10.1177/0044118x97028004001

Motti-Stefanidi, F. (2018). Resilience among immigrant youth: The role of culture, development and acculturation. *Developmental Review, 50*, 99–109. https://doi.org/10.1016/j.dr.2018.04.002

Phinney, J. S. (1992). The multigroup ethnic identity measure: A new scale for use with diverse groups. *Journal of Adolescent Research, 7*, 156–176. https://doi.org/10.1177/074355489272003

Portes, A., & Zhou, M. (1993). The new second-generation: Segmented assimilation and its variants. *The Annals of the American Academy of Political and Social Science, 530*(1), 74–96. https://doi.org/10.1177/0002716293530001006

Raudenbush, S. W., & Bryk, A. S. (2002). *Hierarchical linear models: Applications and data analysis methods*. Sage.

Riechard, D. E., & McGarrity, J. (1994). Early adolescents' perceptions of relative risk from 10 societal and environmental hazards. *The Journal of Environmental Education, 26*(1), 16–23. https://doi.org/10.1080/00958964.1994.9941429

Roberts, R. E., Phinney, J. S., Masse, L. C., Chen, Y. R., Roberts, C. R., & Romero, A. (1999). The structure of ethnic identity of young adolescents from diverse ethnocultural groups. *The Journal of Early Adolescence, 19*(3), 301–322. https://doi.org/10.1177/0272431699019003001

Rumbaut, R. G. (1999). Assimilation and its discontents: Ironies and paradoxes. In *The handbook of international migration* (pp. 172–195). Sage.

Salganik, M. J. (2019). *Prediction in the service of explanation*. Paper presented at the American Sociological Association, New York, NY.

Salganik, M. J., Lundberg, I., Kindel, A. T., Ahearn, C. E., Al-Ghoneim, K., Almaatouq, A., & McLanahan, S. (2020). Measuring the predictability of life outcomes with a scientific mass collaboration. *Proceedings of the National Academy of Sciences of the United States of America, 117*(15), 8398–8403. https://doi.org/10.1073/pnas.1915006117

Sam, D. L., & Berry, J. (2016). *Cambridge handbook of acculturation psychology* (2nd ed.). Cambridge University Press.

Schildkraut, D. J. (2002). The more things change...American identity and mass and elite responses to 9/11. *Political Psychology, 23*(3), 511–535. https://doi.org/10.1111/0162-895x.00296

Shoemaker, A. L. (1980). Construct validity of area specific self-esteem: The Hare self-esteem scale. *Educational and Psychological Measurement, 40*, 495–501. https://doi.org/10.1177/001316448004000231

Small, M. (2019). *Qualitative methods as indispensable to big data research*. Paper presented at the American Sociological Association, New York, NY.

Spencer, M. B. (1995). Old issues and new theorizing about African American youth: A phenomenological variant of ecological systems theory. In R. L. Taylor (Ed.), *African-American youth: Their social and economic status in the United States* (pp. 37–69). Praeger.

Spencer, M. B. (2006). Phenomenology and ecological systems theory: Development of diverse groups. In W. D. R. M. Lerner (Ed.), *Handbook of child psychology: Theoretical models of human development* (pp. 829–893). Wiley.

Spencer, M. B. (2008). Phenomenology and ecological systems theory: Development of diverse groups. In W. Damon & R. M. Lerner (Eds.), *Child and adolescent development: An advanced course* (pp. 696–735). Wiley.

Spencer, M. B., Fegley, S. G., & Harpalani, V. (2003). A theoretical and empirical examination of identity as coping: Linking coping resources to the self processes of African American Youth. *Applied Developmental Science, 7*(3), 181–188. https://doi.org/10.1207/S1532480XADS0703_9

Spencer, M. B., Lodato, B. N., Spencer, C., Rich, L., Graziul, C., & English-Clarke, T. (2019). Innovating resilience promotion: Integrating cultural practices, social ecologies and development-sensitive conceptual strategies for advancing child well-being. In D. A. Henry, E. Votruba-Drzal, & P. Miller (Eds.), *Advances in child development and behavior* (Vol. 57, pp. 101–148). https://doi.org/10.1016/bs.acdb.2019.05.005

Spencer, M. B., McDermott, P. A., Burton, L. M., & Kochman, T. J. (1997). An alternative approach to assessing neighborhood effects on early adolescent achievement and problem behavior. In J. Brooks-Gunn, G. Duncan, & J. Aber (Eds.), *Neighborhood poverty* (*Policy implications in studying neighborhoods*) (Vol. 2, pp. 145–163). Russell Sage Foundation. http://www.jstor.org/stable/10.7758/9781610440868.12

Spencer, M. B., Noll, E., & Cassidy, E. (2005). Monetary incentives in support of academic achievement: Results of a randomized field trial involving high-achieving, low-resource, ethnically diverse urban adolescents. *Evaluation Research, 29*(3), 199–222. https://doi.org/10.1177/0193841x04273329

Spencer, M. B., Tinsley, B., Dupree, D., Fegley, S., Tate, W., & Yeakey, C. (2012). Maximizing culturally and contextually sensitive assessment strategies in developmental and educational research. In W. F. Tate IV (Ed.), *Research on schools, neighborhoods, and communities: Toward civic responsibility* (pp. 299–325). Rowman & Littlefield.

Waters, M. C. (1994). Ethnic and racial identities of second-generation Black immigrants in New York City. *International Migration Review, 28*(4), 795–820. https://doi.org/10.2307/2547158

Chapter 17
Pathways of Positive Development Among Muslim-American Emerging Adults: Not an Oxymoron

Mona M. Abo-Zena

> Rashid discloses, "I've been stopped by the police once, and that was really quite unnerving because he asked me what religion I was.... I told them I was Muslim, and basically the police officer was concerned because someone called the police because I was praying near a train station." (Sirin & Fine, 2008, p. 154)

Nearly a century ago, DuBois outlined the problem of being born into a parallel world that requires Black people to navigate the invisible color line in U.S. society while answering the unasked question, "how does it feel to be a problem?" (DuBois, 1903/1994). Contemporary scholars compare how social issues in U.S. society relate to the plight of an additional societal "problem"—Arab youth growing up post 9/11 (Bayoumi, 2008). While DuBois described the color line as being the problem of the twentieth century, the faith line outlining Muslim youth may mark the challenge of the twenty-first century (Patel, 2010). In the current socio-political climate, Muslims are de-Americanized and "religionized" because of their affiliation with Islam (Mir, 2014, p. 6). Still, they negotiate the re/formation goal of integrating identities in the "third space," a hybrid space in which individuals and groups create alternative social representations between dominant and unequal cultural models (Bhabha & Rutherford, 2006). On individual, community, and social levels, religious and other identities related to race, gender, and socio-economic status intersect (Collins & Bilge, 2016; Crenshaw, 1989).

M. M. Abo-Zena (✉)
University of Massachusetts Boston, Boston, MA, USA
e-mail: Mona.abozena@umb.edu

D. J. Johnson et al. (eds.), *Re/Formation and Identity*, Advances in Immigrant Family Research, https://doi.org/10.1007/978-3-030-86426-2_17

Developing a Religious Identity in a Context Suspicious of the Religion

Discriminatory encounters have been described as knit into the everyday lives of Muslim-Americans (Abu-Ras et al., 2018). Muslim-American youth endure different levels of minority stress in light of within-group variations related to racial, ethnic, socio-economic, and immigrant status (Sirin & Fine, 2008). Muslim children at the time of 9/11 are now emerging adults coming of age in a context where merely fulfilling religious obligations attracts police attention, as evidenced by Rashid's experience. Given such negative social mirrors, how can individuals learn to transform what some consider an oxymoron, being Muslim-American, into a positive developmental trajectory? In particular, how do Muslim youth in a context of heightened Islamophobia navigate religious and identity issues during emerging adulthood, especially when this period marks a period of both transformative and self-defining personal and social development?

Religion and spirituality are important socialization forces, particularly for emerging adults (Smith & Snell, 2009). Arnett contends that while few people have a clearly defined worldview at 18, few leave their 20s without one (2015). For many, this coincides with entrance into college, a largely parentless context where many live and learn in campus contexts likely to be much more religiously diverse than their homes and home communities (Jacobsen & Jacobsen, 2008). In contrast to prior secular descriptions, today's campuses are increasingly religiously diverse and engaged (Calhoun et al., 2007). Conversely, some emerging adults are thought to be lost in transition, resulting in perilous developmental outcomes such as engagement in risky behavior (Smith, 2011).

Muslim emerging adults in the United States may be at a heightened risk for negative outcomes given challenges to their psychological and sometimes even physical well-being. Media's regular projections describe Islam as "a political problem to be solved" (Morey & Yaqin, 2011, p. 2) and imply that Islam is inherently violent, inferring that even moderate Muslims practicing Islam are themselves "extremists," "fundamentalists," or "Islamists."[1] The percentage of Muslims who report discriminatory experiences has remained consistent from 2007 until 2017 and includes being singled out by airport security or law enforcement and being threatened or treated with suspicion (Pew, 2011, 2017). Like many immigrant-origin youth facing a ubiquitous sense of insecurity, Americans of Middle Eastern and North African descent navigate historical trauma, pervasive institutional stress, and hostile institutional and interpersonal interactions (Awad et al., 2019).

Whether an individual Muslim has directly experienced discrimination, implicit or subtle bias may have a deleterious effect on racially minoritized individuals (Okazaki, 2009), and similarly on religiously minoritized ones. Muslim-American adolescents and emerging adults are increasingly seeking mental health services

[1] Words like "Islamist" or "fundamentalist" have no commonly agreed upon meaning, but nonetheless function as code words for "extremist' or "radical."

(Ahmed, 2011). In an attempt to integrate their religious identity and cope with the stressors of daily life, Muslim youth may fall between two extremes: misunderstanding aspects of Islam and adopting radical measures presumably to "protect" their religion, or abandoning public and private practices associated with Islam (Ramadan, 2004). Applied psychological and political science research seeks to understand and curtail the roots of faith-based terrorism (Kressel, 2007). Comparatively less developmental research has explored the positive pathways and coping strategies of maligned social groups (Syed et al., 2018).

Developing Despite Challenges

From a developmental systems theoretical perspective, positive development may be promoted when individuals successfully manage the mutually influential relations between themselves and their context (Lerner et al., 2015). Unfortunately, these contexts include stratified processes such as racism, discrimination, and oppression that challenge the development of diverse children and families and overlook their strengths (García Coll et al., 1996; Spencer 2017). Despite contextual challenges facing Muslim youth, many have contributed successfully to their own development, and to their communities. For parallel samples of Muslim youth ages 12–22 years in New York and Boston, positive pathways were defined largely by the identification with one's religion and with religiosity as a critical factor in the formation of positive values and identities, particularly volunteering and civic engagement (Sirin et al., 2011). Using mixed methods including focus groups, interviews, and identity maps, Sirin and Fine (2008) provide nuanced descriptions of how Muslim students aged 12–18 years fit (or do not fit) the different aspects of their ascribed identities (e.g., gender, ethnic, religious).

While such studies include participants who self-identify as being Muslim, to our knowledge, there has been no examination of the experiences of Muslim participants who are dedicated to actively developing their religious commitment within a "moderate" Islamic framework. Within any religious group, Smith describes a subset of individuals who are "devoted" and may differ qualitatively or quantitatively from other less religiously committed adherents (Smith, 2005). In particular, there have been limited developmental and psychological approaches to describe how "devoted" Muslim young adults may orient to the task of religious identification given the specific social context of Islam in the United States post 9/11.

Religion and its Variation as an Underexplored Asset

Given that scholars recognize that religion is a central, yet overlooked aspect of human behavior (Holden & Vittrup, 2009; Tarakeshwar et al., 2003), there is less research on how its role may vary given contextual, cultural, and religious considerations (Cohen, 2009). Relatedly, some ethnic group members are more likely to embrace criteria for adulthood that favor obligations to others instead of individualist criteria. Further, identity tasks for African American and other minoritized emerging adults are complicated by the need to overcome negative assumptions and experiences of discrimination (Arnett, 2003). In a religiously diverse sample of mostly Catholic and Mormon college students, religious differences were found in the criteria and practices both about adulthood and spiritual beliefs, suggesting that theoretical perspectives and empirical findings associated with one religious group may not accurately be generalized to other religious or cultural groups (Barry & Nelson, 2005).

Given this, research on the role of religion for immigrant-origin and other minoritized youth should reflect a broadening of methodological and sampling approaches of youth and their communities (Abo-Zena & Barry, 2013; Suárez-Orozco et al., 2011). Large, quantitative studies have documented broad effects often by comparing various adherents of religious groups to one another (see Paloutzian & Park, 2005). Certain research designs, though, risk essentializing groups and decontextualizing particular aspects of religious life (e.g., religious practice) from personal and shared meaning (Dollahite & Marks, 2009). Regarding sampling, scholars acknowledge a Protestant-bias that infuses the psychology of religion literature (Hill & Pargament, 2003). Research on diverse samples may help illustrate how developmental pathways and processes are similar and differ across religious groups. For example, a qualitative study of Bangladeshi Muslim-American adolescents found evidence for two distinct groups, internal and external seekers, with respect to religious identity formation (Chaudhry & Miller, 2008). It is unclear what particular developmental challenges religiously seeking immigrant-origin Muslim-American emerging adults of varied ethnic backgrounds face, and what assets they bring to re/forming their development in context.

The Current Study

Through data gathered in focus groups, this study provides person-centered perspectives on identifying the key challenges and resources for Muslim emerging adults in the United States who are actively involved in programs to promote their religious socialization. In particular, this study explores factors that influence identity re/formation processes and experiences.

Method

This chapter is derived from a larger study of spirituality conducted in the Greater Boston area; for a report of the research methodology in the overall study see Lerner et al., 2008.

Sample and Procedures

Participants were recruited because they attended a religious school or youth group, indicating they participated in congregational worship at least weekly. Although the full study included participants with a range of religious affiliations, the present analysis focused on those who self-identified as Muslim. The focus groups ranged from 3 to 15 participants per group; a total of 6 focus groups were conducted with young adults drawn from 4 research sites. The size and the gender composition of the focus groups depended on the nature of the site. The sample included 32 participants and was 20% female with a mean age of 19.2 years ($SD = 2.71$). The self-identified ethnicity of participants was 25.8% other, 16.1% Asian or Pacific Islander, 38.7% Arab, 16.1% Multiethnic, and 3.2% Asian Indian or South Asian, all immigrant-origin.[2]

Borrowing from marketing research (see Greenbaum, 1997), these group interviews sought to provide a more efficient way rather than individual interviews of collecting in-depth, qualitative data (Morgan, 1997) that focus on underrepresented youth perspectives. Although focus groups do not provide the detailed responses of individual interviews, they amplify participants' different insights from those afforded by quantitative data (Krueger & Casey, 2000). Because focus groups are inherently social (Stewart & Shamdasani, 2014), they lend themselves to studying religion, which usually involves a collective experience.

Focus Group Questions

The study constructs focused on religion and spirituality, ethnicity and identity, community service, and positive development and challenges to it. Although the entire focus group transcripts were coded, the analyses focused on the questions: How would you describe someone who is developing in a positive way? What characteristics would they have? Is there a part of <Islam> that is most important to you? Is there a particular way in which being <Muslim> has been hard or challenging? Who or what has had the most influence on your spiritual or religious life?

Data Analysis

Although survey data were collected, this analysis employed only focus group data. While qualitative data analysis may include considerations such as the amount of time spent on an issue or the intensity of expression of various individuals or the

[2] While recognizing a range of reasons for migration and varying legal and generational statuses, in this study we operationalize immigrant-origin as participants having at least one parent born outside the U.S.

group as a whole (Stewart & Shamdasani, 2014), this analysis explored thematic responses to the central research constructs: perspectives on positive development and approaches to integrating an Islamic identity given the broader challenges and assets in the social context. Two coders independently coded transcripts for inductive categories, concepts, and themes (Miles & Huberman, 1994). Then, they jointly discussed codes and organized them into narrative summaries to assist in identifying emerging themes (Maxwell, 2005). Demographic descriptors related to the participant or context are listed when available in order to mark aspects of positionality given an assumption that one's social position may affect and intersect with how one engages (Crenshaw, 1989).

Results

Results about the challenges and resources related to the positive development of Muslim-American participants aligned with categories that reflected the nature of their source: externally or internally derived. The external and internal challenges are presented, followed by the resources.

Challenges and External Social Mirrors

Participants reported negative perspectives from broader society about being religious generally, and about being Muslim in particular.

Being Religious in America

Participants reflected on what they considered mixed messages about the social desirability of being perceived as religious. Managing conflicting messages from public settings and private religious enclaves may contribute to what one female participant described as "leading a double life." A male college participant who is active in an Islamic campus group described how non-religious individuals may evaluate the performance of sacred practices:

> They say that we're just doing – say we're praying, we're just praying for the sake of praying. When we pray, we pray for the sake of Allah. It's not just something we do. There is something behind it. When they say that you're religious, it's just saying that you have this routine you go through every day, and you do it mindlessly.

Participants stressed that their religious practices carry meaning beyond rote routines. Meanwhile, individuals managed what they perceive as broader society's negative evaluation of religion and religiousness. One male college student in an

Islamic student organization reflected on what he considers society's double speak about being religious:

> I think that in our society that word—religious—has a negative spin to it and negative connotations. Living in the United States especially, we have freedoms, freedom of religion and this and that. What it comes down to is our society is highly secular. Right? I think that when someone calls you religious, maybe they're kind of complimenting you, but usually they're kind of looking down on you.

Being Muslim in America

Participants widely described challenges of being Muslim in the United States. Many challenges were associated with public reactions to the events of 9/11, which occurred when these participants, who are currently emerging adults, were children. A male college youth group member recalled, "Basically, people were asking me these questions I couldn't answer, because like I didn't really know much about Islam. I mean, I was only like in first grade." Recalling his reaction to 9/11, one male high school youth group member shared his fear, "don't make fun of me, I think the scariest moment of my life was I was young, I was foolish, when 9/11 happened, I was scared to death that I was going to get deported."

A male high school student recalled the injustice of generalized blame:

> Well, one thing that I'm pretty sure all of us can share would be that after 9/11 happened we experienced the taking away of our pride, basically, because we were blamed for something that allegedly even if a Muslim did it and he's not just claiming he did it, even if he did it, just one person, maybe one extremist if you want to say, it doesn't reflect a whole community...You can't take one person and take the whole community based on that person.

Participants recalled other poignant memories from childhood and adolescence. One male high school participant reflected on observing his father's treatment in public:

> At age 12,...my dad used to wear the turban and stuff...[W]hen I used to be with him, especially in public...they're like "Osama! What's up, Osama?" They called him Osama because I guess he looks like Osama bin Laden...And then in the beginning I was scared because I didn't know what might happen, because what they're trying to do to Osama, I don't know if they might hurt my dad in some way.

Another male participant recalled watching his father manage public affronts with different strategies:

> I was 11, 12. And when I saw my dad, in the beginning my dad used to stay quiet, but then after that he [answered] back...so that they can stop it. I still remember once he was working in that store, and then the guy came and was like, "Yo, Osama, can you give me cigarettes." And he's like, "Okay." Like, he responds in a jolly way so then he makes the other quiet. And there were other scary times when my dad, like once Customs arrested him when we were coming back from Canada. And the way they handled it was shocking, especially in the end when they said, "We're sorry, we do this with almost every person that comes by. Of course, we know that they don't do it, otherwise nobody would be coming back into the U.S." We didn't [raise it]... in the court, we didn't do that, but even though I was still horrified and that kind of made me try not to show my Islam in public. Then after I went to [a

> predominantly Muslim country] and I had my experience there and I lived there and I came back, now I just don't mind it more. Like what are you going to do?...So, I stopped trying to hide my Islam.

One female participant from a co-educational Islamic college student group considered the mistreatment of Muslims before and after 9/11 as consistent, but more sanctioned:

> [A]fter September 11, in my opinion, it didn't change much in terms of how people perceive Muslims. Whatever stereotypes or discrimination already existed beforehand. Maybe September 11 reinforced it into a policy kind of way. Now you see it more in airports, traveling and maybe in jokes that are more acceptable now about terrorism and terrorists. I think all these things stereotyping or discrimination against Muslim was already in American society and already a part of it. It just became more legal.

Other participants considered the poor treatment of Muslims internationally:

> I know what's going on in the world, like in Chechnya and Bosnia and Iraq, Afghanistan, Palestine. I mean, it hurts me so much to see my Muslim brothers and sisters…being killed like without any real purpose. Actually, like, for money, for oil, and whatever the heck it is. Like they're being killed, and like I think that's really wrong. And like I don't think people should be killed for money. You know what I mean? Or should be killed for, like, you know, territorial expansion or something like that.

Some participants described general social experiences of feeling marginalized because of being Muslim. For example, one male Muslim high school student recalled that, "because in public school, once they found out I was Muslim, that's when I used to get picked on." A female college participant who described herself as being "oblivious to most people" recounted a difficult reaction to a common occurrence in her life:

> It wasn't anything in particular, but I was just doing my own thing—like washing up before prayers, and the girl asked, "What are you doing?" or "Your feet are in the sink. What are you doing?" and I said, "I'm Muslim. I'm praying." And she said, "Oh. You're Muslim? I didn't know that." It wasn't really out there.

Feeling Out of Place

Participants reported how conflicting social pressures made it difficult to integrate different aspects of their identities across contexts.

Identity: When You Can't Blend

Across sites, participants described ways their appearance and religiously informed behaviors made it difficult to integrate themselves with their peers. One female participant describes that:

> I had a whole 5 or 6 years of discrimination, which I don't like to talk about….All I can say is that it affected me in very negative way. I had to work on it for a couple of years to take that out of my system.

Others described the clash of trying to combine aspects of religious values with expectations within broader U.S. society, such as this male college participant:

> I am a social person. I did go to dances and whatnot, but I didn't dance. The girl would say, "Why aren't you dancing?" I would say, "Because I was taught it was against my religion." That just goes into a whole awkward kind of thing. By the sheer awkwardness of it, and the halting of the socialization atmosphere just to explain the entire reason behind not dancing with girls and whatnot. It makes you reconsider the actual reasons behind things. That's how it's been hard for me.

Several female participants described the multiple challenges of donning the head scarf:

> Well, obviously, it's hard, especially because we wear *hijab,* when you go outside, people do stare and stuff. But I've been wearing *hijab*, and, yeah, there have been points where I wanted to take it off. But, for the most part, I've been wearing it. And people do stare. And I think I just got used to it because now it doesn't matter to me anymore. People will always stare, and people will always be ignorant, so there's no way you can change everyone. In the beginning it was hard when I would wear hijab and people would look. But, like I said, now I'm just getting used to it, so it's become easier for me to just do whatever I want because it doesn't bother me anymore.

Non-Muslim service providers may try to support Muslim youth to integrate their religious and Western identities, but inadvertently complicate things. One female participant recalled observing a religious fast during a basketball tournament:

> My coach kept forcing me to be like, "I'm not going to let you play unless you're drinking water." I told him I can handle it and that I'd been doing like [that] when I was in Karachi. Everyone can fast when they're playing, but they were just like, "No, you're going to collapse." He was literally forcing me to have water. I told him, "I'm totally fine." Honestly, I wasn't exhausted or slowing down or anything. He just didn't understand why I couldn't just have a sip and then go back to play in the game. So that was interesting and I think my mom had to explain stuff.

Another female participant described the issue generally:

> I think that there was probably the acceptance problem, too. They didn't really see me as an American teenager. They kind of just saw me as this mini-version adult person who's really serious. Who's probably going to get married after high school, or something. So that's why I think that was the acceptance problem. But now, in college, I'm feeling better because I have more room to hang out with different Muslims.

Balancing Act

Participants described various ways of responding to the negative social mirrors including leaving Islam, compartmentalizing their life, shielding themselves, and integrating Islam into their identity. Across focus groups, participants talked regularly about the perils of being "two-faced" where they "pretended to be Muslim" in some contexts, but shed this identity in more public spaces. One male participant from a youth group explained:

> I went on through public school. I kind of started to lose my Islam, to a point where I completely—where I like almost completely lost it—like it was almost like only Muslim in name. And like I believe, but I don't practice. And then I came to…Islamic private school,

> and I kind of got that back. And I feel that if I went back into—that I'm actually ready to go into the real world and not lose my Islam.

Another male participant described maintaining the difficult balance:

> if you're a Muslim, you don't want them thinking that you're a terrorist or this and that, you try your hardest to not make people think that you're a real Muslim and stuff like that. I'm not saying that's happened to me, but like I see some people who do that just so they don't get made fun of.

A female high school student described pros and cons of negotiating social messages:

> I'm sure it can be hard when there are these negative images or negative ideas. But I guess there might be a lot of responses to it. People can just kind of feel really bad…, or people can kind of decide to be really private about their religion or maybe not even to be religious. Or they can kind of retaliate somehow, trying to change, [there] could be a lot of ways. Or just to be angry.

Female participants went beyond discussions of negative social mirrors to discuss particular challenges of managing gender dynamics within an Islamic context that highly regulates interaction between sexes. In an all-female focus group, one participant described that "there needs to be a balance," but that, "I live in a bubble; I always have." The participants discussed parents' desire to "shield" them from social contexts that allow significant interactions between males and females and the relative merits of such shielding.

In one focus group, when the participants were asked if there were additional points they want to share, they outlined deep questions and doubts in their own religious life. For nearly 10 min, participants discussed what sins are forgiven, how to maintain concentration during prayer, how to balance maintaining both a love and fear for Allah, how to manage feelings such as worry about life events with faith that things are in Allah's hands.

Positive Social Mirrors

Participants described how non-Muslim and Muslim social contexts provided positive representations of their identity, beliefs, and behaviors as Muslims.

External Validation

Youth identified affirming feedback from others.

Coherence of Faith

In a social context where Islam and Muslims may be scrutinized, participants identified how reflections by others about the faith system itself helped them feel validated. For example, one female Muslim high school student described reactions when she explains Islamic practices to others outside the faith tradition:

> Even people who aren't Muslim—because I have a lot of non-Muslim friends—they just say, "Oh, that makes sense." One thing that really helps me is the fact that nothing's been changed. People have different interpretations, but the Qur'an hasn't been changed....

A female classmate from the same Islamic high school reflected on feeling validated by what she described as a coherent and comprehensive nature of the religion:

> Personally, I always felt like Islam was very encompassing of all aspects of life. It doesn't just focus on the praying or the ritual acts or even focus only on the self-discipline or your heart. It involves every aspect of your life. So it's not just a weekend kind of thing, or it's not just an hour out of your day. But it's like every minute or every second. It's part of [you] and part of being a Muslim. I don't know. I always felt that that was something that really differentiated Islam from other religions.

Male participants in a youth group referenced feeling validated by the divine origin of a belief system that encompasses all aspects of daily life as a source of guidance:

> To add on to...you could call them divine laws, laws from heaven. They're perfect. And they—because they're perfect, that's how you live your daily life. Like all civilizations in the past have failed. They were all on top. And then they crumbled because of social problems. Now, laws from heaven and divine laws aren't like all these countries with social problems, they made up their own laws. They made up their own system. Whereas our system isn't made up from—we didn't make it up. God made it up and sent it down to us. So...it's beyond our understanding.

Volunteering as a Pathway to Personal and Social Good

Participants discussed multiple ways in which volunteering or civic engagement could help promote their own identity development, contribute to the community, and add to the possibility that they would be accepted by others. Participants indicated that volunteering could help others evaluate the faith/faithful positively, as described by this male participant from a college Islamic organization:

> In our eyes I think that by volunteering and helping other people that are non-Muslims, I think you lead them towards Islam, and maybe like not as like converted, but they can sort of respect you and they don't think of you as what I think the media has said.

In addition to volunteering as individuals, participants recognized benefits when volunteering as part of an organization, as illustrated by this male college student:

> Yeah, I volunteered quite a bit before I came to <University>. Here, it's usually through <University Islamic Organization> (UIO). I think it's important to promote the fact that although we're a separate part of society, we're a segment as Muslims. We still want to show that we want to give back to the society because obviously we all live in the same place. We want to better it, just like all other groups do. In terms of that, if you do it through UIO, that's probably one of the reasons because you want to show that you want to give back to people who are less fortunate.

Social Support and Role Models

Participants sought support from a range of individuals and groups, including parents, close friends, and formal mentors and institutional resources. Participants described how their parents "instilled the basics" when they were younger and that the support continues. One female college student from a co-educational campus group described that despite her parents' desire to support her, their approach did not always meet her personal needs, which prompted her to seek support from a youth mentor:

> My parents did help me. They have influenced me a lot…Islam-wise, but I feel like the youth leader helped me apply it to America, because whenever I come to my parents and say, "I have this problem. It's so hard here," they said, "Just [follow] Islam and make *taubah* [repent], make prayer." I say, "You guys, you don't get it. I'm not doing this to try to be bad. It's really hard for me to do it." Then they act like they got it, but they didn't. They lived in a totally different country. They immigrated here. They came here and went to school and went to work. They don't really understand how it felt to grow up here. So I talked to the youth leader. She said, "I totally get you." She'd be able to help me apply what I need [to America].

The theme of approaching religious education from a dialectic not a didactic approach came up in multiple focus groups. Some participants discussed how being exposed to different religious sources and interpretations "opened my eyes" and encouraged perspective taking. Others discussed how such approaches allowed them to feel more engaged with religious content, its application to history and contemporary society, and its ability to help them connect with other people seeking religious knowledge.

Participants mentioned contemporary religious models (e.g., leaders, teachers, mentors), as well as historical religious figures or resources (e.g., prophets, holy texts, Islamic schools, or social groups). In some cases, participants described the importance of mentors who they perceived as having achieved the balance of integrating the strengths of Islam with U.S. culture. Male participants at one college campus reflected on a particular mentor they considered very well rounded because he "memorized half the Qur'an" and he was "jacked…he was physically accomplished" as well as being "a very motivational speaker" and a "Harvard professional student." Participants concluded that "his social skills were the best" and that "just the way he acts, like…it makes you want to be just like him."

Participants reflected on the importance of strong and regular social support, given the college context and social influences. As one male participant described, "You can't go home at the end of the day like you would in high school, go home to your family." One male Muslim participant recalled his initial reluctance early in college to get involved in a campus Islamic group, followed by a clear decision to deepen his connection:

> At first, I was like – I had my own stereotypes – I was like "This Islamic society is going to be kind of extreme, hardcore." I was kind of turned off by the idea of it. Then I started going to the events. It was basically at that point in my life my freshman year I was going to make a choice. I was either going to make friends with these people, or I was going to make friends elsewhere, like with people on my floor or something like that and hang out with them and go partying with them or whatever. I realized that this was what I wanted and not that. That's what attracted me to Islam.

Participants described how friendships served as personal, social, and religious support and found benefit "just being in the same space as him [a religious friend] and sharing knowledge." In describing seeing each other at a *qyiam* (optional late night prayer), male participants spoke of how the prayer and social program promoted faith and fellowship:

> I mean, those three guys, I mean, they're all about having fun, being, like, meeting new Muslim friends, but they're also about keeping your religion and staying intact with your religion. And it sort of is like—it's sort of like a refueling station, when, like, you sort of get to, like, renew your religion. Like you get to discuss it with your friends and you get to practice it with your friends. It sort of keeps you in high spirits. And it also has like a very positive impact on you for like for quite a while…I think they had the biggest impact on me, because I learned like so many things from it. And, well, I can actually like trace back like—like I can actually like tell like, you know, identify a big difference from like myself a few years ago before I was introduced to [Islamic youth organization] and myself, which is like my current self, and there's a big difference.

Activating Internal Resources for Positive Development

Participants discussed the personal indicators that they are coping successfully with their developmental challenges, primarily their self-regulation, inner-spiritual life, and achieving a sense of balance.

The Ultimate Connection

In addition to a sense of being personally grounded and connected to a spiritual community, participants frequently discussed their direct relationship with God as a source of strength and central to their religious life. The relationship with the Creator was described in various ways, such as the male high school youth group student describing the "constant worship and obviously believing in one God and His message" and the continual search for knowledge to keep up the faith. For many, the five daily prayers provided an anchor. One male participant in a high school youth group shared, "My favorite parts of religion is largely related to prayer and its many aspect – the incorporation of Qur'anic verses and reflecting on them, the direct link to God, not through another [person]." Another explained:

> I think prayer is the most important, because it's the only part of the day when you're isolated by yourself and it's when you talk with Allah, and it basically prostrating down on your knees, you're very low, and it's only—there's no—the only thing between you and Allah is nothing, so you basically—say anything you want, like think about anything and ask Allah for anything. I think it's very important.

Participants described how worship provided a sense of purpose, accomplishment, and satisfaction, as did this female participant from a co-educational college group:

> It makes you take time out of the day and you realize how thankful you should be for what you have. If you think that things aren't going the way they should be going. I don't know, you just need to stop and think. I like that. It's like you don't take things for granted.

Through collective reflection, participants develop both a framework and a communal structure to help put challenges in perspective, as indicated by a male participant in a college student group; "I think Islam helps you to be satisfied with what you've got and what you've achieved. That's how it encourages your own positive development in that sense." Another male member of a Muslim student association reflected:

> The two things that come to my mind are the community aspect—because there's a very unique and strong emphasis on community and the way our communities run and the type of support—and then also the motivation for the idea of heaven and hell that we have. It really not only motivates you, but it also keeps you positive that your life here is not everything. You don't get depressed over little things.

Self-Regulation

Participants discussed how religion in their lives served a regulatory role. For example, in reflecting on how the five compulsory prayers affect behavior, a female participant from a co-educational college Islamic student group shared:

> I think prayer really keeps you in check too….I mean, seriously, you can't go out doing drugs and partying all night and then go back and pray. It really keeps you in check. You can't do all those things. It keeps you sane. It keeps you a good person. That's what I really like about prayer.

In addition, participants discussed how practicing Islam helped promote ongoing learning and development. A female participant from an all-female focus group within a co-educational Islamic high school discussed the importance of critical self-evaluation:

> And I don't feel that Muslims should really think "I'm religious" just because you still have so much you can learn and so much you have to study for what you have to understand. It's just that saying "I'm religious" kind of puts it…like…"Yeah, you've reached the goal." And I don't feel that I did reach the goal.

Feeling Balanced and Integrated

Although mostly talking about future aspirations, many participants described positive development as feeling balanced and able to integrate or apply Islamic teachings to their life generally. Some participants discussed an Islamic concept of seeking knowledge and self-improvement, such as a male college student:

> The more knowledgeable you become in Islam…the more responsible you become… That knowledge, that is a privilege. It is something to be cherished. I think Islam is lessons on self-improvement. I don't think there's any possible end all. It's a constant cycle of self-improvement to get better and better. You gain more knowledge. You become more responsible. You're going to be held responsible for that knowledge because you know what's right and what's wrong. On the Day of Judgment, you will be held accountable. "Why didn't you do this? You had this knowledge."

Another male college student described a Muslim developing positively as:

> I would say someone that is maybe close to their religion, really close with their *deen*. *Deen* means religion. It's kind of a synonym for Islam. Someone that's really close to their religion, applies it to their life and applies it more and more and grows in that way, usually kind

> of ends up succeeding. Those are the people that ultimately develop in a positive manner. I mean, not necessarily from a financial standpoint. Religion helps as far as motivation goes in terms of academics and stuff like that. Those are the people that end up being really content further down the line.

For these youth, identity re/formation integrates Islam across all public and private life domains.

Discussion

Muslim emerging adults discussed their challenges in coping with negative social mirrors at multiple levels: from the media and anonymous people in public contexts to well-intentioned but perhaps culturally or religiously insensitive teachers, coaches, and peers. In addition, they strived to integrate different aspects of their identity, particularly as members of stigmatized religious group, along with other intersectionally minoritized identities.

Participants reported a range of social supports at various individual and institutional levels. Across focus groups, civic involvement and volunteering served multiple purposes including helping bridge to broader society which allows a positive reframing of Muslims. A range of mentors served as role models and helped to address sources of stress. Participation in the religious community, as individual believers and through fellowship, provided an introspective lens to help provide meaning for disappointments and a sense of purpose.

Challenges

The discussion of challenges to being Muslim included the general relegation of religion in U.S. society, and the prevalence of negative social images about Islam and Muslims (Amer & Bagasra, 2013). While the United States is one of the most religious industrialized countries, scientific influences have contributed to a perception that religious commitment is antithetical to reason (Dowling & Scarlett, 2006). While a negative portrayal of Islam affects all Muslims, the participants discussed how negative experiences left them feeling socially strained or personally afraid or vulnerable, challenging their identity re/formation and ability to integrate religious and other identity dimensions. Drawing from the racial socialization literature (e.g., Hughes, 2003; Spencer, 1983), caregivers may prepare their children for bias, promote pride in their group, and warn against trusting broader society. Being religiously minoritized (and possibly also racially, culturally, or linguistically minoritized), Muslim youth may have been prepared for bias similarly, but with a focus that includes building religious identity (Patel, 2010). Like racially marginalized youth, Muslim youth must try to develop a positive self-image as they observe the discrimination encountered by their families and begin their own apprenticeship in how to maintain internal composure in the face of repeated mistreatment (Coard & Sellers, 2005).

While the preponderance of the discussion of challenges to religious development dealt with responding to negative social mirrors, one group of participants delved into detail about seeking to improve their level of religious commitment. Within theological circles and in studies of the psychology of religion, topics such as religious doubt and the elusive desire for transcendence are regular issues for individuals across faith communities (Exline & Rose, 2005). Because these issues were not represented across focus groups, it is unclear how prevalent they are across the sample or population. This absence may be explained in numerous ways: the focus group protocol or its execution (i.e., facilitator effects) did not fully elicit issues of key concern; that participants generally were consumed fielding external questions, so there was less opportunity to identify internal struggles; that discussion of issues of religious doubts and questions may be less prevalent with Muslim participants given the religion-specific limitations to exploring doubt; or that the internal faith issues identified by participants of one focus group are indeed limited to only some of the "devoted" youth population. The other notable absence from participants' responses was a focus on the roles of religion and ethnicity in navigating identity. Particularly for immigrant groups, culture, ethnicity, and religion are often coupled, and these factors are also clustered within Muslim communities (Patel, 2010). Given the focus on the "devoted," it is possible that cultural and ethnic influences may be less salient in favor of religious identity (Levitt et al., 2011).

Resources

Nearly all participants discussed how their religious faith was a source of strength in their life, often discussing a particular aspect of faith (e.g., daily prayer, comprehensive nature of the faith) that guided them. While scholars of religion and positive youth development have discussed ways in which religion is fertile ground for youth development (e.g., King, 2008; King & Furrow, 2004), this study illustrated specific nuances related to the positive development of Muslim youth. The coping methods were similar to other Muslims sampled and facing stressful interpersonal events (Abu-Raiya et al., 2011). Within a religiously pluralistic society, Muslim-American participants focused on defining their own identities partly by putting their faith into action at personal and social levels. Like other youth, participants described how role models provided specific support (Martin et al., 2003). While the benefits of non-parental role models have been documented for youth generally (Schwartz et al., 2006), such modeling may be particularly important for youth for whom parents may not embody the requisite skills and attitudes to which they aspire (e.g., immigrant-origin youth who wish to acculturate, but feel their parents lack a cultural understanding of their current context). Developing a positive relationship with a role model often occurs in the context of building other peer relations, in which case the youth group provides multiple types of companionship, increased religious knowledge, and a lived opportunity to express religious identity. While developing support within ethnic enclaves describes the experiences of racially and ethnically minoritized youth, there may be parallels for religiously minoritized youth and intersections in re/forming ethno-religious identities.

Part of the identity struggle for Muslim young adults is to help them reconcile contradictions: even as immigrant-origin native sons/daughters/people, they struggle to belong as they experience daily guilt by association in a land that professes freedom and liberty for all. Participants described how civic engagement is a particular vehicle to reach multiple destinations. At its essence, volunteering is an expression of religious commitment given the value of performing good in Islam and other world religions (Smith, 1991). Further, civic engagement as a representative of a religious group is seen as a positive and legitimate way to become integrated (Jensen, 2008; Sherrod & Lauckhardt, 2009). Finally, in light of the myriad of negative images about Muslims, public examples of civic engagement may provide specific ways for Muslim youth to rename themselves and reframe their public narratives (Nelson, 2001).

Limitations

While this manuscript contributes to understanding the resources and challenges of Muslim-Americans seeking to develop their religious identity, it has notable limitations. This study used a snowball sample of relatively religiously homogenous Muslim youth in a specific geographic area. The sample itself is not representative of the ethnic, racial, socio-economic, or theological diversity within the Muslim population and is not balanced in terms of sex. Because participants' recruitment was based on involvement in a religious socialization program, it is difficult to determine variation due to program effects. As such, these findings are not necessarily generalizable to other religious groups, or even within the subset of the Muslim population characterized as being "devoted." Further, this study focused on data gathered in focus groups. While the method is conducive to discussing perspectives within a group setting, group data did not provide the depth of exploring individual variation that may have been yielded if followed up by individual interviews, in-depth surveys that included quantitative responses to constructs of interest, or other research designs (e.g., diary, narrative, autoethnography).

Conclusions, Implications, and Future Directions

Positive youth development is important to civic society generally, and particularly to individuals, families, applied organizations, and scholars who seek to better understand and promote development. The development of individuals within marginalized social groups is of particular import because the pathways and the optimal resources required to support their development are less known (Spencer, 2017). Given that all individuals likely hold privileged and marginalized identities, we urgently need to prioritize organizing collective action to reduce Islamophobia and other intersectional forms of structural oppression including Xenophobia, racism, and sexism (Abu-Ras et al., 2018). Societal level change may include psychology scholars, particularly within developmental and social psychology, collaborating with groups within law and social policy to reduce discrimination and increase education about how to partner effectively with minoritized communities. This work should also address intergenerational trauma and internalized oppression perpetuated within Muslim communities given colonial histories (Awad et al., 2019). Currently, Muslim youth's identity re/forming necessitates responding to and coping with Islamophobia and other ethno-religious biases in everyday life. While context contributes to the adjustment process for all youth, the persistence of biased contexts circumscribes a response identity for minoritized Muslims. Some youth develop a support network primarily within their faith and broader community, while responding to ridicule and varied types and levels of violence. While religion is generally considered a factor to support positive development, given the sociopolitical climate, Muslim youth are at risk of not fulfilling their potential given the disaffirming influences of negative social mirrors (Abo-Zena & Barry, 2013). Nonetheless, many devoted youth describe meaning making around these experiences to re/form a situated public-private identity where their religious identity positively anchors their worldview.

This study points to important conclusions and implications to better understand and support the development of Muslim youth, other religiously minoritized youth, and other youth who share multiple minoritized identities. First, negative social mirrors that are propagated by the media and other socialization contexts systems need to be challenged. For example, within schools, there should be policies requiring teacher education programs to increase educators' religious literacy and integrate this lens with other intersectional, anti-bias curricular approaches starting in early childhood education (Moore, 2016; Nimmo et al., 2019). Mental health providers who work with Muslim clients should be aware of the underlying issues and be able to provide culturally sensitive support, particularly given stigma in accessing mental health services. In addition, interventions can be designed to build a network of critical care providers who are positioned to advocate for accommodating ways to support Muslim youth (i.e., organizing anh all-female prom), to avoid inadvertent omissions or problems and help educators, coaches, and youth workers better support all youth navigating difference and inequality. Future research should explore the results of these interventions and partner with youth in action research that more thoroughly maps the challenges and resources and co-designs multi-level interventions that engage a range of stakeholders.

Although this study did not focus on parent-child relations, its findings reinforce the challenges and strengths of communication patterns in highly religious families, namely that youth prefer youth-centered discussions (Dollahite & Thatcher, 2008). The religious community may help introduce caregivers to learning engaging communication strategies, particularly if families have prescriptive discourse styles

(Dollahite et al., 2018). On a family and a youth level, opportunities for socialization in contexts to promote the acquisition of content, beliefs, and attitudes help develop internal resources about one's religious identity and positive value and membership within the religious group. Volunteering and other types of civic engagement may activate one's religious identity, provide a positive opportunity to fulfill religious obligations, and co-author alternative narratives to society about the role of a young, practicing Muslim. Instead of being seen as a problem, young Muslim adults seek resources and opportunities to be and be seen as being part of a solution.

References

Abo-Zena, M. M., & Barry, C. M. (2013). Religion and immigrant-origin youth: A resource and a challenge. *Research in Human Development, 10*(4), 353–371. https://doi.org/10.1080/15427609.2013.846052

Abu-Raiya, H., Pargament, K. I., & Mahoney, A. (2011). Examining coping methods with stressful interpersonal events experienced by Muslims living in the United States following the 9/11 attacks. *Psychology of Religion and Spirituality, 3*(1), 1–14. https://doi.org/10.1037/a0020034

Abu-Ras, W., Suárez, Z. E., & Abu-Bader, S. (2018). Muslim Americans' safety and well-being in the wake of Trump: A public health and social justice crisis. *American Journal of Orthopsychiatry, 88*(5), 503–515. https://doi.org/10.1037/ort0000321

Ahmed, S. (2011). Adolescent and emerging adults. In S. Ahmed & M. Amer (Eds.), *Counseling Muslims: Handbook of mental health issues and interventions* (pp. 251–280). Routledge.

Amer, M. M., & Bagasra, A. (2013). Psychological research with Muslim Americans in the age of Islamophobia: Trends, challenges, and recommendations. *American Psychologist, 68*(3), 134. https://doi.org/10.1037/a0032167

Arnett, J. J. (2003). Conceptions of the transition to adulthood among emerging adults in American ethnic groups. *New Directions in Child and Adolescent Development, 100*, 63–75. https://doi.org/10.1002/cd.75

Arnett, J. J. (2015). *Emerging adulthood: The winding road from the late teens through the twenties* (2nd ed.). Oxford University Press.

Awad, G. H., Kia-Keating, M., & Amer, M. M. (2019). A model of cumulative racial–ethnic trauma among Americans of Middle Eastern and North African (MENA) descent. *American Psychologist, 74*(1), 76–87. https://doi.org/10.1037/amp0000344

Barry, C. M., & Nelson, L. J. (2005). The role of religion in the transition to adulthood for young emerging adults. *Journal of Youth and Adolescence, 34*, 245–255. https://doi.org/10.1007/s10964-005-4308-1

Bayoumi, M. (2008). *How does it feel to be a problem?: Being young and Arab in America*. Penguin Press.

Bhabha, H. K., & Rutherford, J. (2006). Third space. *Multitudes, 3*(3), 95–107. https://doi.org/10.3917/mult.026.0095

Calhoun, C., Aronczyk, M., Mayrl, D., & Van Antwerpen, J. (2007). *The religious engagements of American undergraduates*. Social Science Research Council. http://religion.ssrc.org/reguide/index10.html

Chaudhry, S. R., & Miller, L. (2008). Religious identity formation among Bangladeshi American Muslim adolescents. *Journal of Adolescent Research, 23*(4), 383–410. https://doi.org/10.1177/0743558407309277

Coard, S. I., & Sellers, R. M. (2005). African American families as a context for racial socialization. In V. C. McLoyd, N. E. Hill, & K. A. Dodge (Eds.), *African American family life* (pp. 264–284). The Guilford Press.

Cohen, A. B. (2009). Many forms of culture. *American Psychologist, 64*(3), 194–204. https://doi.org/10.1037/a0015308
Collins, P. H., & Bilge, S. (2016). *Intersectionality.* Polity Press.
Crenshaw, K. (1989). *Demarginalizing the intersection of race and sex: A Black feminist critique of antidiscrimination doctrine, feminist theory and antiracist politics* (pp. 139–168). University of Chicago Legal Forum. https://chicagounbound.uchicago.edu/uclf/vol1989/iss1/8
Dollahite, D. C., & Marks, L. D. (2009). A conceptual model of family and religious processes in highly religious families. *Review of Religious Research, 50*(4), 373–391. https://www.jstor.org/stable/25593754
Dollahite, D. C., Marks, L. D., & Dalton, H. (2018). Why religion helps and harms families: A conceptual model of a system of dualities at the nexus of faith and family life. *Journal of Family Theory and Review, 10*, 219–241. https://doi.org/10.1111/jftr.12242
Dollahite, D. C., & Thatcher, J. Y. (2008). Talking about religion: How religious youth and parents discuss their faith. *Journal of Adolescent Research, 23*, 611–641. https://doi.org/10.1177/0743558408322141
Dowling, E. M., & Scarlett, W. G. (Eds.). (2006). *Encyclopedia of religious and spiritual development.* Sage.
DuBois, W. E. B. (1994). *Souls of Black folk.* Dover Press.
Exline, J. J., & Rose, E. (2005). Religious and spiritual struggles. In R. F. Paloutzian & C. L. Park (Eds.), *Handbook of the psychology of religion and spirituality* (pp. 315–330). The Guilford Press.
García Coll, C., Lamberty, G., Jenkins, R., McAdoo, H. P., Crnic, K., Wasik, B. H., & Vázquez García, H. (1996). An integrative model for the study of developmental competencies in minority children. *Child Development, 67*(5), 1891–1914. https://doi.org/10.2307/1131600
Greenbaum, T. C. (1997). *The handbook for focus group research* (2nd ed.). Sage.
Hill, P. C., & Pargament, K. (2003). Advances in the conceptualization and measurement of religion and spirituality. *American Psychologist, 58*, 64–74. https://doi.org/10.1037/0003-066X.58.1.64
Holden, G. W., & Vittrup, B. (2009). Religion. In M. H. Bornstein (Ed.), *Handbook of cultural developmental science* (pp. 279–294). Routledge.
Hughes, D. (2003). Correlates of African American and Latino parents' messages to children about ethnicity and race: A comparative study of racial socialization. *American Journal of Community Psychology, 31*(1/2), 15–33. https://doi.org/10.1023/A:1023066418688
Jacobsen, D. G., & Jacobsen, R. H. (2008). *The American university in a postsecular age.* Oxford University Press.
Jensen, L. A. (2008). Immigrants' cultural identities as sources of civic engagement. *Applied Developmental Science, 12*(2), 74–83. https://doi.org/10.1080/10888690801997069
King, P. E. (2008). Spirituality as fertile ground for positive youth development. In R. M. Lerner, R. W. Roeser, & E. Phelps (Eds.), *Positive youth development and spirituality: From theory to research* (pp. 55–73). Templeton Foundation Press.
King, P. E., & Furrow, J. L. (2004). Religion as a resource for positive youth development: Religion, social capital, and moral outcomes. *Developmental Psychology, 40*(5), 703–713. https://doi.org/10.1037/0012-1649.40.5.703
Kressel, N. J. (2007). *Bad faith: The danger of religious extremism.* Prometheus Books.
Krueger, R. A., & Casey, M. A. (2000). *Focus groups. A practical guide for applied research* (3rd ed.). Sage.
Lerner, R. M., Lerner, J. V., Bowers, E., & Geldhof, G. J. (2015). Positive youth development: A relational developmental systems model. In W. Overton, P. Molenaar, & R. M. Lerner (Eds.), *Handbook of child psychology and developmental science* (Theory and Method) (Vol. 1, 7th ed., pp. 607–651). Wiley.
Lerner, R. M., Roeser, R. W., & Phelps, E. (2008). *Positive youth development and spirituality: From theory to research.* Templeton Foundation Press.
Levitt, P., Barnett, M., & Khalil, N. A. (2011). Learning to pray: Religious socialization across generations and borders. In K. F. Olwig & M. Rytter (Eds.), *Mobile bodies, mobile souls* (pp. 139–159). Aarhus University Press.

Martin, T. F., White, J. M., & Perlman, D. (2003). Religious socialization: A test of the channeling hypothesis of parental influence on adolescent faith maturity. *Journal of Adolescent Research, 18*(2), 169–187. https://doi.org/10.1177/0743558402250349

Maxwell, J. (2005). *Qualitative research design: An interactive approach* (2nd ed.). Sage.

Miles, M. B., & Huberman, A. M. (1994). *Qualitative data analysis: A sourcebook of new methods* (2nd ed.). Sage.

Mir, S. (2014). *Muslim American women on campus: Undergraduate social Life and identity*. University of North Carolina Press.

Moore, D. L. (2016). Diminishing religious literacy: Methodological assumptions and analytical frameworks for promoting the public understanding of religion. In A. Dinham & M. Francis (Eds.), *Religious literacy in policy and practice* (pp. 27–38). University of Bristol Policy Press.

Morey, P., & Yaqin, A. (2011). *Framing Muslims: Stereotyping and representation after 9/11*. Harvard University Press.

Morgan, D. C. (1997). *Focus groups as qualitative research* (2nd ed.). Sage.

Nelson, H. (2001). *Damaged identities, narrative repair*. Cornell University Press.

Nimmo, J., Abo-Zena, M. M., & LeeKeenan, D. (2019). Finding a place in early childhood programs for the religious and spiritual lives of young children and their families: Taking an anti-bias approach. *Young Children, 74*(5), 37–45. https://www.jstor.org/stable/26842304

Okazaki, S. (2009). Impact of racism on ethnic minority health. *Perspectives on Psychological Sciences, 4*(1), 103–107. https://doi.org/10.1111/j.1745-6924.2009.01099.x

Paloutzian, R. F., & Park, C. L. (Eds.). (2005). *Handbook of the psychology of religion and spirituality*. The Guilford Press.

Patel, E. (2010). *Acts of faith: The story of an American Muslim in the struggle for the soul of a generation*. Beacon Press Inc.

Peek, L. (2005). Becoming Muslim: The development of a religious identity. *Sociology of Religion, 66*(3), 215–242. https://doi.org/10.2307/4153097

Pew Research Center's Forum on Religion & Public Life. (2011). *Muslim Americans: No signs of growth in alienation or support for extremism*. http://www.people-press.org/2011/08/30/muslim-americans-no-signs-of-growth-in-alienation-or-support-for-extremism/

Pew Research Center's Forum on Religion & Public Life. (2017). *U.S. Muslims concerned about their Place in society, but continue to believe in the American dream*. https://www.pewforum.org/2017/07/26/findings-from-pew-research-centers-2017-survey-of-us-muslims/.

Ramadan, T. (2004). *Western Muslims and the future of Islam*. Oxford University Press.

Schwartz, K. D., Bukowski, W. M., & Aoki, W. T. (2006). Mentors, friends, and gurus: Peer and nonparent influences on spiritual development. In E. C. Roehlkepartain, P. E. King, L. Wagener, & P. L. Benson (Eds.), *The handbook of spiritual development in childhood and adolescence* (pp. 310–323). Sage.

Sherrod, L. R., & Lauckhardt, J. (2009). The development of citizenship. In R. M. Lerner & L. D. Steinberg (Eds.), *Handbook of adolescent psychology: Contextual influences on adolescent development* (pp. 372–408). Wiley.

Sirin, S., Abo-Zena, M. M., & Shehadeh, H. (2011). Contributions despite challenges: Exploring positive youth development among Muslim American Youth. In A. E. A. Warren, R. M. Lerner, & E. Phelps (Eds.), *Thriving and spirituality among youth: Research perspectives and future possibilities* (pp. 233–254). Wiley.

Sirin, S. R., & Fine, M. (2008). *Muslim American youth: Understanding hyphenated identities through multiple methods*. New York University Press.

Smith, C. (2011). *Lost in transition: The dark side of emerging adulthood*. Oxford University Press.

Smith, C. & Denton, M. L. (2005). *Soul searching: the religious and spiritual lives of American teenagers*. Oxford University Press.

Smith, C., & Snell, P. (2009). *Souls in transition: The religious and spiritual lives of emerging adults*. Oxford University Press.

Smith, H. (1991). *The world's religions*. Harper Collins.

Spencer, M. B. (1983). Children's cultural values and parental child rearing strategies. *Developmental Review, 3*, 351–370. https://doi.org/10.1016/0273-2297(83)90020-5

Spencer, M. B. (2017). Privilege and critical race perspectives' intersectional contributions to a systems theory of human development. In N. Budwig, E. Turiel, & P. D. Zelazo (Eds.), *New perspectives on human development* (pp. 287–312). Cambridge University Press.

Stewart, D. W., & Shamdasani, P. N. (2014). *Focus groups: Theory and practice* (Vol. 20). Sage.

Suárez-Orozco, C., Singh, S., Abo-Zena, M. M., Du, D., & Roeser, R. W. (2011). The role of religion and religious organizations in the positive youth development of immigrant youth. In A. E. A. Warren, R. M. Lerner, & E. Phelps (Eds.), *Thriving and spirituality among youth: Research perspectives and future possibilities* (pp. 255–288). Wiley.

Syed, M., Santos, C., Yoo, H. C., & Juang, L. P. (2018). Invisibility of racial/ethnic minorities in developmental science: Implications for research and institutional practices. *American Psychologist, 73*(6), 812–826. https://doi.org/10.1037/amp0000294

Tarakeshwar, N., Stanton, J., & Pargament, K. I. (2003). Religion: An overlooked dimension in cross-cultural psychology. *Journal of Cross-Cultural Psychology, 34*(4), 377–394. https://doi.org/10.1177/0022022103034004001

Chapter 18
Identity Development Among African Refugee Immigrant Youth in Australia

Elias Mpofu, Hadidja Nyiransekuye, and Lisa López Levers

Cross-national refugee experiences define a person's identity in many ways important for adult development. Persons with "refugee" status are those who reside outside of their country of nationality or those seeking habitual residence, based on a well-founded fear of persecution due to race, religion, nationality, membership in a particular social group, or any combination thereof. The refugees may be those with certain political opinions and lacking in protection of the country of origin or in danger in returning there for fear of persecution (United Nations High Commission on Refugees: UNHCR, 2011). Refugee immigrants are constantly engaging with the new host culture worlds unravelling before them and having to "think-on-their-feet." Nothing would have prepared them for the cultural uprooting from the culture of origin and transplantation to a foreign, host culture. For this reason, refugee immigrants continuously evolve in their identities to accommodate the rapid cultural learning derived from each new experience they encounter in the host culture.

As emerging adults, young refugee immigrants are struggling with their sense of cultural estrangement, while constructing a sense of personal and social coherence for everyday functioning (Rosenthal et al., 1981). They engage in developmental tasks that are associated with identity formation, including schooling, bonding with peers, family, and forming intimate and loving relationships with potential life partners (Dune et al., 2019; Mpofu et al., 2012). Generally, refugee immigrants achieve these cultural-competency reconfigurations, unconsciously and automatically. However, under certain circumstances, as when the host culture is Western European

E. Mpofu (✉)
University of North Texas, Denton, TX, USA

University of Sydney, Sydney, Australia

University of the Witwatersrand, Johannesburg, South Africa
e-mail: Elias.Mpofu@unt.edu

H. Nyiransekuye
University of North Texas, Denton, TX, USA
e-mail: Hadidja.nyiransekuye@unt.edu

L. L. Levers
Duquesne University, Pittsburgh, PA, USA
e-mail: levers@duq.edu

D. J. Johnson et al. (eds.), *Re/Formation and Identity*, Advances in Immigrant Family Research, https://doi.org/10.1007/978-3-030-86426-2_18

versus African culture of origin, this process becomes conscious and more deliberate, as refugee immigrants try to interpret core meanings of the host culture important for their daily living activities (Berry, 1997). Under circumstances where a large cultural gap exists between the culture of origin and the host culture, refugee immigrants experience what Märtsin (2010) termed as an encounter with "rupturing otherness" (p. 68), or the sense of self-estrangement from pressures to live and learn in the moment, with considerable uncertainty and unpredictability of social outcomes. Moreover, experiences of "ruptured otherness" or self-estrangement are disorienting to identity development and might lead to a sense of self-disavowal (Wolf, 1988), in which emotionally threatening experiences are averted within conscious awareness to protect the self from being overwhelmed. "Otherness" experiences also might result in identity delocalization (Giddens, 1991) or "deterritorialization" (Tomlinson, 1999), in which the youth have a pervading sense of alienation from the host country setting, uncertainty about their future, and a diminished sense of belongingness. These disavowal and delocalization experiences, if accentuated and chronic, might result in an emerging adulthood characterized by a sense of identity confusion arising from a cumulative sense of uncertainty, impermanence, and disorientation. It also might result in identity diffusion (Marcia, 1994), in which the immigrant refugee youth engage in low self-exploration and low commitment to their emerging adulthood in the Australian cultural context.

During the period of adolescence through young adulthood, ethnic and racial identity formation also emerge (Umaña-Taylor et al., 2014). Immigrant refugee youth not only grapple with the normal adolescent pubescent changes but also with adaptation to the host community culture (Abrokwaa et al., 2011). As they mature into adulthood, they must grapple with the possible selves arising from resettlement experiences in the new nation. These processes relate both to the life goals that were out of reach pre-settlement and to new goals the youth need to consider for successful adjustment to the host culture.

Our chapter considers how African youth refugee immigrants to Australia may construct their identities, drawing on their culture-of-origin experiences, and while facing on-going demands to rapidly adapt to the demands of the Australian host country context. First, we provide a broad overview of African refugee immigrant youths' cultures of origin to contextualize their experiences, including a discussion on their resources and liabilities for identity development in the Australian host culture. A brief overview of Australian refugee immigrant policies and regulations provides the backdrop for how these elements influence refugees' adaptation to the host culture setting. Next, we focus on African refugee immigrant youth and their immigrant-oriented identity development statuses of social affiliation, rule, role, and value orientation. We present some illustrative cases on identity development issues, including some background for culture-of-origin expectations, social-integration opportunities, and the impact of visa issuance type on self-perceived adaptation to the Australian host culture. Finally, we consider some issues for research and other forms of scholarship addressing the transient and emergent nature of identity development, gendered identity development, influences of informational technology literacy, and the importance of participatory action research in constructing identities among immigrant refugee youth.

African Refugee Immigrant Youth Cultures and Identities

Africa is a continent comprising 54 nation states. African-origin refugees to Australia are mostly from the war-torn North-Eastern African countries of the Sudan and Somalia. Others are from the Central African countries of Democratic Republic of the Congo, Rwanda, Uganda, and Western African nations of Senegal and Liberia. Australia is ranked 46th, with respect to admission and resettlement of refugees, by the United Nations High Commission on Refugees (UNHCR, 2017).

The vast majority of refugees from Africa are from three countries: Somalia (1,012,323), South Sudan (1,436,651), and Sudan (646,036) (UNHCR, 2017). Africa is a multicultural continent with thousands of subcultures. For instance, the predominantly Arab-language cultures of North and North-Eastern Africa present a distinctly different cultural orientation compared to the "bantu-language" cultures of sub-Saharan Africa. Even within Arab and Bantu culture regions, there are numerous subethnic groups, adding to the cultural diversity among "would-be" refugees to other nations. Regardless of the relative diversity among African heritage cultures, the Arab and sub-Saharan subcultures are collectivist in cultural orientation, placing a higher value on the needs of the social collective over those of the individual. This is unlike in the ethos of individualist cultures, in which the needs of the individual take precedence over those of the social collective (Greif, 1994). Many of the countries of Africa include ethnic communities with distinct cultural identities, even as they share a collectivist culture orientation. Specifically, there are hundreds of ethnic groups in Africa, each with its own cultural mores. Often these ethnic identities determine how an individual self identifies, perhaps more than their African continental affiliation. These ethnic cultural identities are the primary lens through which African refugee immigrant youth (ARIY) may perceive their social worlds and through which they would interpret their refugee experiences.

Nearly the entire African continent, with the exception of Ethiopia, shares a history of colonization by Western countries. Consequently, citizens of many African countries have adopted both Western and indigenous cultural ethos, important for their orientation and ability to adapt and be received in a Western cultural host setting. For instance, most of the countries in the South, East, and West of Africa have an Anglophone colonial heritage, so that the official language in these countries is English. The African countries from Central to North Africa have a predominantly Francophone colonial heritage. Australia has an Anglophone colonial history, and for this reason, the country of origin of African immigrant refugee youth matters to their ease of settlement in Anglophone Australia from having prior exposure to the English language. However, many of the refugee immigrants from Africa admitted to Australia on refugee or humanitarian visas typically have disrupted schooling so that they need initial school enrolment in Australia to acquire basic spoken English (Matthews, 2008).

For the typical African refugee youth, transitional markers associated with emerging adulthood may be altered significantly, following immigration; these include age-based seniority, clan identity, mastery of a subsistence-economy trade or occupation, and getting married or starting a family young. Among indigenous African subsistence-culture communities, by mid- to late adolescence, 13–17 years of age, adult responsibilities are well established. This is because subsistence economies encourage involvement by minors in basic livelihood activities of farming and/or local trading. In these economies, children assume adult-life roles of caring for each other materially and socially (Serpell, 2011). By contrast, in industrialized societies, emerging adulthood is an extended process (18–30 years of age) from the longer period of preparation to assume occupational roles and responsibilities of information-based economies (Arnett, 2002). Thus, for African youth refugees to industrialized Australia, the uprooting from culture of origin and the demands on them to function in a new culture extends their emerging adulthood, by at least a decade, compared to peers in their home-culture setting.

Before discussing the issues of identity reformation development, it is important to understand the Australian immigrant policies and instruments to contextualize youths' experiences. Australia has a variety of refugee immigration admission policies. We discuss some of the country's key refugee immigrant admission instruments next.

Australian Refugee Immigration Policies and Instruments

Host country immigration policy instruments and their implementation influence the extent to which refugee immigrants achieve a healthy sense of personal identity, post-settlement (Matthews, 2008). Australia has long-term refugee detention centers, some of which have been in the news for various reasons (Briskman et al., 2008; Mares & Mares, 2001). However, landed or on-shore refugee immigrants to Australia are typically on one or more of the following visas: refugee visa, global special humanitarian visa, woman at risk visa, or temporary (humanitarian concern) visa. Some young immigrants are on an orphan relative visa, prospective marriage visa, partner visa, or temporary visa. The majority of refugee immigrants settled in Australia are on humanitarian visas.

On average, Australia issues 13,750 humanitarian visa places per year. Visa "places" and allowable slots for visa issuance are predetermined by type, for a set period. Of these places, about 6000 (44 percent) places per year are for refugee resettlement. An additional 5000 special humanitarian visa program are specified for immediate family members of refugees to resettle in Australia.

Our discussion of influences on the identity development of ARIY is with the caveat that the type of visa under which individual refugees are admitted would influence their desire and motivation to adapt to Australian culture values. For instance, those with temporary admission visas may be less invested in developing Australian social affiliations and values orientations in comparison to those under long-stay visa admittance.

Influences of Identity Development of African Youth Refugee Immigrants to Australia

Australia is a culturally pluralistic country, but the mainstream culture is predominantly individualistic. Thus, ARIY must acquire self-identities to function in a predominantly individualistic cultural setting, while also maintaining, as well as maturing in, the collectivistic cultural values of their culture of origin. Navigating both the collectivistic origin culture and the individualistic host culture can be a daunting task and a major stress on ARIY. Moreover, Australian public service agencies are very underprepared to provide support to refugee immigrant youth from Africa (Gatt, 2011; Matthews, 2008), thus increasing the risk for negative social outcomes for the youth, post-settlement.

Identity development is, in part, a product of the "dynamic, irreducible tension" between individual functioning and the sociocultural context (Penuel & Wertsch, 1995, p. 84; see also Wertsch, 1991), and as individuals develop new ways of thinking, feeling, and behaving, they evolve their functional identity for their host community (Märtsin, 2008). This sense of emergent self occurs, in part, because "life structures are in constant motion, changing with time and evolving as new life circumstances unfold" (Hutchison, 2015, p. 277). Moreover, during young adulthood, a majority of individuals acquire a sense of personhood, defined by an awareness of the normative social role affiliations they assume. These social role affiliations include an understanding of the responsibilities to the social collective (Crocetti et al., 2014). An understanding of how one's social world functions is premised on mastery of the rules, roles, and value orientations of the lived or host community (Kelman, 2006). From a self-psychological perspective, individual and group identity evolves from the fact that all humans seek a sense of a functional cohesive self within their social affiliation networks.

Social Affiliation Identity

To achieve aspired identities in a new cultural setting, the refugee immigrant youth must construct a social affiliation identity with networked others who also are of refugee background as well as with host culture others (Abrokwaa et al., 2011). Social affiliation networks are value-laden sociocultural environments comprising individuals' constructed or "figured worlds" with emergent social selves (Holland et al., 2001, p. 52). "Figured world" refers to the worldviews derived from meaning-making of lived experiences within an affiliation social collective (e.g., family, peer group) (Holland et al., 2001). The processes and outcomes of making sense of social sense influence one's sense of selfhood.

ARIY may not be accompanied by immediate family and quickly learn to forge social affiliations and identities that they construct post-settlement. In this regard, the immigrant youth must efficiently acquire social affiliation identities that would enable them to relate to both home-culture and host-culture communities. Specifically, the refugee immigrant youths' sense of social affiliation would include the present refugee community members as well as those who had passed away or been left behind in the countries of origin. Often, the refugee immigrant youth need to construct a working bicultural or hybrid identity "on-the-go" and with no prior preparation as they evolve their social selves.

ARIY are a visible racial minority in Australia, in which the Black population comprises less than one percent of the national population census. For this reason, ARIY to Australia might quickly realize that they are nested within social affiliations that are primarily based on race and secondarily, on their developmental stage as adolescents or young adults. These racialized self-identities may be reinforced by mainstream media reporting on "gang-like" actions by ARIY or on an assumptive "emerging problem population" (Wahlquist, 2018; Windle, 2008). Although many ARIY may have excellent social affiliations with same-age peers of diverse racial backgrounds, perceptions by mainstream Australian culture seem to cast them as would-be gangsters, thus constructing a climate that would constrain social affiliation identities and create social marginalization, alienation, and exclusion (Matthews, 2008).

African refugee immigrants comprise less than one percent of the offender population serving time within Australian correctional facilities (Crime Statistics Agency, 2018). Nonetheless, the risk to African refugee immigrants for a fragile identity development is elevated by being discriminated against, primarily for their skin color (Matthews, 2008; Mansouri & Johns, 2017). Thus, in their identity development, post-settlement in Australia, ARIY must confront the possibility of a sense of self-loss from implicit social affiliation restrictions placed upon them owing to the double jeopardy of being both refugees and youth of Black African heritage.

The typical refugee immigrant youth rely on community services of the host country, including schooling, work-placement, housing, and subsistence. However, they may lack a working knowledge and experiences of the host country's social policy provisions and/or access to mentoring services to assist them in gaining access to the social services they need for successful post-settlement living (Levers et al., 2015; Levers & Mancilla, 2013; Mpofu et al., 2015a, 2015b). Host country social services personnel may be unable to provide timely and appropriate assistance to refugee immigrants, due to a lack of cross-cultural literacy as well as being uninformed on pertinent pre-settlement issues among the refugees, which are important for their post-settlement living (Matthews, 2008).

For ARIY, their assumed identities evolve from efforts to resolve individual and environmental demands and pressures on them. In their identity development, they must address the tensions of competing African culture of origin and Australian mainstream culture. Some ARIY are able to achieve various degrees of integration into the identity aspects of their culture of origin, without subjugating their own emerging values and beliefs (Hatoss, 2012). Others may experience significant acculturative social affiliation stress from navigating host country culture and country-of-origin culture (Berry, 1998). Acculturative stress may be overwhelming to some youths experiencing conflict in the value orientation between their culture of origin and mainstream Australian culture (Australian Human Rights Commission, 2010).

With involuntary migration comes major disruptions to socio-historic identities that are aligned to rule, roles, and values of the home culture (Adepoju, 2005). Thus, ARIY must acquire the functional rule and role identity to successfully manage their activities of daily living in the new cultural setting.

Rule Identity

Rule orientation refers to understanding the affordances and penalties for quality of participation in a particular environment (Kelman, 2006). By thinking, feeling, and behaving consistent with a culture's implicit cultural beliefs, ARIY acculturate to the practices and behaviors of the Australian host culture. They also figure out how to comply with the functional identities for their activities of daily living, post-settlement in Australia. For instance, pertinent rule functions in the broader Australian society may prioritize punctuality based on the 24-hour clock, whereas a flexi-time, socially sensitive personal demeanor, may be prioritized in the collectivist African culture. In African heritage culture, how well one carries out a social activity for or with others matters more than whether it is completed on time by the clock (Mpofu, 1994; Mpofu et al., 1996). Moreover, for whom one performs an activity may matter more in collectivist African culture than the technical skills for that activity (Diale et al., 2019; Grigorenko et al., 2001; Nsamenang, 2006; Serpell, 2011). This is unlike the individualistic Western culture in which one is socialized to an identity prioritizing time-sensitive routines and tasks aimed at meeting inanimate, non-social schedules as befitting the industrial habits of a modern economy (Mpofu, 1994; Mpofu et al., 1996). Conceivably, in an attempt to find a cultural home, ARIY find themselves at a crossroad between maintaining loyalty to their collectivistic culture of origin and adaptation to individualistic mainstream Australian culture (see also Ferguson & Bornstein, 2014). In creating a cultural home in their new environment, ARIY will undergo cultural shedding (Berry, 1997), in which they shed from their behavioral repertoire aspects of culture of origin for adaptive role identity functioning in the Australian setting.

Role Identity

Role identity orientation is about understanding what it takes to implement the obligatory tasks for a social role. For ARIY, role identity development involves rapid acquisition of a dependable knowledge of essential role functions for Australian mainstream culture. These youth may not have the luxury of the period of "waithood" (Honwana, 2013), or a period in which youth in typical developmental settings are searching for ways to facilitate the development of the self. Moreover, the only "waithood" time for refugee immigrants is their extended stay in holding camps pending admittance to the host country. This is potentially a harmful "waithood" time for ARIY, restricting them from important life-planning activities within the confines of a refugee detention camp.

For young people, role identity development is about mastering personal and occupational competences for participation in the host culture such as appropriate self-care and grooming, learning and knowledge acquisition, employment, community living, and interpersonal relationships (Kelman, 2006; Mpofu et al., 2015a, 2015b). While these roles may be generic across cultures, the specific demands on ARIY in their role identity may vary by their pre-settlement experiences. For instance, many ARIY had their schooling disrupted in their home countries or previous countries of asylum. Consequently, with the disruption of their education may require ARIY to re-adjust their role as a full-time student in Australia. For some of the ARIY, the school attendance requirements in Australia would make it impossible to fulfill roles they held previously in their country of origin, causing disruption to the family system. For example, Kango (pseudonym), a 14-year-old immigrant Sudanese African refugee observed the following:

> Sometimes [my uncle] gets upset with me and then not [want to] talk to me … Like, when we were in Africa, [I was] working, washing … doing the housework [for him]… Not like that anymore … [Now] I have to go to school.

In the case of Kango, her culture-of-origin role within the family system changed with immigration to Australia and she was looking forward to school and the opportunity to complete academic degrees rather than home-minding. She needed to self-identify as a student rather than a caretaker for her uncle. For young immigrants, school is one of the most important social contexts for promoting well-being and resettlement in a new country (Gifford et al., 2009; Matthews, 2008).

Higher levels of identification with a student role have been associated with lower levels of depression and higher levels of self-efficacy in young refugee immigrants (Kia-Keating & Ellis, 2007). Schooling may also serve the function of buffering against feelings of otherness in the wider Australian social context, with young people sharing at least the two social identities of a student and of an immigrant.

Similarly, language-use role-function shifts may occur within the family, with younger members of the family translating spoken and written English for the older family members with less schooling in English. As an example, Namupali (16-year Sudanese refugee) said:

> Because [my elder sister/guardian] can't speak English … when she has a problem, she talks to me and if there is no solution, I have to talk to others for ideas, … then I would just come and talk to her like my own idea … like when I first came to [our settlement] school for interview and I talked to the [teachers] for her and … the teachers … could understand.

Cultural translational roles by younger children for their older family members may arise from the fact that younger people are more open and able to new learning (Hatoss, 2012). They are more capable of adaptive learning than older generation family, expanding their "social awareness, responsibility, and obligation" (Hutchison, 2015, p. 270).

However, a young refugee immigrant also may experience social role challenges with home-culture expectations for minors, which may be at odds with the host culture environment. For example, African culture-of-origin expectations for a minor to assume home care for younger siblings and older family members (Serpell, 2011) may be contrary to Australian child protection laws. Thus, an unbridled cultural translation success characterization of immigrant youth may obscure the difficulties of immigrant youth with added family responsibilities; this may lead to failure, depression, and resentment in the absence of diversity friendly support systems.

Value Identity

Value orientation refers to the extent to which an individual identifies with the underlying beliefs behind the workings of a social system. Australia has a pluralistic culture, although the mainstream culture is predominantly individualistic. For this reason, ARIY must acquire value centered identities to enable them to function in a predominantly individualistic cultural setting, while also maintaining, as well as maturing in, the collectivistic cultural values of their culture of origin. They typically seek to achieve value orientation with acculturation, which refers to the adoption of the ideals of an "*other*" culture along with the retention of own culture values (Berry, 1997; Schwartz et al., 2006). Specific threats to the young refugee immigrants' value-identity development include (a) concerns for loss of respect for the cultural heritage and values, (b) difficulty maintaining the culture and value of the country of origin, (c) negative gender relations, (d) intergenerational issues, and (e) social exclusion (Australian Human Rights Commission, 2010). Navigating both collectivistic and individualistic cultures can be a daunting task and a major stress on ARIY.

For ARIY, their native cultural values promote being hardworking, resilient individuals and strong survivors who work to adapt to their host culture (Matthews, 2008). Moreover, many ARIY have a deep sense of compassion and kindness, stemming from the influence of a lost parent, sibling, or other relative as well as from their reception by the Australian host country (Märtsin, 2008, 2010). ARIY who have keen memories of the compassion and kindness of others may aspire to integrate those memories into their individual identity and may believe that they are carrying forward the values and meanings that connect them to the broader humanity (Abrokwaa et al., 2011). Thus, ARIY, who are in the process of developing identity, are likely to draw on previously internalized voices and dialogues to determine how their experiences fit with their self-constructions (Märtsin, 2008, 2010).

For ARIY, their settlement in Australia may be after many cycles of internal displacement within their country of origin as well as external displacement to neighboring countries (Abrokwaa et al., 2011). These within-country or region displacements often accompany feelings of deprivation and recurrent trauma from failed, under-resourced refugee-care systems (Levers et al., 2015). As a result, recurrent trauma from displacement, resettlement, and stigmatization experiences from being treated as needy outsiders influence their identity construction (Adepoju, 2005; Boikanyo & Levers, 2017; Levers & Hyatt-Burkhart, 2011). For instance, depending upon how the individual youth refugee acts in and reacts to all of the environmental layers of the new location, a chain of person-environment interactions is likely to set up a more or less functional adaptation process. Depending upon the quality of the interactions, new or amplified eco-systemic risk or protective factors may arise, thus affecting the adaptation process. New circumstances may unfold rapidly with involuntary migration, and how the individual youth adapts to these circumstances defines his or her evolving identity. Hatoss (2012) reported that young Sudanese refugee immigrants to Australia have a positive sense of self that may have been the result of a strong cultural identity and their immigrant community social capital, thereby reinforcing protective factors in their new social environment.

Many of the ARIY may have a history of repeat displacements and of subjection to ethnic and racial discrimination in previous countries of asylum, perhaps due to remnants of tribal or colonial structures still embedded in or transported to asylum cultures. Moreover, ARIY likely have histories that lack formal education, reliable means of subsistence, and secure livelihoods (Honwana, 2013; Porter & Haslam, 2005). These experiences of displacement, discrimination, and deprivation are likely to trigger feelings of disavowal or "otherness" from chronic awareness of social vulnerability, making for fragile identity development. For instance, Abdul, an 18-year-old Somali refugee on a temporary settlement visa, expressed his fears of not being able to stay in Australia as follows: "I don't think about the future. I just think about today and tomorrow. Nothing more than that." This statement from Abdul suggests a pervading sense of disavowal from a lack of certainty of his own future in Australia. Abdul's fears of deportation are based on the reality that youth refugee immigrants on bridging visas may be technically removed from Australia and deported back to their places-of-origin, where they likely would suffer resumption of persecution, and are particularly vulnerable to a sense of disavowal or identity-development paralysis.

A common theme among identity theories is that of exploration and integration of multiple cultural perspectives, with greater exploration and integration leading to a more balanced and cohesive self-concept (Kunst et al., 2015; Märtsin, 2008, 2010). However, ARIY who experience racism may not self-explore sufficiently, from fear about speaking out against racist actions toward them, believing that there was no trustworthy host-country person with whom to share violations of their personhood (Matthews, 2008). For this reason, young refugee immigrants with chronic exposure to race-based social exclusion and discrimination adopt a social compliance value orientation to accept unequal treatment, passively, in classrooms, playgrounds, shopping malls, and public transportation. The sense of disavowal or self-estrangement from being a social refugee from the host country population would make for atypical identity development in which youth refugees would not internalize the full cultural values of the host country.

Issues for Research and Other Forms of Scholarship

Identity "has to be created and recreated on a more active basis than before" (Giddens, 2000, p. 65). This certainly is the case with ARIY to Australia, especially regarding the need to construct identities that bridge their experiences prior to and after settlement. The greater the cultural distance between their culture of origin and the host culture, the more they have to engage in adaptive learning to foster functional identities in the host culture (Berry, 1997). Pre- and post-settlement encounter experiences rather than "refugee-ness" or "African-ness" may determine the aspired and achieved identity development of ARIY (Porter & Haslam, 2005; Rutter, 2006). Unfortunately, there has been limited attention to how pre- and post-settlement factors influence identity development among ARIY. Moreover, young refugee immigrants with the same settlement outcomes might vary in their identity development outcomes by gender, personal histories, sociocultural backgrounds, level of pre-migration schooling, literacy in mother tongue, acculturation, and personal resilience. Thus, there is a need for research on the varied resettlement pathways and how these pathways influence identity development among refugee immigrant youth.

Technological advances have played a role in shaping and changing identities, from increasing cultural exchanges on cyberspace to creating global citizens, and this particularly occurs among the younger generation of millenials across the world (Arnett, 2002; Barker, 2009; Jensen, 2007). However, there is a gap in knowledge on how information technology literacy influences identity development among ARIY. For instance, stressors from navigating both the collectivistic cultural norms and the individualistic norms would vary among young refugee immigrants, based on their access to and familiarity with information technology and social media tools (Idang, 2015). It may be possible that those with information technology and social media literacy prior to resettlement would have a relatively global culture orientation that, in turn, may enhance their adaptation to Australian culture by developing a compatible identity. In contrast, those with minimal or limited information technology and social media access or literacy may be more constrained in their adaptation to the host culture, hindering their host culture identity development.

Research on identity development among young refugee immigrants poses a number of ethical and methodological challenges, including the increased potential for disempowerment and misrepresentation (Barker & Weller, 2003; Block et al., 2013) and difficulties associated with cross-cultural communication and limited English language abilities (Gifford et al., 2007). Participatory action research approaches with refugee immigrant communities might obviate some of these limitations. Tree- of- Life is one example of a participatory research approach (Denborough, 2008; Ncube, 2006; Schweitzer et al., 2014). The ToL is a narrative method, with demonstrated evidence as a sound research tool for use with culturally and linguistically diverse children and young people to study their social self-constructions. It utilizes the metaphor of a tree to map out aspects of a young person's life. In the ToL, the roots signify a person's family, community, and/or place-of-origin, and the ground is representative of the person's current day-to-day routine (e.g., school, work, living arrangements). At the trunk of the ToL, individuals reflect on their personal strengths, including any particular skills, avoiding a deficit view of the self. At the branches of the ToL, individuals are empowered to reflect on their hopes and dreams for the future, rather than dwelling on past limitations. The ToL leaves are for narratives about people with whom individuals share a significant relationship. The fruits depicted in the ToL represent gifts to the person by others, such as love and support or shared values. The ToL approach has the advantage to reliably map the social affiliation, rule, role, and value orientation identities among young refugee immigrants on their own terms rather than those of the research leaders.

Conclusion

Identity development among ARIY to Australia tends to be of a hybrid type, combining home and host culture. ARIY develop cultural adaptation from operating within a constructed post-host country identity for a fresh start in life while a visible racial minority person. Moreover, identity development outcomes may vary between and among these youth, depending on their sociocultural background, pre-and post-settlement pathways, the school and community supports to which they have access, and opportunity for self-development. ARIY with stronger social affiliations along with opportunities to learn the rules, roles, and values for successful living in Australia likely will develop a healthier, bicultural identity. Those with chronic social marginalization and poorer support for education to acquire requisite rule, role, and value functionality in host Australia likely will develop more depraved personal identities from being day-to-day social refugees in their country of sojourn. Thus, implementation of policy instruments that are sensitive to the refugee immigrant sociocultural background, as well as pre- and post-settlement experiences by national and international agencies, would support healthy identity development among youth immigrants of refugee background.

References

Abrokwaa, C., Ngoma, M. S., Shizha, E., & Mpofu, E. (2011). African refugees: Challenges and prospects. In E. Mpofu (Ed.), *Counseling people of African ancestry* (pp. 166–179). Cambridge University Press. https://doi.org/10.1017/CBO9780511977350

Adepoju, A. (2005, September). Migration in West Africa. *Global Commission on International Migration.* http://www.iom.int/jahia/webdav/site/myjahiasite/shared/shared/mainsite/policy_and_research/gcim/rs/RS8.pdf

Arnett, J. J. (2002). The psychology of globalization. *American Psychologist, 57*, 774–783. https://doi.org/10.1037/0003-066X.57.10.774

Australian Human Rights Commission. (2010). *In our own words: African Australians: A review of human rights and social inclusion issues.* https://www.humanrights.gov.au/sites/default/files/content/africanaus/review/in_our_own_words.pdf

Barker, J., & Weller, D. (2003). "Is it fun?": Developing children centred research methods. *International Journal of Sociology and Social Policy, 23*, 33–58. https://doi.org/10.1108/01443330310790435

Barker, V. (2009). Older adolescents' motivations for social network site use: The influence of gender, group identity, and collective self-esteem. *Cyber Psychology and Behavior, 12*, 209–213. https://doi.org/10.1089/cpb.2008.0228

Berry, J. W. (1997). Immigration, acculturation and adaptation. *International Journal of Applied Psychology, 46*, 5–34. https://doi.org/10.1111/j.1464-0597.1997.tb01087.x

Berry, J. W. (1998). Acculturative stress. In P. B. Organista, K. M. Chun, & G. Marin (Eds.), *Readings in ethnic psychology* (pp. 117–122). State University of New York Press. https://psycnet.apa.org/record/1998-06033-001

Block, K., Warr, D., Gibbs, L., & Riggs, E. (2013). Addressing ethical and methodological challenges in research with refugee-background young people: Reflections from the field. *Journal of Refugee Studies, 26*, 69–87. https://doi.org/10.1093/jrs/fes002

Boikanyo, M. N., & Levers, L. L. (2017). The effects of climate change on water insecurity in Botswana: Links to the criminal justice system. In G. Magill & K. Aramesh (Eds.), *The urgency of climate change: Pivotal perspectives* (pp. 434–449). Cambridge Scholars Publishing. https://www.duq.edu/events/integrity-of-creation/2015-conference

Briskman, L., Latham, S., Goddard, C., Street, K., & Bentley, P. (2008). *Human rights overboard.* Scribe Publications.

Crime Statistics Agency (2018). *Correction of country of birth data incorrectly reported and attributed to the crime statistics agency.* https://www.crimestatistics.vic.gov.au/media-centre/news/correction-of-country-of-birth-data-incorrectly-reported-and-attributed-to-the-crime-statistics-agency

Crocetti, E., Erentaite, R., & Zukauskiene, R. (2014). Identity styles, positive youth development, and civic engagement in adolescence. *Journal of Youth and Adolescence, 43*(11), 1818–1828. https://doi.org/10.1007/s10946-014-0100-4

Denborough, D. (2008). *Collective narrative practice: Responding to individuals, groups, and communities who have experienced trauma.* Dulwich Centre Publications.

Diale, B. M., Mpofu, E., Fatima, A., Nadia, A., Dunbar-Krige, H., Ntinda, K., & Pillay, J. (2019). Social function acquisition among sub-Saharan Africa culture children and teenagers with orphanhood. *International Society for the Study of Behavioral Development Bulletin, 2*(76), 8–9. https://issbd.org/resources/files/JBD_Nov11.pdf

Dune, T., Mpofu, E., Hossain, Z., Mupedzanhamo, V., & Liamputtong, P. (2019). *Migrant and refugee youths' sexual and reproductive health and rights.* Western Sydney University.

Ferguson, G. M., & Bornstein, M. H. (2014). Tridimensional (3D) acculturation: Culture and adaptation of Black Caribbean immigrants in the USA. In R. Dimitrova, M. Bender, & F. Van de Vijver (Eds.), *Global perspectives on well-being in immigrant families* (pp. 31–51). Springer. https://doi.org/10.1007/978-1-4614-9129-3_3

Gatt, K. (2011). Sudanese refugees in Victoria: An analysis of their treatment by the Australian Government. *International Journal of Comparative and Applied Criminal Justice, 35*, 207–219. https://doi.org/10.1080/01924036.2011.591904

Giddens, A. (1991). *Modernity and self-identity: Self and society in the late modern age*. Polity Press. https://www.sup.org/books/title/?id=2660

Giddens, A. (2000). *Runaway world: How globalization is reshaping our lives*. Routledge. https://www.routledge.com/Runaway-World-How-Globalization-is-Reshaping-Our-Lives/

Gifford, S. M., Bakopanos, C., Kaplan, I., & Correa-Velez, I. (2007). Meaning or measurement? Researching the social contexts of health and settlement among newly arrived refugee youth in Melbourne, Australia. *Journal of Refugee Studies, 20*, 414–440. https://doi.org/10.1093/jrs/fem004

Gifford, S. M., Correa-Velez, I., & Sampson, R. (2009). *Goods starts for recently arrived youth with refugee backgrounds: Promoting wellbeing in the first three years of settlement in Melbourne, Australia*. La Trobe Refugee Research Centre. https://eprints.qut.edu.au/69057/

Greif, A. (1994). Cultural beliefs and the organization of society: A historical and theoretical reflection on collectivist and individualist societies. *The Journal of Political Economy, 5*, 912–950. https://www.jstor.org/stable/2138652?seq=1

Grigorenko, E. L., Geissler, P. W., Prince, R., Okatcha, F., Nokes, C., Kenny, D. A., Bundy, D. A., & Sternberg, R. J. (2001). The organization of Luo conceptions of intelligence: A study of implicit theories in a Kenyan village. *International Journal of Behavioral Development, 25*, 367–378. https://doi.org/10.1080/01650250042000348

Hatoss, A. (2012). Where are you from? Identity construction and experiences of 'othering' in narratives of Sudanese refugee-background Australians. *Discourse and Society, 23*, 47–68. https://doi.org/10.1177/0957926511419925

Holland, D., Lachicotte, W. S., Skinner, D., & Cain, C. (2001). *Identity and agency in cultural worlds*. Harvard University Press. https://www.hup.harvard.edu/catalog.php?isbn=9780674005624

Honwana, A. (2013). *Waithood, restricted futures and youth protests*. http://africanarguments.org/2013/08/12/youth-waithood-and-protest-movements-in-africa-by-alcinda-honwana/

Hutchison, E. D. (2015). *Dimensions of human behavior: The changing life course* (5th ed.). Sage. https://edge.sagepub.com/hutchisonclc6e

Idang, G. E. (2015). *African culture and values* (Vol. 16, p. 97). UNISA Press. S1561-40182015000200006.

Jensen, B. T. (2007). Understanding immigration and psychological development: A multilevel ecological approach. *Journal of Immigrant & Refugee Studies, 5*, 27–48. https://doi.org/10.1300/J500v05n04_02

Kelman, H. C. (2006). Interests, relationships, identities: Three central issues for individuals and groups in negotiating their social environment. *Annual Review of Psychology, 57*, 1–26. https://doi.org/10.1146/annurev.psych.57.102904.190156

Kia-Keating, M., & Ellis, B. H. (2007). Belonging and connection to school in resettlement: Young refugees, school belonging, and psychosocial adjustment. *Clinical Child Psychology and Psychiatry, 12*, 29–43. https://doi.org/10.1177/1359104507071052

Kunst, J. R., Thomsen, L., Sam, D. L., & Berry, J. W. (2015). "We are in this together": Common group identity predicts majority members' active acculturation efforts to integrate immigrants. *Personality and Social Psychology Bulletin, 41*, 1438–1453. https://doi.org/10.1177/0146167215599349

Levers, L. L., Biggs, B. A., & Strickler, A. (2015). International implications for addressing immigrant and refugee health issues. In E. Mpofu (Ed.), *Community-oriented health services: Practices across disciplines* (pp. 271–292). Springer. https://www.springerpub.com/community-oriented-health-services-9780826198174.html

Levers, L. L., & Hyatt-Burkhart, D. (2011). Immigration reform and the potential for psychosocial trauma: The missing link of lived human experience. *Analyses of Social Issues and Public Policy, 12*, 68–77. https://doi.org/10.1111/j.1530-2415.2011.01254.x

Levers, L. L., & Mancilla, R. (2013). Educating the children of undocumented immigrant parents: The impact of trauma on citizen children's development. In F. E. McCarthy, M. H. Vickers, & E. Brown (Eds.), *International advances in education: Global initiatives for equity and social justice* (Migrants and refugees: Equitable education for displaced populations) (Vol. 6, pp. 51–72). Information Age Publishing. https://www.abebooks.com/9781623964665/Migrants-Refugees-Equitable-Education-Displaced-1623964660/plp

Mansouri, F., & Johns, A. (2017). Social networks and perceptions of intergenerational difference among migrant youth in Australia. *Journal of Sociology, 53*, 127–144. https://doi.org/10.1177/1440783316636244

Marcia, J. E. (1994). The empirical study of ego identity. In H. A. Bosma & L. G. Tobi (Eds.), *Identity development: An interdisciplinary approach* (pp. 67–80). Sage. https://books.google.com/books/about/Identity_and_Development.html?id=ni5-AAAAMAAJ

Mares, P., & Mares, P. (2001). *Borderline: Australia's treatment of refugees and asylum seekers* (Vol. 204). UNSW Press. https://www.google.com/books/edition/Borderline/

Märtsin, M. (2008). *Identity construction as a personal sense-making process: a case study of Estonian students in the United Kingdom* (PhD Thesis, University of Bath). http://opus.bath.ac.uk/

Märtsin, M. (2010). Rupturing otherness: Becoming Estonian in the context of contemporary Britain. *Journal of Integrative Physiological and Behavioral Science, 44*, 65–81. https://doi.org/10.1007/s12124-009-9109-8

Matthews, J. (2008). Schooling and settlement: Refugee education in Australia. *International Studies in Sociology of Education, 18*, 31–45. https://doi.org/10.1080/09620210802195947

Mpofu, E. (1994). Exploring the self-concept in an African culture. *Journal of Genetic Psychology, 155*, 341–354. https://doi.org/10.1080/00221325.1994.9914784

Mpofu, E., D'Amico, M., & Cleghorn, A. (1996). Time management in an African culture: Correlates with college academic grades. *Canadian Journal of Behavioural Science, 28*, 102–112. https://doi.org/10.1037/0008-400X.28.2.102

Mpofu, E., Levers, L. L., Mpofu, K., Tanui, P., & Hossain, Z. S. (2015a). Family assessments in rehabilitation service provision. In M. Millington & I. Marini (Eds.), *Families in rehabilitation counseling: A community-based rehabilitation approach* (pp. 251–266). Springer. https://www.google.com/books/edition/Families_in_Rehabilitation_Counseling/

Mpofu, E., Ruhode, N., Mhaka-Mutepfa, M., January, J., & Mapfumo, J. (2015b). Resilience among Zimbabwean youths with orphanhood. In L. Theron, L. Liebenberg, & M. Ungar (Eds.), *Youth resilience and culture* (pp. 67–80). Springer. https://www.google.com/books/edition/Youth_Resilience_and_Culture/

Mpofu, E., Stevens, C., Biggs, H. C., & Johnson, E. T. (2012). Sociocultural influences on the work participation of refugees: An exploratory mixed systematic mixed studies review. *Vulnerable Groups & Inclusion, 5*(4), 1–12. https://doi.org/10.3402/vgi.v310.16066

Ncube, N. (2006). The Tree of Life Project: Using narrative ideas in work with vulnerable children in Southern Africa. *International Journal of Narrative Therapy and Community Work, 1*, 3–16. dn=020925808336979.

Nsamenang, A. B. (2006). Human ontogenesis: An indigenous African view on development and intelligence. *International Journal of Psychology, 41*, 293–297. https://doi.org/10.1080/00207590544000077

Penuel, W. R., & Wertsch, J. V. (1995). Vygotsky and identity formation: A sociocultural approach. *Educational Psychologist, 30*, 83–92. https://doi.org/10.1207/s15326985ep3002_5

Porter, M., & Haslam, N. (2005). Predisplacement and postdisplacement factors associated with mental health of refugees and internally displaced persons: A meta-analysis. *Journal of the American Medical Association, 294*, 602–612. https://doi.org/10.1001/jama.294.5.602

Rosenthal, D. A., Gurney, R. M., & Moore, S. M. (1981). From trust to intimacy: A new inventory for examining the Erikson's stages of psychological development. *Journal of Youth and Adolescence, 10*(6), 525–537. https://doi.org/10.1007/BF02087944

Rutter, J. (2006). *Refugee children in the UK*. Open University Press.

Schwartz, S. J., Montgomery, M. J., & Briones, E. (2006). The role of identity in acculturation among immigrant people: Theoretical propositions, empirical questions, and applied recommendations. *Human Development, 49*, 1–30. https://doi.org/10.1159/000090300

Schweitzer, R. D., Vromans, L., Ranke, G., & Griffin, J. (2014). Narratives of healing: A case study of a young Liberian refugee settled in Australia. *The Arts in Psychotherapy, 41*, 98–106. https://doi.org/10.1016/j.aip.2013.10.006

Serpell, R. (2011). Social responsibility as a dimension of intelligence, and as an educational goal: Insights from programmatic research in an African society. *Child Development Perspectives, 5*, 126–133. https://doi.org/10.1111/j.1750-8606.2011.00167.x

Tomlinson, J. B. (1999). *Globalization and culture*. University of Chicago Press.

Umaña-Taylor, A. J., Quintana, S. M., Lee, R. M., Cross, W. E., Rivas-Drake, D., Schwartz, S. J., Syed, M., Yip, T., Seaton, E., & Ethnic and Racial Identity in the 21st Century Study Group. (2014). Ethnic and racial identity during adolescence and into young adulthood: An integrated conceptualization. *Child Development, 85*, 21–39. https://doi.org/10.1111/cdev.12196

UNHCR. (2017). *Global trends forced displacement in 2016*. http://www.unhcr.org/5943e8a34

United Nations High Commissioner for Refugees (UNHCR). (2011). *Handbook and guidelines on procedures and criteria for determining refugee status under the 1951 convention and the 1967 protocol relating to the status of refugees*. http://www.refworld.org/docid/4f33c8d92.html

Wahlquist, C. (2018, January). Is Melbourne in the grip of African crime gangs? The facts behind the lurid headlines. *The Guardian*. https://www.theguardian.com/australia-news/2018/jan/03/is-melbourne-in-the-grip-of-african-gangs-the-facts-behind-the-lurid-headlines

Wertsch, J. V. (1991). Voices of the mind: A sociocultural approach to mediated action. *Harvester Wheatsheaf*. https://www.google.com/books/edition/Voices_of_the_Mind/

Windle, J. (2008). The racialisation of African youth in Australia. *Journal for the Study of Race, Nation and Culture, 14*, 553–556. https://doi.org/10.1080/13504630802343382

Wolf, E. S. (1988). *Treating the self: Elements of clinical self-psychology*. The Guilford Press. https://www.google.com/books/edition/Treating_the_Self/

Index

D. J. Johnson et al. (eds.), *Re/Formation and Identity*, Advances in Immigrant Family Research, https://doi.org/10.1007/978-3-030-86426-2